Human Genetics: possibilities and realities

The Ciba Foundation for the promotion of international cooperation in medical and chemical research is a scientific and educational charity established by CIBA Limited–now CIBA-GEIGY Limited–of Basle. The Foundation operates independently in London under English trust law.

Ciba Foundation Symposia are published in collaboration with Excerpta Medica in Amsterdam.

Excerpta Medica, P.O. Box 211, Amsterdam

Frontispiece
Sir Gordon Wolstenholme, Director of the Ciba Foundation, 1949–1978.
(Portrait by June Mendoza, owned by the Royal Society of Medicine, London.)

Human Genetics: possibilities and realities

Ciba Foundation Symposium 66 (new series)

In honour of Sir Gordon Wolstenholme

1979

Excerpta Medica

Amsterdam • Oxford • New York

ISBN Excerpta Medica 90 219 4072 8
ISBN Elsevier/North-Holland 0 444 90064 0

Published in June 1979 by Excerpta Medica, P.O. Box 211, Amsterdam and Elsevier/North-Holland, Inc., 52 Vanderbilt Avenue, New York, N.Y. 10017.

Suggested series entry for library catalogues: Ciba Foundation Symposia.
Suggested publisher's entry for library catalogues: Excerpta Medica.

Ciba Foundation Symposium 66 (new series)
437 pages, 58 figures, 42 tables.

Library of Congress Cataloging in Publication Data

Symposium on Genetics and Human Biology: Possibilities and Realities, London, 1978.
Human genetics.

(Ciba Foundation symposium; 66 (new ser.) "Held at the Ciba Foundation, London, 27th-29th June 1978."
Includes indexes.
1. Human genetics–Congresses. 2. Medical genetics–Congresses. I. Title. II. Series: Ciba Foundation. Symposium; new ser. 66.
QH431.S98265 1978 573.2'1 79–10949
ISBN 0–444–90064–0

Printed in The Netherlands by Casparie, Heerhugowaard.

Contents

S. BRENNER Introduction 1

L. L. CAVALLI-SFORZA Cultural change and its relevance for human genetics 5
Discussion 16

P. M. B. WALKER Genes and non-coding DNA sequences 25
Discussion 39

C. WEISSMANN, H. WEBER, T. TANIGUCHI, W. MÜLLER and F. MEYER Reversed genetics: a new approach to the elucidation of structure–function relationship 47
Discussion 58

J. B. GURDON, D. A. MELTON and E. M. DE ROBERTIS Genetics in an oocyte 63
Discussion 73

P. E. POLANI DNA repair defects and chromosome instability disorders 81
Discussion 127

General Discussion I 135

D. J. WEATHERALL, J. B. CLEGG, W. G. WOOD and G. PASVOL Human haemoglobin genetics 147
Discussion 174

HARRY HARRIS Multilocus enzymes in man 187
Discussion 199

W. F. BODMER Gene clusters and the HLA system 205
Discussion 223

P. J. LACHMANN and M. J. HOBART The genetics of the complement system 231
Discussion 246

C. MILSTEIN, G. GALFRE, D. S. SECHER and T. SPRINGER Monoclonal antibodies and cell surface antigens 251
Discussion 266

General Discussion II 277

M. SINISCALCO Human gene mapping and cancer biology 283
Discussion 308

HENRY HARRIS Some recent progress in the analysis of malignancy by cell fusion 311
Discussion 327

G. KLEIN The role of viral transformation and cytogenetic changes in viral oncogenesis 335
Discussion 349

B. CHILDS and J. M. FINUCCI The genetics of learning disabilities 359
Discussion 373

C. R. SCRIVER On being an individual, or: the man in the red hat 377
Discussion 388

W. BODMER Molecular and genetic organization: the future 395

Final General Discussion 401

S. BRENNER Chairman's closing remarks 413

Index of contributors 415

Subject index 417

Participants

Symposium on Genetics and Human Biology: possibilities and realities, held at the Ciba Foundation, London, 27th–29th June 1978

S. BRENNER *(Chairman)* MRC Laboratory of Molecular Biology, University Postgraduate Medical School, Hills Road, Cambridge CB2 2QH, UK

W. F. BODMER Genetics Laboratory, Department of Biochemistry, University of Oxford, South Parks Road, Oxford OX1 3QU, UK

L. L. CAVALLI-SFORZA Department of Genetics, Stanford University School of Medicine, Stanford, California 94305, USA

B. CHILDS Department of Pediatrics, Johns Hopkins Hospital, Baltimore, Maryland 21205, USA

F. H. C. CRICK The Salk Institute, P.O. Box 1809, San Diego, California 92112, USA

J. R. S. FINCHAM Department of Genetics, University of Edinburgh, King's Buildings, West Mains Road, Edinburgh EH9 3JN, UK

R. L. GARDNER Department of Zoology, University of Oxford, South Parks Road, Oxford, OX1 3PS, UK

J. B. GURDON MRC Laboratory of Molecular Biology, University Postgraduate Medical School, Hills Road, Cambridge CB2 2QH, UK

HARRY HARRIS Department of Medical Genetics, Richard Buildings G-4, University of Pennsylvania, Philadelphia, Pennsylvania 19174, USA

HENRY HARRIS Sir William Dunn School of Pathology, University of Oxford, South Parks Road, Oxford OX1 3RE, UK

F. JACOB Institut Pasteur, 25 Rue du Dr Roux, 75015 Paris, France

G. KLEIN Department of Tumour Biology, Karolinska Institutet, S-104 01 Stockholm, Sweden

H. KOPROWSKI The Wistar Institute, 36th and Spruce Streets, Philadelphia, Pennsylvania 19105, USA

P. J. LACHMANN MRC Group on Mechanisms in Tumour Immunity, The Laboratory of Molecular Biology, The Medical School, Hills Road, Cambridge CB2 2QH, UK

J. V. NEEL Department of Human Genetics, University of Michigan Medical School, 1137 Catherine Street, Ann Arbor, Michigan 48104, USA

SIR GUSTAV NOSSAL The Walter & Eliza Hall Institute of Medical Research, Royal Melbourne Hospital Post Office, Victoria 3050, Australia

R. PETO Department of the Regius Professor of Medicine, Radcliffe Infirmary, Oxford OX2 6HE, UK

P. POLANI Paediatric Research Unit, Guy's Hospital Medical School, Guy's Tower, London Bridge, London SE1 9RT, UK

G. PONTECORVO Imperial Cancer Research Fund Laboratories, P.O. Box 123, Lincoln's Inn Fields, London WC2A 3PX, UK

F. H. RUDDLE Departments of Human Genetics and Biology, Yale University, New Haven, Connecticut 06520, USA

C. R. SCRIVER The McGill University–Montreal Children's Hospital Research Institute, 2300 Tupper Street, Montreal H3H 1P3, Quebec, Canada

M. SINISCALCO Sloan Kettering Institute for Cancer Research, 424 East 68th Street, New York, NY 10021, USA

P. M. B. WALKER MRC Mammalian Genome Unit, Department of Zoology, University of Edinburgh, West Mains Road, Edinburgh EH9 3JT, UK

D. J. WEATHERALL Nuffield Department of Clinical Medicine, The Radcliffe Infirmary, Oxford OX2 6HE, UK

C. WEISSMANN Institut für Molekularbiologie 1, University of Zürich, 8093 Zürich, Switzerland

SIR GORDON WOLSTENHOLME The Ciba Foundation, 41 Portland Place, London W1N 4BN, UK

Editors: RUTH PORTER *(Organizer)* and MAEVE O'CONNOR

Introduction

SYDNEY BRENNER

MRC Laboratory of Molecular Biology, University Medical School, Cambridge

I would like to repeat Dr Porter's welcome to you all, and particularly to those who attended the Ciba Foundation symposium on the *Biochemistry of Human Genetics* nearly 20 years ago. Our subject has undergone enormous development since that time, with the introduction of a range of new methods and new concepts.

To introduce the symposium, I should first note that it is a particularly appropriate time to discuss genetics and human biology. The great technical advances of the past few years, not the least of which has been the ability to clone and analyse DNA sequences, have begun to throw new light on the organization of genes in higher organisms. The discovery of several instances of what might be called spaced-out genes, that is coding sequences which are not continuous but are interrupted by other sequences, offers the first clear insight into the relationship between primary RNA transcripts in the nucleus and the messenger RNAs into which they are processed. We now know that this processing involves the precise excision of the intervening region. There are already cases where a single DNA sequence can ultimately produce more than one protein by different excision events. Whether this novel step in information transfer is used for control is still an open question, and one which is bound to attract wide attention in the coming years. No doubt there will be discussion on this question during the meeting.

The possession of these very powerful methods for molecular analysis of gene sequences will be crucial for the development of the genetics of higher organisms, particularly man. Together with somatic cell genetics, we now have the tools for tackling many problems in human genetics which have so far been totally intractable. But, as some of the papers will show, there is still more to come. Cloned DNA sequences can be inserted into cells and the value of the sequences assayed. Furthermore, since the sequences can be modified in a directed manner, recloned and assayed again, this will allow the analysis of a function of any particular region of the DNA sequence. Genetics will, in principle, be freed from the search for rare mutational events, although I for one will regret the passing of that exciting laboratory pastime—the great mutant hunt! And that other great blood sport, the vast compendium of human haemoglobin variants, will be minuscule compared with the variants that will be produced in the test tube in the next few years.

The great ease with which molecular information can be collected on the genomes of higher organisms will tempt many. We can inevitably expect vast compendia of sequences but, without functional reference, these compendia will be uninterpretable, like an undeciphered ancient language. Many people and many computers will play games with these sequences, but we will have to find out by experiment what the sequences do and how the products they make participate in the physiology and development of the organism. Thus, although the analysis of the genotype has been taken care of, we still need better ways of analysing phenotypes. Many of us are ultimately interested in the causal analysis of development and the reduction of the complex phenotypes of higher organisms to the level of gene products. This is still the major problem of biology. We must understand what cells can do because all of what we are is generated by cells growing, moving, and differentiating. Here too, powerful tools are becoming available, among which the capacity to produce monoclonal antibodies is certain to become a major method for the analysis of the phenotypes of cells in higher organisms.

These experimental forays have naturally led to speculation on whether anything could be done to manipulate the genomes of the somatic cells of man or, in the extreme case, to alter the genetic constitution of the human population. As many of us know, the study of human genetics is opposed by some because of the fear that information acquired and techniques developed for what might be the purest of scientific motives may come to be misapplied in the future. Many of the present controls of genetic research are partly based on these fears. I can only make one or two comments on this subject today.

First, genetic engineering as a true technology in the sense that we can design an organism and implement that design does not exist. Nor will it exist until we know much more about the biology of cells and organisms than we know now. All we can do is a little 'tinkering', but that, as François Jacob (1977) has pointed out, is nature's way and not ours.

The second point is that much of human activity depends not on our biology but on our culture, and it is much cheaper and easier to change people by talking to them than by altering their genes. In one sense, 'Brave New World' has been with us now for a million years, based on cultural changes rather than on changes in genetic polymorphisms. This area of human biology—the old argument about nature versus nurture—particularly where it touches on human behaviour, is controversial. We shall no doubt hear views on this controversy at this meeting. These are difficult and fascinating questions but they do not match the fascination and challenge of discovering how a fertilized egg with 10^9 nucleotide pairs of DNA can make a human being.

References

CIBA FOUNDATION (1959) *Biochemistry of Human Genetics (Ciba Found. Symp.),* Churchill, London

JACOB, F. (1977) Evolution and tinkering. *Science (Wash. D.C.) 196,* 1161-1166

Cultural change and its relevance for human genetics

L. L. CAVALLI-SFORZA

Department of Genetics, Stanford University School of Medicine, Stanford, California

Abstract The first part of this paper summarizes conclusions drawn from theoretical analysis of cultural change, as appeared in various papers (published and unpublished) by the author in collaboration with Marc Feldman. Among conclusions emphasized are the tendency to homogeneity of cultural traits with most mechanisms of cultural transmission, the great variation in rates of change and conditions determining them, and the major factors responsible for change. The possibility of genetic variation in learning ability adds considerable complications and determines joint biological and cultural evolution.

In the second part of the paper, one very specific example of biological and cultural coevolution is outlined. Archaeological information shows that agriculture spread slowly from a Near East area of origin of domestication of plants and animals. The spread towards Europe is particularly well mapped. There are good reasons why the spread of agriculture may have been accompanied by a spread of farmers from the area of origin. It turns out that synthetic gene maps of Europe showing such a spread of farmers would be an excellent explanation for the geographic distribution of genes in Europe.

About ten years ago I became interested in the variation in, and evolution of, cultural traits. To a biologist like myself, one reason for excitement was the discovery that much of the conceptual framework which has been developed for biological evolution could also be usefully applied in the context of cultural change. In other words, the evolutionary factors which we identify with respect to genes, such as mutation, selection, drift and migration, translate into cultural equivalents when, instead of the evolution of genes, we consider the evolution of customs, languages, technology or any other aspect of culture.

Somewhat surprisingly, however, the literature of cultural evolution is very sparse. How is it possible that linguists, anthropologists and so on have had so little interest in the evolutionary aspects of culture? Certainly there were attempts to study these aspects, but most of them aborted, some so badly that they generated widespread taboos. Thus there is a tendency among some scholars to avoid even the use of the expression 'cultural evolution'; others limit the term to very specific problems of sociopolitical change. Most of the treatments in existence are typically macroevolutionary. The micro-evolutionary approach, which was so fruitful in biology in providing opportunities for the verification and extension of theory, is almost absent.

What seems to have been neglected beyond belief is the study of cultural transmission. It is true that in biology it was possible to understand, before specific knowledge of biological transmission was available, how adaptation could evolve under natural selection. The full understanding and verification of Mendelian principles, which are at the basis of biological transmission, developed later. But it was only with the help of knowledge of transmission that it became possible to fill the evolutionary picture with the other factors: mutation, migration and drift. In fact evolution can be viewed as the sum of exceptions to a perfect transmission so that an understanding of transmission rules is a necessary complement to the study of evolution.

One cannot fail to be struck by the enormous difference between biological and cultural transmission. Even if almost no scientific work is available on the latter, everyday experience provides some insight. By contrast, biological transmission has been the object of innumerable and often very sophisticated studies and we know that, with few exceptions, it takes place exclusively from parent to offspring. According to the well-known Mendelian probabilities, with one gene and two alleles (three genotypes) there are nine possible matings, each of which gives rise to the three genotypes with fixed probabilities (for instance 1,0,0 for genotypes AA, Aa, aa in mating AA × AA, 1/2:1/2:0 in matings Aa × AA or AA × Aa, etc.).

In lieu of this set of probabilities, a cultural trait existing in three types could have any set of values leading to potentially very complex rules of transmission. In practice there will be eighteen parameters to specify parent–offspring transmission. In addition there is no guarantee that these parameters will be constant from generation to generation. Even more important, there will be transmission not only from parent to offspring but also between sibs, other types of relatives, friends, especially age peers, and from chiefs, teachers and through public media, etc. (Cavalli-Sforza 1971; Cavalli-Sforza & Feldman 1973*a,b,* 1976 and unpublished; Feldman & Cavalli-Sforza 1975, 1977 and unpublished).

The task of formalizing this protean set of rules may seem absolutely hopeless. But there are bound to be obvious simplifications. My Stanford collegue, Marc Feldman, and I have done the spadework and have started considering the generalizations which emerge. It is of interest to contrast these generalizations with the expectations in biological evolution. In the latter, transmission is such that in the absence of evolutionary factors such as mutation etc., there is no change in populations. This static condition (otherwise known as neutral, or indifferent, equilibrium) is rarely to be expected for a cultural trait. Transmission rules for determining neutral equilibria may exist, but only exceptionally; more often, unlike the biological counterpart, transmission of a cultural trait is likely to change the frequency of the trait. This does not deny the possibility of stable equilibria; cultural transmission rules will indeed lead to stable equilibrium frequencies for a trait, but these will often be frequencies of zero or one. However, internal stable equilibria may exist, especially if parent–offspring transmission is also involved. In sum, even in the absence of the classical evolutionary factors, such as mutation etc., a great variety of types of evolution may be generated through transmission. For instance, a cultural trait (which can exist in three phenotypes) may take Hardy-Weinberg proportions at every generation (M. Feldman & L. L. Cavalli-Sforza, unpublished work), but the frequency of the trait will be unstable and change all the time until it hits a stable equilibrium. Rather subtle changes in the transmission rules may make proportions widely different from Hardy-Weinberg. Superficial imitation of Mendelian transmission and even of sex linkage is possible, and has in fact been observed in the case of kuru, an infectious transmission determined by cultural customs.

There is one important correlate of the fact that stable cultural equilibria are very frequently found at the boundaries, for trait frequencies of zero or one. Extinction or fixation of a cultural trait means one thing: *cultural homogeneity*. The trend towards homogeneity is even clearer if we study a continuous rather than a discrete trait. In the latter, in fact, homogeneity involves the rather extreme conditions of fixation or extinction, but in the former one can estimate the variance of the trait at every generation.

The interest of these conclusions is strengthened if one considers that stabilization of the variance of a trait to a sufficiently low value, that is a high degree of cultural homogeneity, is a prerequisite for the existence of culture and society. Homogeneity of the language spoken by all individuals belonging to the same linguistic group, for example, must be very high for effective communication to be possible. One conception of culture is that of biological adaptation making it possible for an individual to learn from the experience of others in ways that will increase the fitness of the individuals. For this aim, some kind of communication is essential; language is only one way, but a very efficient one. It is not, however, the only part of culture which requires a high degree of homogeneity for full effectiveness. Clearly, those behavioural rules that fall in the domain of ethics must be standardized to a large extent to make social life possible. Indeed, the very essence of ethics for Immanuel Kant was that the same moral rules should be valid for everybody, that is they should reach absolute homogeneity. Also behavioural rules without specific ethical content, which nevertheless govern much of our daily behaviour and are sometimes labelled as *etiquette,* are highly homogeneous. It has been stated that homogeneity of rules of etiquette is essential for social life in that they guarantee that a person we meet belongs to our group and is not a foreign—perhaps hostile—being. Etiquette thus helps in distinguishing between 'we' and 'they' (White 1959).

The rules of cultural transmission thus seem to favour homogeneity and social life seems to need it. There are, however, at least three conditions of non-transient departure from cultural homogeneity: (i) the division of labour; (ii) the establishment of classes and especially castes; (iii) the diversification of groups that are separated, geographically or otherwise, and do not communicate. In the first case there is clearly an advantage in diversification. In the third case both adaptation to different environmental conditions and drift create differences. This differentiation may, and generally will, proceed to the point where mutual understanding becomes impossible. Mathematically it corresponds to the case of a reducible transmission matrix, in which there are groups of individuals who do not communicate at all. They will then tend to diverge and their difference is expected to increase indefinitely with time.

It is a commonplace that a major difference between biology and culture is the rate of evolutionary change. In biology rates are low. Natural selection is the most important factor which can cause rapid change, but only at the expense of great loss of life or fertility. This can hardly be the case for Man, an organism in which the maximum possible number of progeny is only a small multiple of that necessary for replacement, which means that mortality rates cannot be high; and also fertility variation is relatively low. But in cultural evolution extremely high rates of change are possible, and an almost instantaneous change in a large mass of individuals is not unthinkable. This does not mean that rates are always necessarily high. On the contrary, one can find groups that have very similar cultures in spite of long separation. Extreme conservation of cultural traits is perhaps likely to be observed when a cultural trait is ordinarily learnt at a very early age, and therefore usually from parents (vertical transmission), but is unlikely to change at a later age. Another powerful agent for cultural conservation is group pressure from the parental generation. We have called transmission from members of the older generation other than parents 'oblique' transmission. When the grandparental generation plays a role the rate of change is further reduced. By contrast, with transmission from age peers, and more generally 'horizontal' transmission, to use the term introduced by epidemiologists, rates will tend to be higher. Highest rates of change are likely to be observed when transmission takes place from one individual (a teacher, or a leader) to all the rest of the group. This is of course greatly facilitated by the mass media. No doubt their increasing importance will bring the rate of cultural evolution close to its maximum. But even before the mass media made it possible to have very large audiences, hierarchically structured societies provided the means for fast cultural change.

Another important difference between biological and cultural evolution is familiar with natural selection which, as Darwin showed, is the automatic increase, generation after generation, of those inherited types which give their carriers the chance to survive and reproduce more efficiently (have higher 'darwinian fitness') than the others. This is darwinian selection, acting typically at the individual level. Now the phenotypes on which natural selection acts can very well be 'cultural' in the sense of having been acquired through the learning of behavioural rules. These may be special skills, like those of making and using weapons and tools of importance for defence or for obtaining food; recognition of dangers (poisonous or otherwise dangerous plants, or animals; the capacity to cross roads is a modern example); hygienic rules and, in general, all behaviour useful for avoiding disease. Specific skills like these may very well confer higher darwinian fitness; but the acquisition of the skill may be a largely cultural phenomenon and follow the rule of cultural inheritance.

A more complex situation arises when there is innate (genetic) variation in the capacity to acquire the skill. Then the transmission is mixed, part biological, part cultural; selection may still be of the strictly darwinian type.

There are two important complications to remember when thinking of the role of selection in cultural change. One is the frequent appearance of 'group selection' due to cultural inheritance. Certain skills if acquired by some (or many) members of a group may provide the whole group with an advantage over other groups. The advantage may be limited to situations of competition such as war but it may also express itself simply through a relative increase in numbers, without direct conflict between the groups. Thus the diffusion of medical knowledge from developed countries to developing ones has caused a rapid decrease of mortality rates in the latter without an adequate response of birth rates. The outcome is a net rate of increase up to and over 3% per year in many developing countries, with consequences that are well known. Another striking example of group selection due to cultural change, determining an earlier demographic transition which had important genetic consequences, will be explained at the end of this paper.

Some anthropologists have been especially interested in the adaptive nature of many social customs and in the selective retention of cultural traits which seem advantageous. But the observations of anthropologists also show that there exist enormous differences among cultures. Are all these traits truly adaptive, or are they 'neutral'? In other words, is the variation between groups at least to some extent the outcome of 'cultural drift'? There clearly exist cultural adaptations: the ways people are dressed or houses are built are obviously the result of careful choices, some of which are dictated by the available material, local climate and other local features. But people (and buildings) also obey fashions, which are hardly adaptive. Other customs are less clearly, if at all, adaptive: what is, for instance, the advantage of being superstitious? There are many difficulties, however, in reconstructing the reasons and events behind choices and decisions made over the centuries by our ancestors—choices and decisions which shaped our customs and beliefs and were passed on to us as part of our cultural inheritance.

Whatever the adaptive significance of social customs or of cultural traits in general, their acceptance by a group can hardly be the result of darwinian selection, or even of group selection in every case. Acceptance—or refusal—must be a matter of cultural selection, a choice made by the subject which may be voluntary or involuntary depending on the age and status of the individual, and on the amount of conditioning or coercion from various sources. Both the involuntary—or subconscious—training which determines the acceptance of a cultural trait and the voluntary component are eventually built into transmission coefficients. These, the probabilities that given individuals will show a certain cultural trait, are in fact the joint outcome of many components:

(1) Teaching drive (if the teacher plays an active part) stimulated by the gratification of accomplished duty, a gain made or a punishment avoided;

(2) The capacity of the teacher to teach, which is to some extent at least independent of drive;

(3) The level of knowledge of the teacher;

(4) The capacity to learn of the pupil;

(5) The pupil's opportunity to retain, and

(6) To make use of the teaching at the right time.

Thus cultural transmission and selection are deeply confounded and in most cases it must be difficult to measure cultural selection separately from transmission.

Much of learnt behaviour is somehow acquired through an internal drive. There is an inner urge to learn language, which is maximal at a certain age. In other cases there is a teaching urge, for instance of parents or individuals taking a parental role. These drives are not necessarily accompanied by easily perceptible gratification of the subject who does the learning or the teaching, although there is probably always some kind of gratification as a result of such actions. Moreover, there are important mechanisms which favour the occurrence of certain types of learning during activities made desirable by their being pleasant *per se*. There is general recognition that play is such a mechanism, for instance. Some kind of gratification (that is, stimulation of pleasure centres, or inhibition of some kind of pain) must be a frequent, if not general, event accompanying and determining cultural selection.

Natural selection helped to develop pleasure and pain (or fear, etc.) centres, which made it possible to direct the behaviour of an individual towards certain goals. Even the major goals of natural selection, survival and reproduction are generally achieved by coupling positive rewards or the avoidance of negative ones with actions that increase darwinian fitness. Naturally, the same reward mechanisms as are useful for innate behaviour may be employed in learnt behaviour, which is one of the basic determinants of culture. While natural selection is measured on the scale of darwinian fitness, cultural selection should be measured, if possible, in terms of gratification, level of stimulation of internal reward centres, or the like. Our present ignorance of the functions of such centres makes it impossible to make satisfactory measurements of cultural selection. Measurement of elicited emotions, as is sometimes done by researchers in marketing or advertising, may be the nearest substitute which one might use today, but the validity and usefulness of measurements of cultural selection made by such means would obviously be limited. The main reason for attempting such a difficult task is that we will remain in doubt about what exactly is meant by cultural selection until we know how to measure it. At the moment one may just note that cultural selection is mostly hidden in coefficients of cultural transmission. At least the task of measuring the latter is not very difficult. What we want to study, quantitatively if at all possible, is who teaches what to whom, and why; this is the central problem in cultural transmission, and it is surprising that it has been largely if not entirely neglected so far.

In the rest of this paper I shall present a short summary of our results in an attempt to distinguish, in one specific case, between cultural and genetic diffusion (or migration, as it could also be called). This research was begun a long time ago but came only recently to a conclusion. It belongs to a field of investigations linking human genetics, history and archaeology, to the development of which a Ciba Foundation symposium *(Medical Biology and Etruscan Origins)* was dedicated in 1958, thanks also to the encouragement and personal interest of Sir Gordon Wolstenholme.

It is well known today that the domestication of plants and animals was a rather recent event in the history of Man, having started about 10 000 years ago. In previous times food was obtained only by gathering, hunting and fishing. The change to an economy of food production started in several nuclear areas, of which the best investigated one is in the Near East, and is known as the fertile crescent. It is here that many cereals such as wheat, barley and oats, and animals such as cattle, sheep, goats and pigs, were first domesticated. From this area the new economy spread, presumably in all directions, thus reaching Europe, Africa and proceeding eastwards in Asia. Other independent centres existed elsewhere but the regions around them are far less well known archaeologically.

Archaeologists did not seem interested in the problem of whether it was the technology, or the people, or both, that spread from the centres of origin of agriculture. Archaeological data alone can hardly be informative on the problem of whether it was farming or farmers that spread. This was a non-problem to archaeologists, although some of them at least implied that culture (farming) rather than people (farmers) diffused.

In general, even major cultural changes are not accompanied by striking changes in numbers of people. Conquerors need not be many to impose their domination, their cultural life and their language. When the numerical ratio of conquerors to the conquered is small, genes will change little, but cultural changes may be very substantial. With agriculture, however, it might have been different. The new economy of food production greatly enhanced the potential population density of an area. Although the density of hunting gatherers can only go up to at most 0.1-0.2/km^2, the density of farmers can be much higher, even with primitive agricultural techniques. There are various other reasons that would favour a population explosion and hence a population spread with the transition to agriculture.

The first step that archaeologist Albert Ammerman and I decided to take in investigating this problem was to measure the rate of spread, by using the available radiocarbon datings of the earliest records of wheat consumption in the archaeologically best-known European sites. Wheat grows wild even today in the Middle East but it was not present in Europe before the arrival of farming; finding of wheat is thus a good marker of the transition to the new economy. This transition is always a sharp one, including in most of Europe the simultaneous arrival of many other features, such as pottery. The new culture is often referred to as neolithic.

Our analysis showed that there was a rather slow and regular spread of early farming to Europe from the area of origin in the Near East, starting about 9000 years ago and finishing about 5000 years ago (Ammerman & Cavalli-

Sforza, 1971, 1973). The radial rate of advance was approximately 1 km/year. As I mentioned, the process of spread might have taken place by diffusion of the farmers themselves ('demic' diffusion), or the technology might have spread to pre-existing populations of hunters-gatherers living in these areas ('cultural' diffusion).

We are interested here in the fact that the genetic consequences of the two modes of diffusion are quite different. The migration of farmers from the Near East towards Europe should spread the farmers' genes to all of Europe. By contrast, with purely cultural diffusion no direct effect on genes is expected (except for selection due to the changed way of life). Finally, if farmers spread and also mixed with pre-existing hunter-gatherers, one should observe in Europe clines, that is gradients, of gene frequencies originating in the Near East and expanding radially from there. Such a gradient is of course expected (i) only for those genes for which there were, before the spread, clear-cut differences in allele frequencies between the European hunter-gatherers and the Near East farmers; (ii) and only if concurrent or subsequent evolutionary events, selective and migrational, were not of such magnitude as to modify the gradient beyond recognition. Data from some single genes (e.g. the Rh negative allele) are in agreement with the idea of an earlier, largely Rh negative population occupying mostly the West of Europe, with which a slow wave coming from the East intermingled gradually. Mourant *et al.* (1976) had already postulated that Basques, with a high Rh^- frequency, are the remnants of a proto-European population. Other genes (especially many HLA-B alleles) show a similar NW-SE gradient across Europe (Ryder *et al.* 1978). One would like, clearly, to summarize the information from all genes for which sufficient data are available, using a suitable composite index (Menozzi *et al.* 1978). Such an index is supplied by a statistical technique known as 'principal components analysis'. This generates 'synthetic' variables, usually linear combinations of the original variables, the gene frequencies:

$$S_\alpha = \alpha_1 p_1 + \alpha_1 p_2 + \ldots + \alpha_k p_k$$

where p_1 is the gene frequency of an allele at locus 1 in a given population, p_2 the gene frequency of another allele at the same or another locus, etc., with a total of k alleles (usually belonging to many loci). The population can then be represented by the new synthetic variable S_α. Using one of various related statistical models, the coefficients α_1, $\alpha_2 \ldots \alpha_k$ can be determined so as to minimize the amount of information lost. The process can then be renewed by computing another set of coefficients $\beta_1 \ldots \beta_k$ (to replace the α's) and thus generate a new synthetic variable S_β which retrieves as much as possible of the information lost by S_α; and so on. The amount of information that can be actually summarized using a small number of variables depends on the data set. Clearly, the procedure will be most useful if much of the information can be summarized with the use of only a few variables. One can also evaluate the amount of information synthesized by each variable. With data from ABO, Rh, MNS, Le, Fy, Hp, PGM_1, HLA-A and HLA-B, 38 independent alleles (counting the number of alleles minus one for each locus) were available—17 non-HLA and 21 HLA. The HLA and non-HLA subsets gave similar results. The first principal component alone explains about 28% of the total variation, the second about 18% and the third 11%.

There is a remarkable similarity between the map of the first principal component and the archaeological map of the advance of early farming. In both cases there is a general gradient originating in the Middle East and spreading to the rest of the land. Both the genetic and the archaeological maps correspond to a series of concentric circles centred in the Middle East which peripherally tend to flatten into ellipses with the major diameter horizontal. Probably the flattening corresponds to the fact that migration by sea was a little easier than in areas of South Europe, which are mostly rugged. Thus this analysis supports the hypothesis that this spread was largely a diffusion of the farmers themselves, a *demic* spread, rather than a purely central diffusion of farming technology, and that this early event has largely conditioned the modern distribution of genes in Europe.

The second principal component shows a marked east-west gradient, with a few irregularities of uncertain meaning. The third principal component shows a gradient NE-SW, almost perpendicular to that of agriculture, with a central area in Ukraine. The gradients may represent other migrations, some of which can be perhaps tentatively identified with a variety of 'barbarian' invasions.

Plotting the first three principal components on a trichromic map, overall similarities between populations are more easily visible, and about 60% of the total information is recovered.

The areas to which expansion of Near East farmers may have spread include, in addition to Europe, North Africa (the Sahara not being a desert at the time), Arabia and East Africa, and Southwest Asia as far as the Indus valley. The archaeological information about these areas is, however, much more limited than that available for Europe. In addition, the genetic information for these areas is quite meagre. Some data on non-European populations from neighbouring regions have been used to improve the construction of map edges but we have basically confined our study to Europe and the Near East. In particular, the maps are especially poor for North Africa, where the poverty of data makes it necessary to employ rather remote populations, thus generating considerable uncertainty about North African estimates. When sufficient data become available it will be of great interest to extend the present analysis to other areas around the Near East as well as to other centres of origin of agriculture.

We have assumed that the gradient we find is due to migration, which is a sound hypothesis, since migration affects all genes equally. The other evolutionary factor of importance, which could cause gradients, is natural selection. We have built our map looking for a specific gradient corresponding to a specific migrational hypothesis and we have found it. There is no simple reason why natural selection should give precisely the observed gradient, but this hypothesis cannot be rigorously excluded. One might instead expect gradients due to the response of natural selection to climate. They should follow approximately the latitude; such gradients, however, were not found.

It is not too surprising that the diffusion pattern connected with the spread of early farming is the dominant one, even if it is relatively ancient, and perhaps older than those connected with the other patterns. The chance of observing a mass migration at the genetic level depends among other things on how massive the migration is in terms of the ratio of the numbers of colonists to the numbers of people who lived earlier in the area. It also depends on patterns of differential reproduction of coexisting groups. Of all known migrations, that connected with early farming is likely to have been among the most important in terms of the ratio of colonizers to colonized, given that the transition to agriculture creates the potential for a major increase in population density (by more than one order of magnitude). By contrast, the invasions of 'barbarians' of Central and Eastern Europe are poorly known and archaeologically difficult to trace. Those that took place in historical times did not usually involve numbers large enough to influence gene frequencies strikingly. Thus, even though it is tempting to assume that the other clines were also associated with demic spreads, it is at the moment difficult to link them unequivocally with any particular demic spread.

References

AMMERMAN, A. & CAVALLI-SFORZA, L. (1971) Measuring the rate of spread of early farming in Europe. *Man (Lond.) 6,* 674-688

AMMERMAN, A. & CAVALLI-SFORZA, L. (1973) A population model for the diffusion of early farming in Europe, in *The Explanation of Culture Change* (Renfrew, C., ed.), Duckworth, London

CAVALLI-SFORZA, L. (1971) Similarities and dissimilarities of sociocultural and biological evolution, in *Mathematics in the Archaeological and Historical Sciences* (Hodson, F. *et al.*, eds.), pp. 535-541, Edinburgh University Press, Edinburgh

CAVALLI-SFORZA, L. & FELDMAN, M. (1973*a*) Models for cultural inheritance. I. Group mean and within group variation. *Theor. Popul. Biol. 4,* 42-55

CAVALLI-SFORZA, L. & FELDMAN, M. (1973*b*) Cultural versus biological inheritance: phenotypic transmission from parent to children (a theory of the effect of parental phenotypes on children's phenotype). *Am. J. Hum. Genet. 25,* 618-637

CAVALLI-SFORZA, L. & FELDMAN, M. (1976) Toward a theory of cultural evolution. *Interdiscip. Sci. Rev. 3* (2), 99-107

CIBA FOUNDATION (1958) *Medical Biology and Etruscan Origins (Ciba Found. Symp.),* Churchill, London

FELDMAN, M. & CAVALLI-SFORZA, L. (1975) Models for cultural inheritance: a general linear model. *Ann. Hum. Biol. 2,* 215-226

FELDMAN, M. & CAVALLI-SFORZA, L. (1977) The evolution of continuous variation II. Complex transmissions and assortative mating. *Theor. Popul. Biol. 11,* 161-181

MENOZZI, P., PIAZZA, A. & CAVALLI-SFORZA, L. (1978) Synthetic maps of human gene frequencies in Europeans. *Science (Wash. D.C.) 201,* 786-792

MOURANT, A., KOPEC, A. & DOMANYIEWSKA-SOBCZAK, K. (1976) *The Distribution of the Human Blood Groups,* 2nd edn., Oxford University Press, Oxford

RYDER, L., ANDERSON, E. & SVEJGAARD, A. (1978) An HLA map of Europe. *Hum. Hered. 28,* 171

WHITE, L. (1959) *Evolution of Culture,* McGraw Publishers, New York

Discussion

Bodmer: It is puzzling that the British and the French, for example, have differences between them that look as if they were inherited. If the teacher/ leader is the main source of cultural transmission, and if what that individual transmits is actually determined by the genotype, a cultural trait that distinguishes two populations may look as though it is genetic. Newton's genius, for example, must undoubtedly have had a genetic component but the spread of Newtonian mathematics can be cultural. The interaction of culture and genotype may explain those differences between populations that look as though they are genetic but which can't be genetic because there hasn't been time for differentiation of such complex characters.

Cavalli-Sforza: Probably part of Newton's genius was genetic in origin but it must also have come from the environment in which he developed and worked. However that is an extreme example of genetic drift, since there was just one man or a few men like that in the whole history of England, and one person or a few people like him in the whole history of France. Some cultural differences between England and France may come from the fact that Newton and Descartes started different lines of thought which others followed up later. The interaction between genetics and cultural events may amplify the effects of a smallish genotypic contribution.

Bodmer: The genetic contribution of the originator may however be a major determinant. If one population has a large proportion of people who know maths because they were taught at school and another population has very few such people, and if one regards the ability to do maths as something that might have a strong genetic component, that would look as though the two populations differed with respect to their genetic ability to do maths, which is clearly nonsensical. The difference lies in the genetic attributes of the originator of the maths.

Cavalli-Sforza: Until there is proof to the contrary, I would assume that the frequency of those genes is very similar in different populations. Two factors determine differences in development. One is that the events we are speaking of are so rare that during a whole century no one with those genetic qualifications may be born in a particular country. That is why I am speaking of 'drift' in the sense that there is a chance element in the occurrence of a particular rare genotype.

The other point is that the environment is also necessary. In China there might have been people who were as mathematically inclined as Newton or Einstein but perhaps there was less interest in mathematics.

Ruddle: You imply that population size must have a profound effect on the frequency with which such innovative clusters of genes arise in a particular person, and that the size must have some effect on the rate of cultural progression.

Cavalli-Sforza: I think it must have such an effect.

Crick: But you also said the converse, namely that a country with a large population like China didn't have a Newton whereas a country like England with a relatively small population did. Both factors have to be weighed up, and I would have thought that a very small size was a handicap. A reasonable size is a necessary condition but it may not be sufficient. The whole question of why science proper didn't arise in China is an unsolved problem, as I understand it.

Peto: You want to infer from analysis of your first and second principal components that you have strong evidence for the spread of genes from the Middle East, Professor Cavalli-Sforza, but I am not sure that this is correct.

The principal components are defined as certain linear combinations of the frequencies of the genes for which you have data. You could, however, have examined the distribution across Europe of any other linear combinations of these frequencies, not necessarily a principal component. If you did so, then since you are probably correct in supposing that the major determinant of gene frequencies is recent migration rather than recent selection, you would probably observe a pattern with respect to the chosen linear combination which made sense in terms of past migrations. For example, the north of

Scotland and Northern Ireland would probably look rather similar to Scandinavia for almost any such combination, unless one deliberately selected a combination that systematically avoided this. The frequencies of different genes should show a strong association with each other because the effects of migration are common to many genes, and the magnitude of the first few principal components confirms that this is indeed the case. However, I am not sure that this implies that the migratory pattern suggested by the first principal component is any more fundamental than that suggested by some other linear combination, and in fact many linear combinations don't exhibit the pattern you want. One very strong rule of statistics is that nothing very sensible comes out of principal components analysis, though this study could be the exception to that rule.

Cavalli-Sforza: I don't agree with that rule. Principal components were not used on a very large scale over a long time until recently, when computer programs became available. Now they are used on a larger scale, although the interpretation of multivariate analysis is still in its infancy. From an evolutionary point of view, I would expect the form that gene maps take to be due basically either to migrational history or to local natural selection. I am pleased with these results because I was expecting gradients starting from a certain point in the map and I find the first principal component gives them exactly as I want them. If the observed gradients were due to natural selection, one type that I would expect to be important is that due to climatic differences, which would go with latitude. But none of the major principal components showed an important correlation with latitude. I take this as evidence that natural selection due to climate was not important for the set of genes considered. But there are good reasons for thinking that agriculture determined a major demographic transition. From hunting-gathering to agriculture there is a factor of 10, 100 or even 1000-fold population increase. Because of that one expects a population explosion, and if there is a population explosion there must be territorial expansion. There is of course no easy way of proving any historical facts rigorously. All I am saying is that all the proofs we look at convince us that this is the right interpretation.

Peto: But the third principal component didn't show this pattern. I am not saying that your data are contradictory but that they really don't demonstrate this spread in any way. We know that there must have been migration to northern Europe because 12 000 years ago the whole area was covered in an ice-sheet, and there definitely weren't any people there then.

Cavalli-Sforza: There were people in England later, and there were people in France at that time.

Peto: Yes, but there must have been a northward migration when the ice-sheets disappeared.

Cavalli-Sforza: There certainly were migrations of all sorts. What I claim is that extensive migrational waves are likely to be fairly accurately portrayed by principal components. Because of the orthogonality of principal components each one may represent a different migrational history. I am not surprised that we find that the major migration as indicated by the first principal component is that due to agriculture. Agriculture is the one major demographic change which took place in this area in the last 10 000 or 100 000 years and it must have changed the numbers of people enormously. We have direct evidence that this is true. And as soon as there are major changes in numbers one expects major migrations.

Bodmer: There are two important points on this. One is the parallelism of results obtained from different genes. That is what is brought out by the principal components analysis and which would not be expected with natural selection. Different genes give the same information, which one would not expect from natural selection alone. The other point relates to migration. This study is not about migration into empty space but about genetic migration, in which there is admixture with a pre-existing population. One would not get this effect without a pre-existing population, since the people migrating would have the same gene frequency as the people who stayed in their place of origin. This study represents an admixture of the migrating people gradually diluted with the pre-existing people.

Crick: In future there will probably be markers with very little selective advantage in the intervening sequences that we will be talking about. It will be surprising if some are not polymorphic. To underline your point, rather than taking mixtures of genes you will probably use pure ones in the future, looking at DNA intervening sequences. When it will become practical to take samples from people I don't know.

Cavalli-Sforza: The third bases of the DNA triplets may also be informative.

Bodmer: The accuracy of the picture you showed depends on the differences in gene frequency between the original population and the migrating population. So the difficulty here is that the markers that are the most informative are the ones where there was originally the biggest difference in gene frequency, and it is a matter of luck how many of those one finds.

Harry Harris: You started off by contrasting migration of ideas with migration of people, and you come out in favour of migration of people.

Cavalli-Sforza: That doesn't exclude the migration of ideas.

Harry Harris: How far would the population densities over the whole area at the beginning, and while it was all happening, affect the issue?

Cavalli-Sforza: There were certainly people in the area. If there was no admixture, which is not likely to be the case, one would have to assume they were somehow killed by the invaders.

Harry Harris: If there were a cline in population density, how would this affect the analysis?

Cavalli-Sforza: We have done some simulations and there are no major problems in that respect. One can make even more definite assumptions than we have made. For instance, the hunter-gatherers of Europe, the so-called mesolithics, may have been 100% Rh negative while the invaders were all Rh positive. The invasion was not a real invasion—there was no war. The people who came in were in very small groups and they probably also lived in different parts of the environment. Hunter-gatherers preferred certain types of soil and forest and the farmers preferred other types so there was not even much contact to begin with. The farmers at first had low population densities—perhaps of the order of five inhabitants/km^2 after saturation in central Europe about 5000–6000 years ago. Hunter-gatherers might have had a density about 10 times lower but the farmers came in little by little so there was some progressive mixture all the way.

Pickering: Do you think that religion is a good example of the way in which culture can mimic genetics? There is even the phenomenon of Mendelian dominance because when a Protestant marries a Catholic the Pope insists that the children should be brought up as Catholics. And the Jews are a splendid example of the descent of a culture rather than the descent of genes.

Cavalli-Sforza: I think you are right on religion. We find it is a good candidate for this kind of simulation of, if not Mendelism, at least biological inheritance by culture. One other set of cultural traits in which vertical transmission from parent to child seems to be very strong is political opinion.

Pickering: That is a variant of religion!

Henry Harris: Walter Bodmer sent me a reprint the other day about HLA polymorphism in people in Israel (Bonné-Tamir *et al.* 1978). It seems fairly clear that, for the Jews, it is not only the descent of culture, but also the descent of a great packet of genes, although, of course, there are admixtures.

Lachmann: One difference between cultural and genetic evolution may be that the gains of cultural evolution can be more easily lost. The Spaniards brought about the destruction of Meso-American culture by killing a relatively small number of people. If there were mathematicians as able as Newton among the Mayas we wouldn't know about them.

Brenner: The Spaniards killed a large number of people by another mechanism. They brought with them viruses for measles and smallpox.

Bodmer: Were genetic differences responsible for those deaths?

Brenner: No, immunological differences; the indigenous population had never been exposed.

Bodmer: That may be genetic susceptibility.

Cavalli-Sforza: The rates of cultural evolution can be very fast not only when there are catastrophes, but also wherever there is a teacher or leader type of cultural transmission. Also, in older times, hierarchical societies were almost as efficient at spreading information and orders as the mass media are today. Under these conditions there could be very fast cultural changes, but culture is not necessarily always very fast in changing: it can also be extremely conservative. Vertical transmission from parent to child is almost indistinguishable from biological transmission. Very good geneticists have been misled into considering as Mendelian the inheritance pattern of a disease due to a virus, kuru, found in a New Guinea tribe, which tends to be transmitted from parent to child or close relative because of cultural customs.

Brenner: In the special case of human evolution, one has to consider changes which are different from the classical genetic changes encountered in other organisms. First, it is interesting to ask in what sense cultural evolution simulates biological evolution and whether we can use the same concepts to look at the mechanisms without necessarily thinking of them as analogous to each other.

The other question is how to assign the factors in a particular instance to biological or cultural mechanisms. Lurking behind this is another issue which Walter Bodmer just touched on: granted that the superstructure of cultural change exists, in the end, is it all conditioned by the genetic constitution of the population? Is there a gene for farming? Did agriculture in the Middle East start because some people were mutants showing farming behaviour?

Cavalli-Sforza: There was nothing in what I said that requires or even indicates genes for farming.

Brenner: You didn't bring that out fully. Let me ask you directly: do you believe that there were mutants for farming?

Cavalli-Sforza: No, but it depends what you call 'mutants for farming'. If this involves, for instance, a greater interest in planning ahead for many years, which is what farming demands compared to hunting-gathering, then maybe the people for whom it is easier to think ahead for many years would go more easily into farming than into hunting-gathering. Hunting-gathering as an activity is much lighter and more pleasant and doesn't require a great deal of long-term planning.

Bodmer: So you are arguing that there was a genetic difference?

Cavalli-Sforza: There could always be differences of those kinds. They are extremely difficult to prove. It is perfectly possible that the introduction of agriculture was entirely a matter of new inventions and everybody had an absolutely equal probability of accepting it or rejecting it.

Crick: I am trying to think of examples where things did *not* happen, which is sometimes helpful. The most striking example that I can think of at the moment is where cultural transmission was stopped by the Japanese. Later, once the cultural barrier was removed, western culture flowed very rapidly into Japan. Internally the culture, including the genetics of the people, was capable of absorbing these external influences rather easily. Is there a case where transmission was easy, in the sense that the culture was available, but where it never actually 'took'?

Cavalli-Sforza: It is very difficult in any case to distinguish whether the make-up that we speak of is genetic or environmental. One of my reasons for being interested in the problem of spread of agriculture was that I have been working with pygmies as an example of hunter-gatherers, one of the few in existence. One gets the impression that pygmies really resist the transition to agriculture as strongly as they can. But on the whole they probably resist agriculture because they are trained to hunt, and live in a hunting society which is very different from a farming society. The transition to agriculture is a complete change from all the values and skills that they have learnt.

Crick: That is a good example of what I was looking for, namely where the activity is available to them and some of them even do it temporarily, but it doesn't take. Whether that is genetic or cultural is a very good question.

Bodmer: In that case the pygmy farmers can leave the pygmy community.

Cavalli-Sforza: They can, but very few do.

Crick: The pattern is usually to mix and then to go back.

Siniscalco: Have you given any thought to why people should abandon the Garden of Eden to come to these northern shores? Might disease have forced them to travel to places with less agreeable kinds of ecological make-up?

Cavalli-Sforza: In the Middle East the population stayed put. Where wheat was available and used as a wild plant people may have thought it was more convenient to have a crop exactly where they wanted it and not just where it grew wild. As they had stone houses and needed storage places it was easier to shift the field near the house and the storage place than the other way round. It might have been as simple as that but other suggestions have been made. There was no major change of habitat that we know of. The climatic changes that happened in Europe between 15 000 and 10 000 years ago also have threatened the hunters and their economy and kept their population down, contributing motives to effect the shift to agriculture.

Brenner: Continuing the analogy between culture and biology, I have often thought it would be worth investigating what might correspond in culture to

DNA. DNA has two functions. It has the property of replication but it also contains a description of the organism in a condensed form. The origin and propagation of formal legal structures may really be our attempt to mimic the role of DNA as a source of codified information.

Jacob: Or the Bible, for example.

Brenner: Exactly. The Bible is a mixture of everything, but parts of it are the formal codification. Then one would be interested in how that is read and interpreted into the phenotype of society at any given time. This may be nothing more than an amusing analogy.

Reference

Bonné-Tamir, B., Bodmer, J. G., Bodmer, W. F., Pickbourne, P., Brautbar, C., Gazit, E., Nevo, S. & Zamir, R. (1978) HLA polymorphism in Israel. *Tissue Antigens 11,* 235-250

Genes and non-coding DNA sequences

P. M. B. WALKER

MRC Mammalian Genome Unit, Department of Zoology, University of Edinburgh

Abstract Many new categories of DNA sequence organization in higher organisms have been discovered in the last 20 years. Most of these are not found in bacteria and each in turn is hailed as a likely candidate for those regulatory elements which are assumed to distinguish higher from lower organisms. This article stresses the similarities between their sequence organizations and the continuing lack of convincing function for any of them. It suggests that these sequences often appear after chance recombinational events and their presence then facilitates further change. Within the constraints of proper function the genome is clearly able to find many organizational solutions of apparently equal advantage to the organism.

Finally, an example of rapid genome reorganization resulting from selection for resistance to the drug methotrexate is considered.

Probably the greatest contribution which molecular biology will ever make to medicine is the understanding that the form of all organisms, including man, is ultimately controlled by the permutations and combinations of the four bases in DNA. But despite our present considerable knowledge of how base sequence specifies the structure and conformation of proteins, we know virtually nothing of how the temporal and spatial regulation of genes, essential for differentiation, is effected.

What is interesting to me in this situation is that during the last 15 years it has been possible to study the organization of DNA in higher organisms in greater and greater detail, and each new discovery has revealed sequence organizations, some exceedingly bizarre and entirely absent from bacteria, but, as far as I know, no convincing function has been demonstrated for any of them; certainly none that throws any light on the control of the expression of genes. Nevertheless, most people remain convinced that the functional and structural peculiarities of the genome of higher organisms must be somehow related. This conviction is reinforced by the finding first that there is in higher organisms a very substantial proportion of non-coding sequences, that is DNA which does not specify proteins, and second, that these novel organizations almost always involve non-coding DNA. This is precisely where we might expect to find those sequences concerned with regulation.

The most striking feature of these sequence organizations and the one which has made their functional analysis immeasurably more difficult has been their extreme variability even between closely related species. That the genome is rather a flexible organization has, of course, been known for a long time. The woolly rhinoceros has 82 very small chromosomes, while the Indian red muntjak has six or seven large ones. A similar variety in DNA content has been recorded in more closely related species, but it is even more striking when we come to consider in detail some of the sequence organizations which newer techniques of analysis have revealed.

In the next section I want to discuss some kinds of sequence in more detail, not only to illustrate this variability but also because they share additional properties which, to my mind, make it likely that they are not directly related to gene regulation but that many of them are the result of evolutionary accidents, and that a singular property of the eukaryote genome is its ability to accommmodate such accidents.

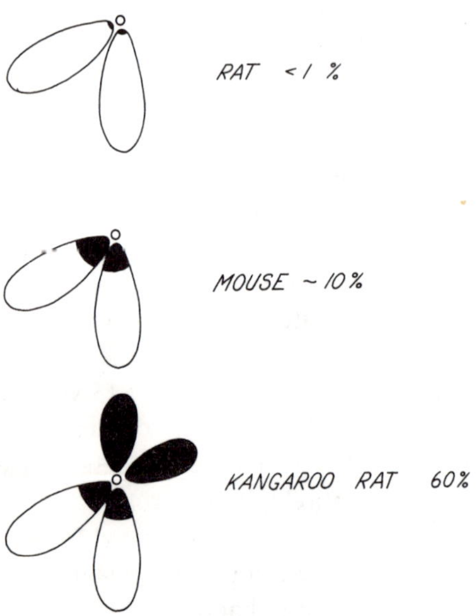

FIG. 1. Content of satellite DNA and heterochromatic regions in mammals.

Chromosomes often have characteristically stainable regions which remain condensed during interphase, the constitutive heterochromatin. Fig. 1 shows diagrams of chromosomes of three species of rodent and it will be seen that the amount of heterochromatin varies from vanishingly small in the common rat to massively large in the kangaroo rat *(Dipodomys ordii)*. For general reviews of this and related topics see, for example, Walker (1971) and Bostock & Sumner (1978). This condensed material is almost entirely composed of special kinds of DNA, containing many millions of repeats of rather short sequences, satellite DNA. Some of these sequences are illustrated in Fig. 2. Again, the sequence varies; from almost pure AT in the crab to a complex long sequence in the wood mouse *(Apodemus agrarius)*. We now know something of the steps by which these sequences multiply. In the mouse, for example, a short sequence has been multiplied, base changes accumulate in the course of time, and a section of this sequence is further amplified, probably by unequal crossing-over. A number of multiplication steps followed by quiescent periods accounts for the present-day patterns of many satellite DNAs.

DNA with rather similar properties occurs elsewhere in the genome; in particular it has been well studied in the spacers which separate repetitive genes such as those that specify the various ribosomal RNA molecules and the five histones (Fig. 3). Again, we see the same pattern of variability. The size of the spacer varies nine-fold in the 5S genes of two *Xenopus* species which interbreed in the laboratory. The spacers themselves often have internal repetition, strikingly reminiscent of satellite DNA, and again the short repeat varies in length, but in the histone genes there is little evidence of internal repeats in the spacer.

1. Crab	-----ATATATAT-----(3% C+G)
2. *Drosophila*	Several e.g. ACAAACT, ATAAACT, ACAAATT } >90% pure Some more complex
3. House mice	Complex, ---GAAAAATGA---common but very variable Major long repeat at 240 bp.
4. Wood mice	Reassociates slowly like middle repetitive DNA

FIG. 2. Representative satellite sequences.

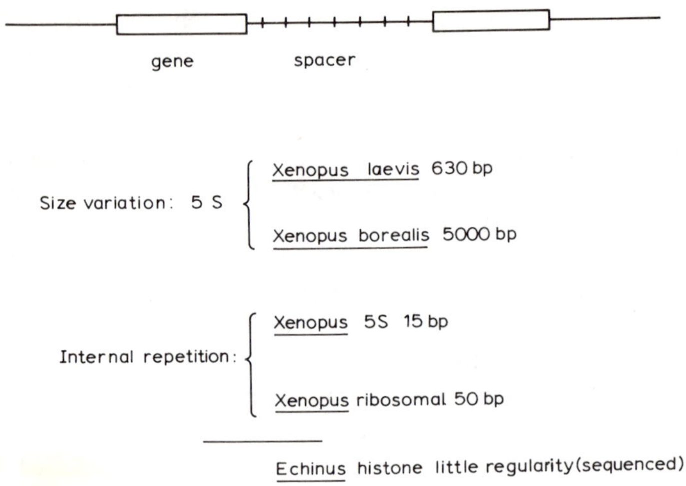

FIG. 3. Spacers between repetitive genes.

To bring these examples more up-to-date, we can consider the remarkable insert sequences recently found in the middle of gene sequences but not in the messenger or processed RNA (Fig. 4). For the *Drosophila* ribosomal example (Glover & Hogness 1977), Glover *et al.* (1978) and R. Appels (unpublished work) have shown that a recombinant DNA clone containing only the insert hybridizes to a number of sites including several salivary bands as well as the chromocentre. Such clones also hybridize to non-ribosomal DNA molecules, including, as Dawid & Wellauer (1978) have shown, a distinct density fraction. On the other hand, some of those discovered in the haemoglobin and ovalbumin genes appear to be unique to these genes, at least at the criteria of stringency used in these experiments.

We now have to consider a whole range of much less well-defined DNA sequences, first discovered by Britten & Kohne (1967), which undoubtedly occur in all eukaryotes and are apparently scattered throughout the genome. These are the middle-repetitive sequences, so called in an attempt to distinguish them from the satellite DNAs on the one hand, and the unique sequences which, by the same methods of analysis, occur only once or a few times in the haploid genome. They are poorly characterized because they are easily analysed only by studies of their reassociation kinetics, that is the rate at which they hybridize together after they have been denatured. While this is a relatively precise and very powerful method when one component such as a well-characterized radioactive probe is used, there are considerable uncertainties when attempts are made to use a mixture of sequences with varying

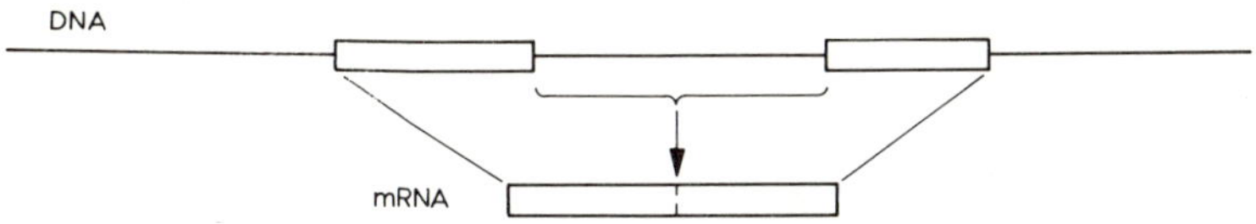

1. *Drosophila* rDNA : insert hybridizes
   - a. other non-ribosomal sequences
   - b. other salivary bands
   - c. chromocentre
2. Haemoglobin and ovalbumin DNA
   some inserts unique to these sequences

FIG. 4. Inserted sequences. rDNA: ribosomal DNA.

degrees of length and mismatch. Nevertheless, some generalizations are possible and these are summarized in Fig. 5. Much the most usual pattern is for repetitive sequences around 300 base pairs long to be interspersed between longer sequences around 1000. In *Drosophila,* as Manning *et al.* (1975) have shown with comparable techniques, the repetitive sequences behave as if they are 5000 bases long and are interspersed between much longer sequences. In addition to these sequences, there are a number of others, less well-characterized, which are listed in Fig. 6. They include inverted repetitious or fold-back DNA, discovered because of the ability of a proportion of DNA made single-stranded to form stable looped structures. They probably do not occur in nature, although a single-stranded DNA transcript may form such loops and hairpins. They appear to have a certain regularity in size but this

1. Characterized only by hybridization kinetics/EM

2. a) General pattern : 50% or more of genome

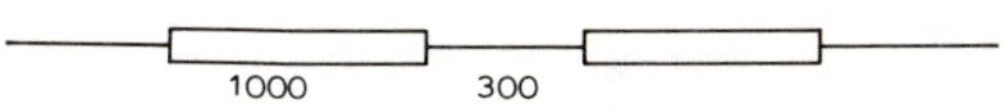

'Average pattern'

b.) *Drosophila* :

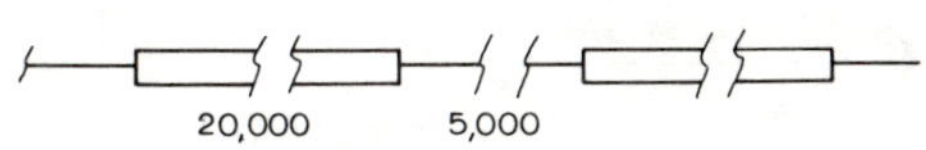

FIG. 5. Middle repetitive DNA.

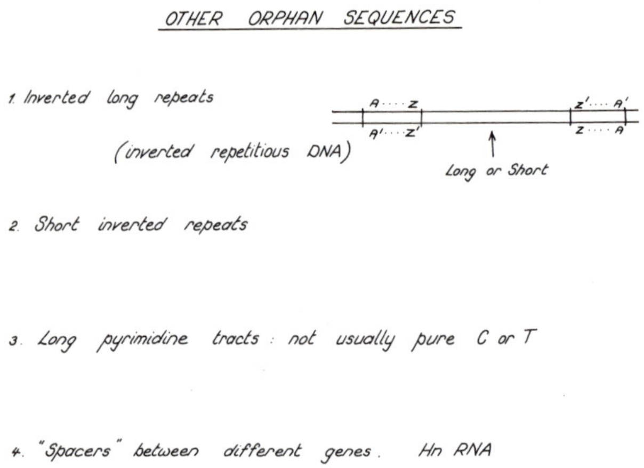

FIG. 6. Additional orphan sequences.

may simply mean that middle-repetitive DNA can become inverted. Pyrimidine-rich tracts of up to about 80 bases can occur on one strand and there is now considerable evidence for non-coding regions at both ends of messenger RNA in the larger RNA molecules which occur in the nucleus.

What nearly all these sequence arrangements share is the property of repetition and it is very difficult to assign a unique distinguishing feature to any of them (Fig. 7). As we have seen, each kind of sequence can have a range of properties, none of which appear to be uniquely characteristic of any of them. It has been argued that satellite sequences are in a different, even an autonomous, class because they seem to be confined to the condensed heterochromatic regions of the chromosome. It is true that wherever large blocks of these kinds of sequences occur, either near the centromere or at the ends of the

General properties

1. Extreme variability

2. No clear distinction between the sequence organization of any non-coding DNA

Attribute, possibly transient

No convincing function

FIG. 7. General properties of non-coding DNA.

chromosome, or even in the middle of an otherwise euchromatic arm, that part of the chromosome is condensed. Because of the method of detection it is very difficult to show that small amounts of similar sequences with short internal repeats do not occur elsewhere. Indeed, such sequences are found in spacers, but they do not appear to be related to the commoner kinds of simple-sequence DNA. On present evidence I would suggest that the heterochromatic regions are the main sites of sequence reduplication, that some of this material can be lost directly from these regions, but that there is also a flow of sequences during evolution to other regions of the chromosomes. This flow is probably mediated by inversions or translocations involving the heterochromatin, and thereafter base substitutions will render the sequence unrecognizable and at the same time destroy whatever property causes constitutive chromatin condensation.

How then can we explain the presence of substantial fractions of DNA with very variable sequence properties, much of which shows evidence of duplications which have occurred both earlier and later in the evolutionary history of the species? Firstly, do we need a sophisticated explanation? By this I mean a hypothesis which explains the special organization of the eukaryote genome in terms of the capacity of this genome to control the orderly differentiation of a multicellular organism. Most people still instinctively feel that it should be possible, but there are alternatives.

An alternative unsophisticated explanation could be one in which chromosomes simply need to have a minimum size in order to function properly. In one instance there seems to be good evidence for this (Cooke 1976); the human Y chromosome is small and yet it contains a high proportion, over 50%, of satellite DNA and there is no evidence for more than very few genetic markers. At the other end of the evolutionary scale there is the ciliate protozoon, *Oxytricha,* studied by Prescott and by Ammerman (see, for example, Prescott & Murti 1974). In this complex unicellular organism there is no mitosis in the macronucleus, where transcription occurs, and this large structure contains many copies of only 5% of the DNA sequences. The discarded 95%, made up of repetitive sequences different from the 5%, is required for the chromosomes of the micronucleus, where alone meiosis and mitosis occur. This suggests again that it is chromosomal mechanics, including the ability to recombine, which demands the excess DNA.

There have been, of course, specific functions suggested in terms of chromosome mechanics. One of the more persuasive suggestions is that put forward by Peacocke and his colleagues in Canberra (Peacocke *et al.* 1977) who have tended to stress the sequence homogeneity of certain satellites. They have found that each chromosome in *Drosophila melanogaster* has a specific

FIG. 8. The eight diploid chromosomes of *Drosophila melanogaster* each have a different pattern of satellite sequence near their centromeres, e.g.

leading to Peacocke's hypothesis that the pattern mediates meiotic pairing. But laboratory-made stocks by Yamamoto and Miklos with different patterns of heterochromatin

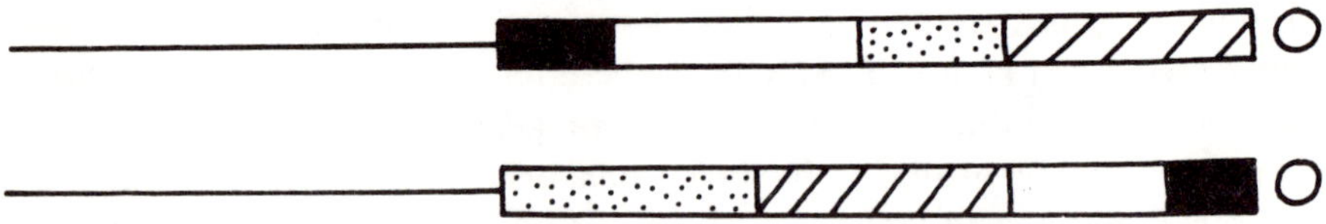

pair according to their euchromatic segment.

pattern of satellite sequences in terms of the amount and order of the several different sequences (Fig. 8). They have therefore suggested that the pattern of satellite DNA in the centric heterochromatin controls the pairing of homologues. In *Drosophila* the number of chromosomes is small, so the patterns may be different by chance. Moreover, the mechanism may be more complicated, as is suggested by Yamamoto & Miklos (1977), who also happen to work in Canberra. They have constructed special chromosomes in which the euchromatin is normal but the heterochromatin regions have been derived from the other chromosomes of the set. These man-made chromosomes pair according to their euchromatin and not according to the pattern of satellite sequences.

Miklos then goes on to suggest that the function of heterochromatin is to suppress recombination in adjacent regions of the chromosome. He and others have certainly demonstrated that this occurs in many organisms, and one adaptive advantage would be to preserve a functional set of genes (a supergene) close together on a chromosome (Fig. 9). Like so many other

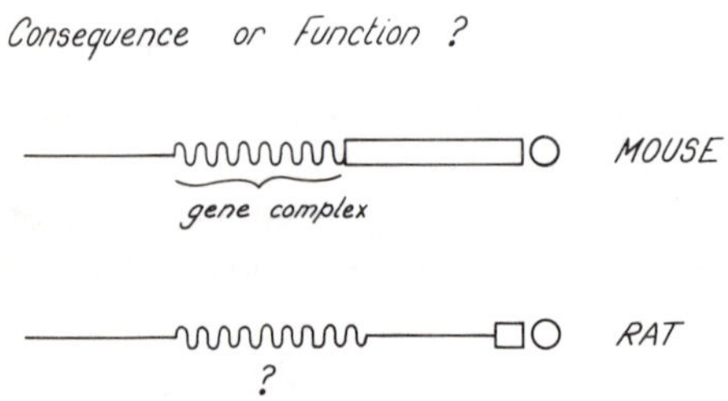

FIG. 9. Recombination suppression: consequence or function?

functions for highly repetitive DNA, this one seems to fall down because closely related species have such varying amounts of this recombination-suppressing material.

Nevertheless, all the evidence suggests first that repetitious DNA has occurred as the result of events involving recombination, such as the unequal crossing-over mechanism proposed by Smith (1976) and, second, that the presence of repeated sequences in the genome is likely to facilitate recombination. These mysterious sequences, widespread as they are in the genome, may just be the leaf-scars of earlier chromosome rearrangements, which in turn will facilitate and even channel future changes, some of which may have adaptive significance.

There is, however, one attractive explanation for extra DNA in terms of the regulation of gene expression, which follows from the recent discoveries on leader sequences in adenovirus (Berget *et al.* 1977; Chow *et al.* 1977). Here, a long region of DNA containing leader sequences, an intervening region and the several coding regions are apparently transcribed into large precursor RNA molecules. This is then processed to give the much shorter mRNA in

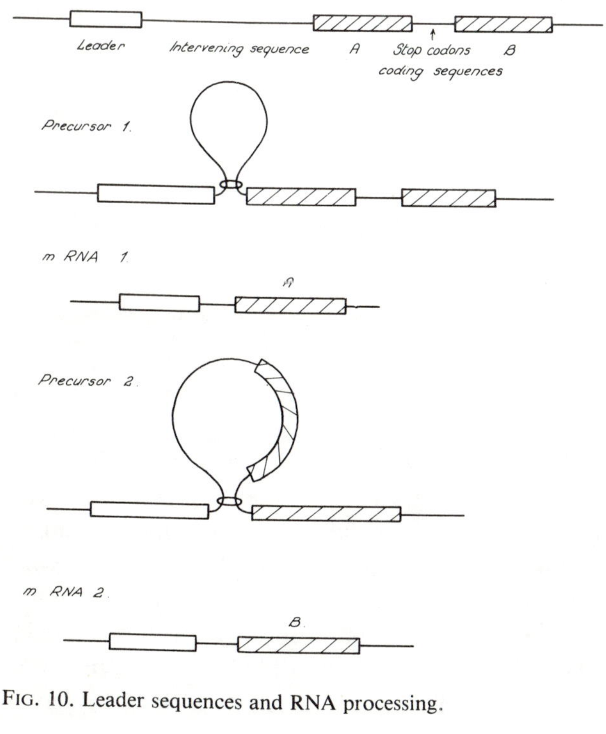

FIG. 10. Leader sequences and RNA processing.

which a leader is covalently linked to one of the coding sequences. It is quite possible that the precursor RNA first loops out and is transiently held by the stem of the loop while the loop is excised (Fig. 10). Because different coding sequences can be joined to different leaders, alternative looping patterns will be needed. The relative frequency of each pattern could be economically governed by the stabilities of the structures which make the stems of the loops stay together for longer or shorter times. This kind of model has a very desirable property apart from accounting for heterogeneous nuclear RNA (HnRNA): the length and sequence of the loops need not be stringently conserved, since longer loops can be compensated by greater stability of the stem-holding sequences. The orphan DNA which I have been discussing could then be the non-specific part of a mechanism which allows the genome to overcome the constraints of linear arrangement and reassort its elements transiently during development, or more permanently during evolution.

Finally, I want to describe some experiments, which appropriately stem from a suggestion made by the chairman of this symposium and look as if they may provide new evidence on repetitive DNA. They concern the resistance of cell lines to the anti-tumour drug, methotrexate, which is often accompanied by the overproduction of the enzyme dihydrofolate reductase. The suggestion was that the kinetics of induction and loss of resistance after withdrawal of the drug was that which might be expected if gene reduplication was the method that these cells adopted to achieve such high levels of resistance. Schimke and his colleagues have recently shown this to be true by reassociation analysis using a radioactive probe derived from the appropriate messenger RNA. They find a 200-fold increase in the number of gene sequences (Alt *et al.* 1978). Dr Veronica van Heyningen and I in the early stages, and Dr Bostock more recently, have selected a highly resistant line derived from mouse PG19/T3 cells (Fig. 11). This line now grows in 10^{-4} M-methotrexate and has an enzyme activity 700 times that of the parent line.

The original karyotype has 72 mouse chromosomes, of which six are metacentrics. It is therefore nearly tetraploid. The resistant line has three to five very large chromosomes (Figs. 11,12). From the G-banding pattern round the centromeres it is apparent that these chromosomes are derived from duplications of two original chromosomes. The resistant chromosomes are all metacentric, but loss of parts of an arm of an original isochromosome is common. It is probable that these giant chromosomes are derived from the original metacentrics. The additional regions are often easily identifiable as homogeneously stained regions, as shown less dramatically by Biedler & Spengler (1976) for methotrexate-resistant Chinese hamster cell lines. Simple measurements of length indicate that about 20% of the total resistant DNA content is

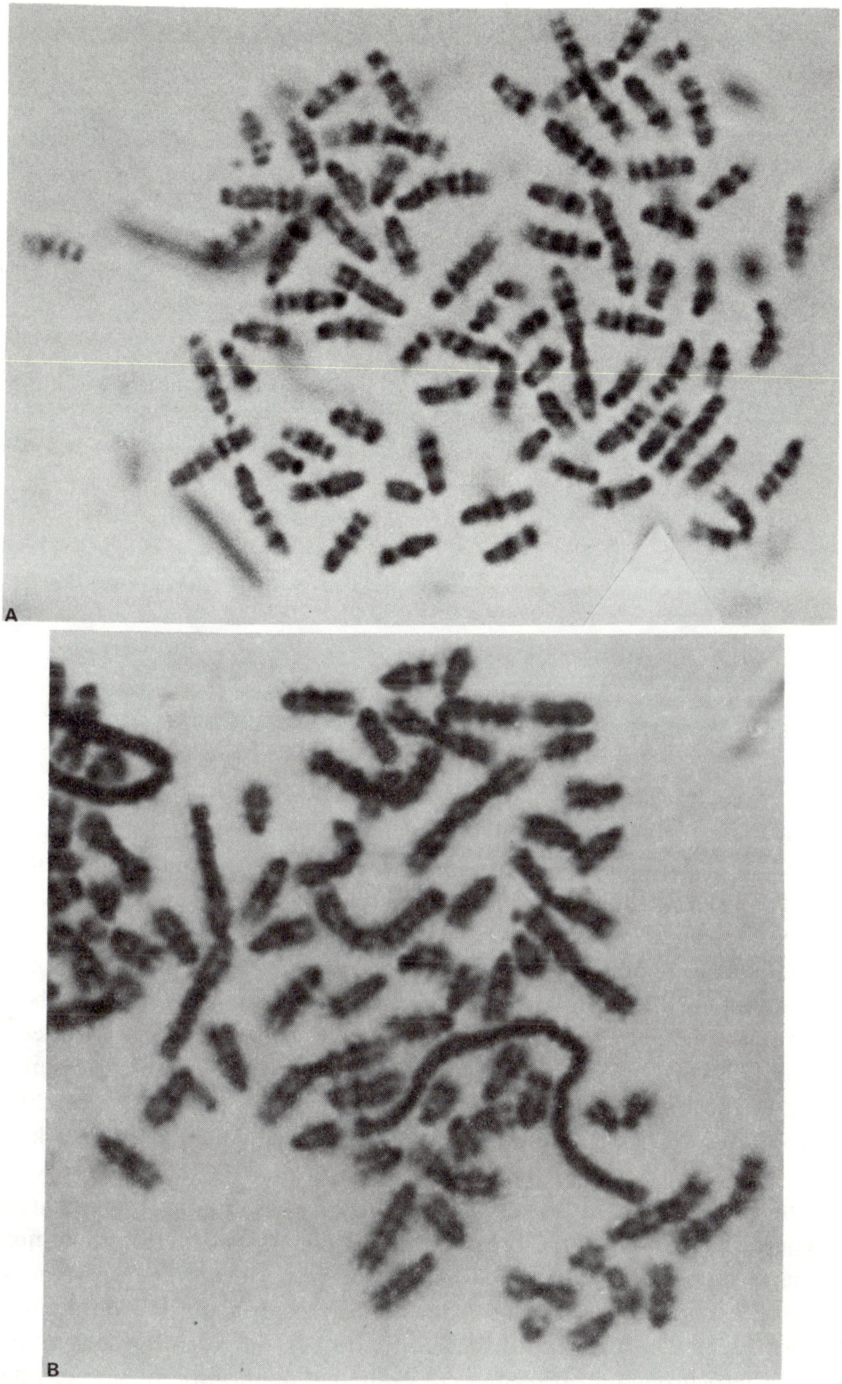

FIG. 11. Comparison of (A) non-resistant and (B) resistant cell lines stained with Giemsa.

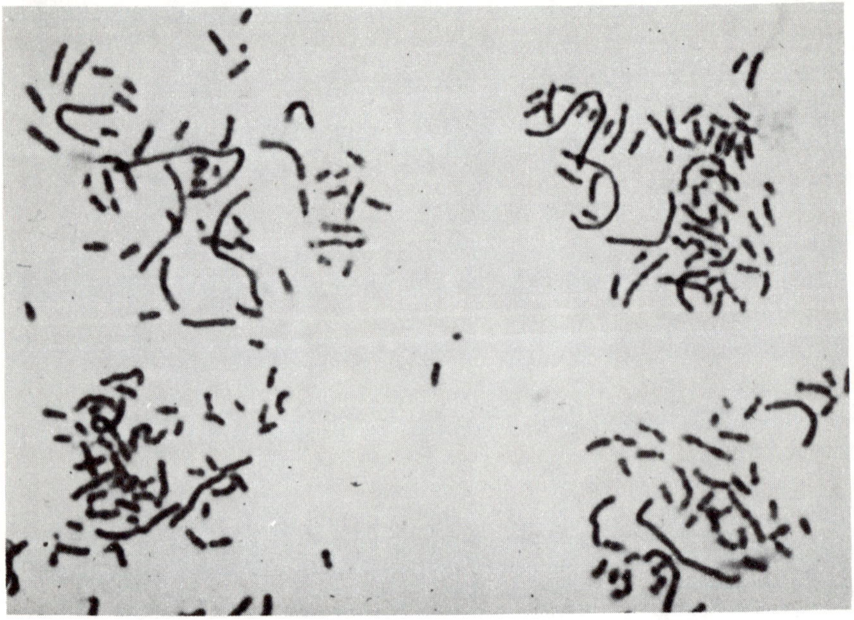

FIG. 12. Four contiguous metaphase plates to show range of size of large chromosomes.

now in these regions, corresponding to over 10^9 base pairs. There is, of course, no suggestion that there are now 10^6 copies of the dihydrofolate reductase gene, which is probably only reduplicated a few hundred times. So what has happened? Either by chance the initial reduplication which permitted subsequent amplification by unequal crossing-over was a large block of genes round the dihydrofolate reductase gene, or alternatively there has been massive concomitant amplification of some other sequence. There is evidence (Baimai 1977) that in Hawaiian *Drosophila* inversions with one break-point in the heterochromatin cause the appearance of new and enlarged heterochromatic regions. That is, chromosome breakages and reunions of the kind which must have occurred during gene amplification in the methotrexate-resistant cell line have, in *Drosophila,* resulted in the additional amplification of sequences involved in the break. Production of new satellite-containing regions is supported by Fig. 13, where it will be seen that the new regions fluoresce as brightly as the centromeric heterochromatin. The stain is Hoechst 33258 which is said to be specific for AT-rich regions, such as the mouse heterochromatin. It seems very likely that duplication of the gene has

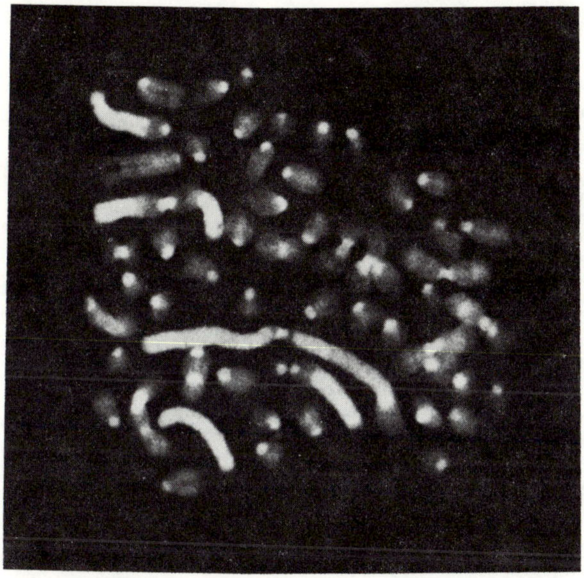

FIG. 13. Resistant cell line stained with Hoechst 33258.

induced the multiplication of some other AT-rich sequence. It does not seem to be the same as mouse satellite, but work is still in progress on this. Such a finding would lend considerable weight to the view that chromosome rearrangements may themselves engender the multiplication of sequences in the genome just as they probably facilitate the occurrence of these arrangements.

In conclusion, I would argue that nearly all non-coding DNA has a rather passive role, specifying, for example, the RNA which is excised during processing, and reflecting chromosome rearrangements in the past. I would caution against trying to fit every new eukaryote sequence anomaly into some regulatory hypothesis. Challenging as they are, inserts and detached leaders can be thought of as simply additions to an already long list of anomalous sequence arrangements in higher organisms. No doubt just as there are short conserved regions in the non-coding parts of messenger RNA molecules, so there may be other sequences elsewhere, which may have some regulatory function. As it becomes increasingly possible to study individual eukaryotic sequences rather than general properties of whole genomes, it will be possible to discover which, if any, of these DNA fractions are really concerned with the control of the expression of genes.

ACKNOWLEDGEMENTS

I would like to thank Dr C. Bostock for the use of the data on the methotrexate-resistant cell line, and for the photographs of these cells.

References

ALT, F. W., KELLEMS, R. E., BERTINO, J. R. & SCHIMKE, R. T. (1978) Selective multiplication of dihydrofolate reductase genes in methotrexate-resistant variants of cultured murine cells. *J. Biol. Chem. 253,* 1357-1570

BAIMAI, V. (1977) Chromosomal polymorphisms of constitutive heterochromatin and inversions in *Drosophila. Genetics 83,* 85-93

BERGET, S. M., MOORE, C. M. & SHARP, P. A. (1977) Spliced segments at the 5′ terminus of adenovirus 2 late mRNA. *Proc. Natl. Acad. Sci. U.S.A. 74,* 3171-3175

BIEDLER, J. L. & SPENGLER, B. A. (1976) Metaphase chromosome anomaly: association with drug resistance and cell-specific products. *Science (Wash. D.C.) 191,* 185-187

BOSTOCK, C. J. & SUMNER, A. T. (1978) *The Eukaryotic Chromosome,* pp. 13-70, North-Holland, Amsterdam

BRITTEN, R. J. & KOHNE, D. E. (1967) Nucleotide sequence repetition in DNA. *Carnegie Inst. Wash. Year Book 65,* 78

CHOW, L. T., GELINAS, R. E., BROKER, T. R. & ROBERTS, R. J. (1977) An amazing sequence arrangement at the 5′ ends of adenovirus 2 messenger RNA. *Cell 12,* 1-8

COOKE, H. J. (1976) A repeated sequence specific to human males. *Nature (Lond.) 262,* 182-186

DAWID, I. B. & WELLAUER, P. K. (1978) Ribosomal DNA and related sequences in *Drosophila melanogaster. Cold Spring Harb. Symp. Quant. Biol. 42,*1185-1194

GLOVER, D. M. & HOGNESS, D. S. (1977) A novel arrangement of the 18S and 28S sequences in a repeating unit of *Drosophila melanogaster* rDNA. *Cell 10,* 167-176

GLOVER, D. M., KIDD, S. J., ROIKA, H. T., JORDAN, B. R., ENDOW, S. & APPELS, R. (1978) Interrupter sequences that are widely distributed in the *Drosophila* genome. *Biochem. Soc. Trans. 6,* 692-696

MANNING, J. E., SCHMID, C. W. & DAVIDSON, N. (1975) Interspersion of repetitive and non-repetitive DNA sequences in the *Drosophila melanogaster* genome. *Cell 4,* 141-157

PEACOCKE, W. J., APPELS, R., DUNSMUIR, P., LOKE, A. R. & GERLACH, W. L. (1977) Highly repeated DNA sequences: chromosomal localization and evolutionary conservation, in *International Cell Biology (*Brinkley, B. R. & Porter, K. R., eds), pp. 494-506, Rockefeller University Press, New York

PRESCOTT, D. M. & MURTI, K. G. (1974) Chromosome structure in ciliated protozoans. *Cold Spring Harb. Symp. Quant. Biol. 38,* 609-618.

SMITH, G. P. (1976) Evolution of repeated DNA sequences by unequal cross-over. *Science (Wash. D.C.) 191,* 528-535

WALKER, P. M. B. (1971) Repetitive DNA in higher organisms, in *Progress in Biophysics and Molecular Biology,* vol. 23 (Butler, J. A. V. & Noble, D., eds.), pp. 145-190, Pergamon Press, Oxford

YAMAMOTO, M. & MIKLOS, G. L. G. (1977) Genetic dissection of heterochromatin in *Drosophila:* the role of basal X heterochromatin in meiotic sex chromosome behaviour. *Chromosoma (Berl.) 60,* 283-296

Discussion

Harry Harris: How do you envisage the unequal crossing-over?

Walker: Sister chromatid exchange.

Harry Harris: Recently in Philadelphia, Balaban-Malenbaum & Gilbert (1977) reported on marker chromosomes containing long homogeneous staining regions (HSRs) in human neuroblastoma cell lines. One line contains two classes of cells: one with an HSR marker chromosome and the other with large numbers of double-minute chromosomes. Each cell had one of these abnormalities, no cell had both, which suggests that the double minutes might have been derived from the HSR. It would be worth looking for this kind of effect in your line as you relax selection.

Walker: We have not so far seen anything like that in the nucleus but then we haven't relaxed selection on it.

Henry Harris: Did you start your selections on the PG19 when it was already tetraploid? PG19 comes from me, and it was originally a pretty good diploid.

Walker: We used PG19 so that we could later select for hybrid lines if we wanted them, but we had great difficulty in establishing the first stages of selection. We passaged it through mice thereafter and got out a melanotic line which grew well. It is this sub-line which we have used. It soon lost the melanotic component on selection. It is tetraploid and has the six isochromosomes.

Henry Harris: Back in the Middle Ages when I was heavily involved in intranuclear RNA I advanced the thesis (Harris 1965) that the stuff that was breaking down in the nucleus was being transcribed from non-coding sequences and that maybe that was where evolution really took place, because most mutational events in the coding sequences wouldn't get you very far very fast. In a recent article in *Nature,* Walter Gilbert (1978) has advanced in a rather more sophisticated way the idea that changes in the non-coding regions and the inserts might be kicking evolution along at a much greater rate. Do you have any views on this? You have been avoiding committing yourself to function.

Walker: Yes, that was intentional. These sequences are presumably the result of earlier duplications, earlier translocations within chromosomes. They are the leaf scars of earlier chromosomal breakages and reunion of one sort or another. I also think that the presence of scattered repeated sequences will facilitate further changes.

Crick: The intervening sequences are complicated and perhaps we'll be discussing them again later. The idea that there is a mechanism which multiplies things, not necessarily by exchange, seems evolutionarily very reasonable.

Obviously in some cases one can imagine that the mechanism is next door to something of which a lot of copies are needed. But I am not sure that applies to the kangaroo rat. But an increase of DNA might occur in many ways, so one has to ask why this additional DNA stays around. Perhaps one ought to think of it in evolutionary terms as a rate of production, which may be by some random events, and a rate of elimination. So if there is extra DNA, how fast is it eliminated during evolution? I suspect that the answer is something along these lines: the usual effect of extra DNA is to put a small burden on the cell. It also makes the nucleus bigger, and if the nucleus is bigger, the cell is bigger; if the cell is bigger—as it apparently is in plants and probably in animals too—things tend to go rather more slowly and are often associated with a longer life cycle. This has been clearly shown in plants. So is the life cycle of the kangaroo rat a little slower than that of the rat or the mouse? Perhaps the rat and the mouse are in such a hurry to get on with their life cycles that there has been a selective advantage in eliminating extra DNA.

Walker: The one thing we know about the kangaroo rat is that it doesn't drink very much. It survives very well in very arid conditions and it recolonizes bomb craters very early. I don't know whether it has a longer life cycle than the rat or mouse.

Crick: The argument about the length of the life cycle is not true of tetraploids. It is the ratio between the effective number of genes and the nonsense DNA that matters. A plant which has more DNA but presumably the same number of genes, that is to say more of this junk, in general has a slower life cycle. It is the ratio of sense to nonsense which gives this effect.

Ruddle: What is the copy number of dihydrofolate reductase genes in your methotrexate-resistant cell?

Walker: We haven't done the copy number. We have an agreement with Sydney Brenner that his laboratory will try and produce a cloned sequence from the message but I don't know whether this has been done yet.

Ruddle: It would be interesting to look at the copy number also in relationship to the length of these long sequences, to see whether there is a regular distribution of the gene along their length. If copies are distributed over all of this material, one would expect that in chromosome-mediated transfer, where one can transfer only subchromosomal pieces into recipient cells, there would be difficulty in transmitting the trait to a recipient cell. We have been trying to do that over the last six months without success. It may be that the copies are not organized in tight packages of repeated units but are spread out.

Polani: If heterochromatic X chromosomes are added to the complement, the cycle is lengthened; by contrast we know that when only one sex chromosome is present (the X), the cell cycle is shortened. So one could predict the amount of lengthening for each extra chromosome.

Crick: Is the cell slightly longer in your cell culture, Peter? And I am sure you are going to set up a culture with the old cells and the new ones and see whether the new ones are selected under non-selective conditions.

Walker: Yes.

Bodmer: I think it is a mistake to look for the primary function of highly repetitive DNA. All these things about the size and length of chromosomes must be quite secondary, because if size of chromosomes was important why not put them all together in bigger chromosomes? It seems to me that repetitive DNA is the only true selfish gene. It is DNA which is essentially parasitic on the cell, and that has its own replicative advantage. This reminds me of the experiments that Sol Spiegelman did with Qβ RNA in the test tube (Spiegelman *et al.* 1965). A sequence could acquire intrinsic replicative advantage using the cellular machinery and the cell can't get rid of the DNA, so it is then used in some secondary way. I think all the explanations for its evolution in terms of a primary functional advantage are wrong. Pairing is totally nonsensical to me, because one can have hybridization between species with totally different repetitive DNA fractions. Suppression of recombination doesn't make sense. Even size is limited because humans after all have a fairly slow metabolic rate compared to mice, yet they have much less satellite DNA. But given that the higher organism can tolerate large amounts of DNA and make use of it, these repetitive sequences have this intrinsic advantage. Secondarily they may have some more general functional advantage. I can see no alternative to looking at this material in that way.

Crick: Why doesn't the kangaroo rat get rid of it? What is the selective advantage?

Bodmer: The selective advantage is the replicative advantage of the material itself.

Crick: Then why are there species that don't have any?

Bodmer: One could develop a theory of natural selection and drift of these organisms within the bigger organisms. I think that these satellite DNA sequences may arise and evolve mainly by saltatory steps of replication outside the chromosome rather than unequal crossing-over. They may be built up, in part, by sister chromatid exchange, but I am sure that replication outside the chromosome must be terribly important.

Fincham: DNA sequences which, by whatever mechanism, can get themselves replicated differentially have fitness at the molecular level and will tend to accumulate within the genome. But some species must have a way of life such that there is strong selection at the organism level against the accumulation of this junk.

Crick: That is what I was trying to say. My question is a very precise one: is the mitotic cycle on average longer in the kangaroo rat than in the rat?

Peto: In what sense do you mean that the gene is managing to be selfish within the organism, Professor Bodmer? What sort of mechanisms of selfishness did you have in mind?

Bodmer: My view is that the expansion mechanisms for those sequences are likely to be outside the chromosomes. They are likely to involve replication in a plasmid-like state that is non-chromosomal, and in those circumstances there may be very rapid replication and later reinsertion. Quite simple sequences may have the necessary properties for this sort of evolution. They can be expanded in the various ways that we know about and then just stick there because the organism doesn't need to get rid of them or can't do so unless there is some serious disadvantage in having that extra material.

Peto: Given that they do expand and sit there, and that secondary advantages can be taken, is there any reason to suppose that these particular sequences would have any such advantage in terms of becoming plasmids and proliferating when they are plasmids?

Bodmer: Other people may be able to answer that better than I can. Sol Spiegelman used Qβ in an *in vitro* system where he made it keep replicating. There was a sequence which was selected out more rapidly. It is a question of the origins of replication, and we know that reiterated sequences are places where homologies would allow unequal crossing-over to take place. In general terms I think it is plausible.

Brenner: We might ask whether higher organisms contain origins of replication in their DNA. From the observed lengths of replicating units one can calculate that a reasonable fraction of randomly cloned DNA fragments might contain origins of replication, especially if eukaryotic replication resembles what is found in eukaryotic viruses and prokaryotes. A very simple question can be asked: does cloned satellite DNA replicate as a plasmid when inserted into the nucleus?

Weissmann: Dr Mantei in our laboratory did one experiment like that which didn't work. We joined plasmid DNA to mouse satellite DNA, cloned it and then attempted to reintroduce it into mouse cells. We found less than one plasmid per cell.

Brenner: But it may have to get into the nucleus of the cell.

Weissmann: Goebel & Schiess (1975) reported that ColE1 moved to the mouse cell nucleus by itself and could be recovered intact after 24 hours. That was our basis for doing the experiment, but it just didn't work.

Bodmer: It has been disappointing to find that there is no obvious evolution of satellite DNA in tissue culture. One might argue that in that system evolution isn't taking place, but one might have wondered about it over the period of time that we have had cultures of mouse cell hybrids available. Maybe we wouldn't have detected a new sequence against the background of what already exists. Isn't the problem of the plasmid in the bacterium and its evolution already an indication that evolution may go on? If cow satellite DNA incorporated into a plasmid is put in a bacterium, what happens over a period of time? Even though that is an alien environment, wouldn't there be quite extensive evolution and expansion?

Walker: There is evidence that suggests that cloned repetitive sequences frequently lose some of their repeats.

Bodmer: I suppose one could argue that that is efficient for the plasmid because it must survive and has to be economical.

Henry Harris: What do you mean when you say the cell couldn't 'get rid' of these things?

Bodmer: I was speaking teleologically—there is not a sufficient advantage to the cell. The disadvantage of having that material there, if there is one, is not strong enough to prevent its selfishness from keeping it there. We are talking about the satellites, which are highly repetitive.

Brenner: For an organism of a given size, what is the difference between making it with a small number of very large cells, or a large number of small cells? Salamanders have very large cells.

Bodmer: The satellite structure in the mouse is not necessarily all that different from the structure in humans. It is just that the mouse has a lot of one satellite and humans have a lot of different ones. So although the two organisms are very different in every other respect, the DNA structure is essentially similar.

Milstein: The problem may be complicated. Each animal has reached a certain equilibrium and the change of equilibrium requires an interplay of many different factors, the amount of DNA being only one of them. Each species must have a well-balanced situation, so that the total DNA is kept constant within that particular species.

Peto: Maybe we are looking about 600 million years too late? Maybe the selfishness and so on got fixed earlier on.

Neel: The wood mouse, *Peromyscus,* has about the same body weight as the house mouse but three times the lifespan. There is the material you are looking for.

Cavalli-Sforza: But the synthetic phase may be short enough in the whole life of the cell for a very small increase in time to be necessary even if the DNA is doubled.

Jacob: Is anything known about the sequence of the mammalian Y chromosome?

Walker: There is a specific sequence for about 50% of that chromosome. It is also very variable, and there are many polymorphisms in human lines.

Henry Harris: I suspect that when one first explants the kangaroo rat cells, they grow very slowly. Then chromosome irregularities occur, such as tetraploidization, and rapid growth ensues.

Weissmann: We have compared the β-globin-specific chromosomal DNA of rabbit and mouse. The β-globin gene of the rabbit has a small invervening sequence and a larger one. Between exactly the same two triplets in the coding region, the mouse also has two intervening sequences—a smaller and a larger one. This suggests that the intervening sequences or introns were present before the divergence of rabbit and mouse. Surprisingly, the nucleotide sequences of the rabbit and mouse introns show little resemblance at the nucleotide sequence level. The only obvious regions of similarity are the ones bordering the junction between the coding and the non-coding sequences. As one moves away from the junctions, the intervening sequences become more and more different. If one accepts that the introns are derived from a common ancestral sequence, then one must attribute the sequence diversity to strong genetic drift, that is, a mutation acceptance rate far higher than that of the coding region of the mRNA leader sequence.

Bodmer: Is there any polymorphism between individuals?

Weissmann: We have only looked at one mouse clone and one rabbit clone so we cannot tell. Tom Maniatis has also isolated a rabbit clone and a comparison with as much as he and his colleagues have sequenced shows a few nucleotide differences; however these results are not yet definitive.

Bodmer: There is a paradox here between rapid evolution and lack of polymorphism.

Weissmann: I can't comment on that until the complete intron sequences are known.

The introns could be an evolutionary relic. At a certain stage of evolution prokaryotes may have come to an impasse when expression of additional genetic information became necessary. The nature of the impasse may have been the following: in prokaryotes there are nucleolytic enzymes which process RNA, for example ribosomal RNA and tRNA precursors. These enzymes recognize specific sequences, so whenever new DNA sequences (arising by gene reduplication or a transduction-like process) were transcribed, the RNA was susceptible to the action of these pre-existing nucleases. At a certain point in evolution RNA ligase was created (maybe as a modified DNA ligase) and joining of RNA ends became possible. This would have laid the foundations of the 'splicing process', in which a processing enzyme clips an RNA in two places and then a ligase joins the ends. Once splicing is introduced, the difficulty of expressing expanded genetic material can be overcome. You can see that any RNA produced has now to go through a sort of sieve consisting of processing and ligating enzymes. Any evolutionarily new RNA that is made will be subjected to this mechanism. Therefore genetic information is selected for at the expression level *after* the DNA transcripts have gone through this net of processing and ligation. Once this system has started evolving it may be as conservative as the genetic code. The intervening sequence cannot be taken out partially because accurate processing of the remainder might be precluded. A complete deletion of an intron must be an extremely rare event, because of the precision required for conserving the translatability of the mRNA. Moreover, even if such a deletion occurs, it may confer little or no advantage on the organism, so that it might be rare in the population. It is of course possible that the presence of the intron leads to some advantages; such speculations abound in published work now.

Walker: That doesn't help over the ribosomal genes in *Drosophila,* does it?

Weissmann: The introns in ribosomal genes are perhaps different since, as you pointed out, they are repetitive and they occur not only in ribosomal genes but in other parts of the genome as well.

Brenner: There is no evidence in *Drosophila* that these are excised or that the genes function in the organism.

Weissmann: No, they are not even transcribed.

References

BALABAN-MALENBAUM, G. & GILBERT, F. (1977) Double minute chromosomes and the homogeneously staining regions in chromosomes of a human neuroblastoma cell line. *Science (Wash. D.C.) 198,* 739-741

GILBERT, W. (1978) Why genes in pieces? *Nature (Lond.) 271,* 501

GOEBEL, W. & SCHIESS, W. (1975) The fate of a bacterial plasmid in mammalian cells. *Mol. Gen. Genet. 138,* 213-223

HARRIS, H. (1965) The short-lived RNA in the cell nucleus and its possible role in evolution, in *Evolving Genes and Proteins* (Bryson, V. & Vogel, H.J., eds.), pp. 469-477, Academic Press, New York

SPIEGELMAN, S., HARUNA, I., HOLLAND, I. B. *et al.* (1965) The synthesis of a self-propagating and infectious nucleic acid with a purified enzyme. *Proc. Natl. Acad. Sci. U.S.A. 54,* 919-927

Reversed genetics: a new approach to the elucidation of structure–function relationship

CHARLES WEISSMANN, HANS WEBER, TADATSUGU TANIGUCHI, WILLI MÜLLER AND FRANÇOIS MEYER

Institut für Molekularbiologie I, Universität Zürich

Abstract Methods for generating point mutations at predetermined sites of RNA or DNA genomes have been developed. With Qβ RNA as a template, minus strands were synthesized *in vitro* in a stepwise, substrate-controlled reaction. The nucleotide analogue N^4-hydroxyCMP was introduced in the desired position, the minus strands were completed with the four standard triphosphates and used as templates to synthesize plus strands; about 30% of the progeny plus strands showed a base transition at the position corresponding to the nucleotide analogue. Two mutant RNAs with extracistronic nucleotide substitutions have been generated; one of these was viable, albeit with a reduced propagation rate, while the other was non-infectious. Furthermore, mutants with changes at the initiation codon of the coat cistron were prepared. An analysis of ribosome binding to such mutant RNAs revealed the importance of the A-U-G region for the formation of the initiation complex.

With a similar approach applied to the β-globin complementary DNA (cDNA) plasmid PβG, point mutations have been introduced at the positions corresponding to amino acids 121 to 123.

The elucidation of the relationship between structure and function of biological macromolecules is greatly aided by the availability of appropriate mutants. The classical genetic approach depends on mutations which arise at random, either spontaneously or as a consequence of mutagenic agents. Mutants are usually selected or screened for by virtue of some desired phenotype, or picked up because of some striking property. Difficulties may arise when the role of non-coding, functionally undefined segments of a genome are to be studied, since one does not know what property to screen or select for. Moreover, in haploid organisms unconditionally lethal mutations cannot be picked up at all. The problems are compounded for higher organisms where recessive mutations are frequently only recognizable when homozygous. In addition, the genome of higher eukaryotes is so large that the probability of finding a mutation in a particular DNA region without a potent selection system is quite low.

The advent of hybrid technology has changed this situation dramatically. It has become possible to integrate *in vitro* any gene of interest, with its neighbouring regions, into a vector, clone and amplify it, and reintroduce it into a eukaryotic cell. This constitutes the basis for what we have called 'reversed genetics' (Weissmann 1978), an approach wherein a nucleic acid is modified at predetermined positions *in vitro* and the effects of these interventions are scored *in vitro* or *in vivo,* in contradistinction to classic genetics, where deviant phenotypes are first isolated and the lesion giving rise to them is identified subsequently. The modifications of the DNA may be gross, such as deletions, insertions or transpositions of DNA segments by the techniques of *in vitro* recombination, or point mutations such as the base substitutions generated by site-directed mutagenesis. In this article we show how site-directed mutagenesis can be applied to generate point mutations at predetermined regions of Qβ RNA as well as of hybrid plasmid DNA.

SITE-DIRECTED MUTAGENESIS IN Qβ RNA

Phage Qβ contains an RNA of about 4500 nucleotides which serves both as genome and messenger RNA. Qβ RNA consists of three translatable and four non-translatable (extracistronic) segments. Whereas the regions immediately preceding the cistrons are involved in the initiation and regulation of protein synthesis, the function of the longer untranslatable segments at the ends of the genome is unknown (Weissmann 1974).

Qβ RNA can be replicated *in vitro* by purified Qβ replicase, in conjunction with a host factor, to yield infectious progeny RNA in large excess over the input template (Spiegelman *et al.* 1965). The product of the first step of synthesis, a single stranded Qβ minus strand, is non-infectious but serves as an excellent template for the synthesis of infectious Qβ RNA (Feix *et al.* 1968).

Site-directed mutagenesis is based on the introduction of a mutagenic nucleotide analogue into a predetermined position of the viral RNA minus (or plus) strand by substrate-limited synthesis. The purified substituted strand is replicated *in vitro* to allow the generation of base substitutions at the site corresponding to the analogue (Flavell *et al.* 1974). To study as yet unknown functions of the RNA phage genome, we have generated Qβ RNAs with point mutations in the 3′ terminal extracistronic region and in the coat initiator region, and have examined their properties.

(A) Qβ RNA with an A→*G substitution in position 40 from the 3′-terminus (Domingo et al. 1976)*

Synthesis of the mutated RNA. Stepwise, substrate-limited synthesis of Qβ *minus strands with* N^4-hydroxyCMP in position 39 is shown schematically in Fig. 1. The purified, substituted minus strands were used as a template for the synthesis of one round of ^{32}P-labelled plus strands. T_1-fingerprinting of the purified product showed that a new large oligonucleotide, designated T-1*, had appeared, while the amount of oligonucleotide T-1 was diminished. The ratio of T-1* to T-1 was 1 : 3. Fig. 2 shows the positions of these two oligonucleotides in the fingerprint. T-1 is derived from positions −63 to −38 at the 3′ end of wild-type Qβ RNA (cf. Fig. 1) and has the sequence

−60 −50 −40
A-A-U-A-A-A-U-U-A-U-C-A-C-A-A-U-U-A-C-U-C-U-U-A-C-Gp.

The structure of T-1* was

A-A-U-A-A-A-U-U-A-U-C-A-C-A-A-U-U-A-C-U-C-U-U-Gp.

Therefore, an A→G transition at position −40 of the wild-type sequence accounts for the appearance of the new oligonucleotide. Since synthesis starts at the penultimate nucleotide of the plus strand, the 40th nucleotide from the 3′ end of the plus strand is complementary to the 39th position from the 5′ end of the minus strand.

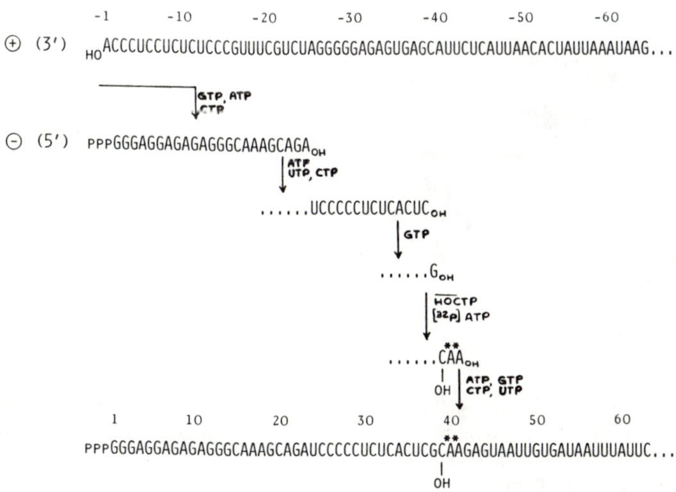

FIG. 1. Scheme for the stepwise synthesis of Qβ minus strands with introduction of $\overline{HOC}$MP in position 39 from the 5′ terminus.

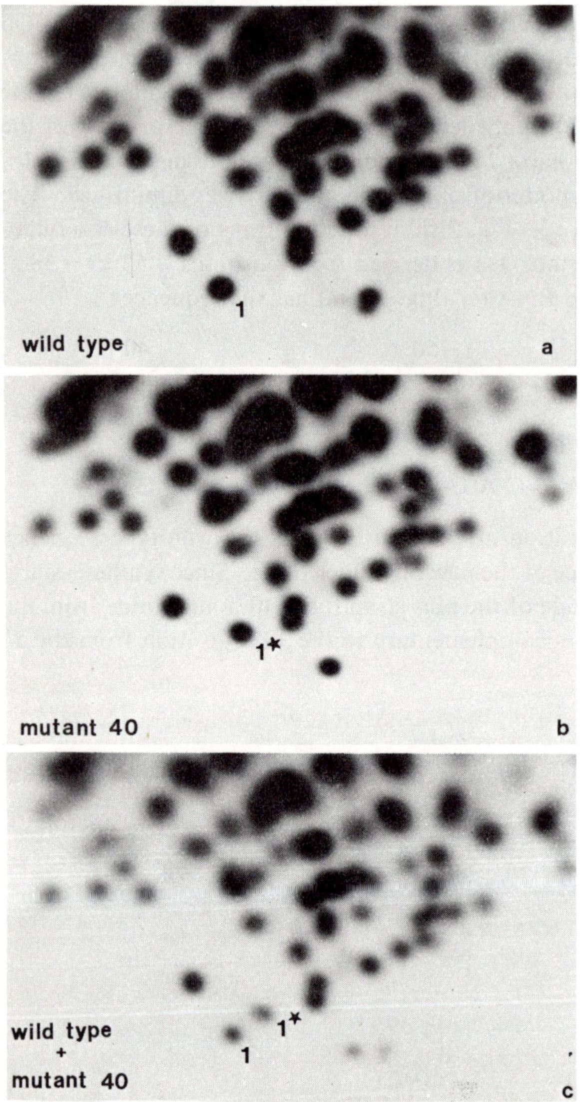

FIG. 2. Two-dimensional polyacrylamide gel electrophoresis of the T_1 oligonucleotides of uniformly ^{32}P-labelled wild-type RNA and mutant ($A_{-40} \rightarrow G$) RNA prepared from cloned phage. (a) Wild-type RNA, (b) mutant ($A_{-40} \rightarrow G$) RNA, (c) a mixture of wild type and mutant ($A_{-40} \rightarrow G$). (From Domingo *et al.* 1976.)

Isolation of phage carrying an A→G transition in position −40 of its RNA. One hundred and twenty nanograms of minus strand substituted with N^4-hydroxyCMP in position 39 were used as template for one round of plus strand synthesis. Nine nanograms of plus strands with a specific infectivity similar to that of wild-type Qβ RNA were formed. Spheroplasts were infected with these plus strands and the ^{32}P-labelled RNA of 18 resultant phage clones was examined by T_1-fingerprinting. Four preparations showed T-1*, diagnostic for the mutant A_{-40}→G, and no significant amounts of T-1; 14 RNAs gave rise to T-1 but not to T-1* (Fig. 2). This proportion of mutant phage, 22%, reflects rather accurately the proportion of mutant RNA, 25%, determined chemically in the preparation used for transfection.

Growth competition in vivo between mutant A_{-40}→G and wild-type phage. *E. coli* Q13 was infected with a 1 : 1 mixture of cloned mutant and wild-type phage and the resulting lysate was used to infect a fresh culture. After four such cycles of infection the mutant content was only about 3% and no mutant was detected after ten cycles. Propagation of plaque-purified mutant phage A_{-40}→G in the absence of added wild-type reproducibility resulted in the appearance of wild-type phage after a few cycles, showing that revertants arose at a substantial rate and outgrew the mutant. We have estimated a reversion rate of 10^{-4} per doubling and a growth rate of 0.25 of the mutant relative to wild-type under competitive conditions (Batschelet et al. 1976).

In vitro competition experiments showed that (A_{-40}→G) RNA is replicated less effectively by Qβ replicase than wild-type RNA (Domingo *et al.* 1976).

It is of interest that the nucleotide in position −40 is part of a sequence (−63 to −38) which binds both host factor and S_1 protein (Senear & Steitz 1976). Host factor is required by Qβ replicase for initiation on plus strands (Franze de Fernandez *et al.* 1972); S_1 is a ribosomal protein, which, after infection, is recruited as the α subunit of Qβ replicase (Kamen 1970; Kondo *et al.* 1970). Goelz & Steitz (1977) found that the mutant oligonucleotide T-1* is bound less efficiently by protein S_1 than its wild-type counterpart, T-1, suggesting that the reduced efficiency of RNA replication could be due to weaker binding of replicase and/or host factor to the mutated binding site.

(B) Qβ RNA with a G→A substitution in position 16 from the 3′ terminus (Flavell et al. 1974)

A procedure similar to the one described above was used to prepare Qβ RNA with a G→A substitution in position 16 from the 3′ terminus. To determine whether this RNA was infectious, spheroplasts were infected with the first generation of plus strands synthesized on minus strands with $\overline{\text{HOCMP}}$-substitution in position −15, as in the previous experiment. One hundred and twenty clones were analysed by T_1 fingerprinting: all were wild type. The plaque formation efficiency of mutant relative to wild-type RNA was estimated to be less than about 0.03 (Sabo *et al.* 1977). The deleterious effect of the ($G_{-16} \rightarrow A$) transition on the infectivity of the RNA does not appear to be due to impaired RNA replication (Flavell *et al.* 1975); its cause is not yet known.

(C) Qβ RNA with mutations in the coat cistron initiator region (Taniguchi & Weissmann 1978)

Under initiation conditions, *E.coli* ribosomes bind to Qβ RNA almost exclusively at the coat cistron initiation site. To determine to what extent the AUG triplet is required for 70S complex formation, we prepared Qβ RNA with G→A transitions of the third and fourth nucleotides of the coat cistron, i.e. with modifications in the third position of the A-U-G codon and the following nucleotide (cf. Fig. 3).

In order to carry out stepwise synthesis in the required region we synchronized minus strand synthesis at a ribosome attached to the coat initiation

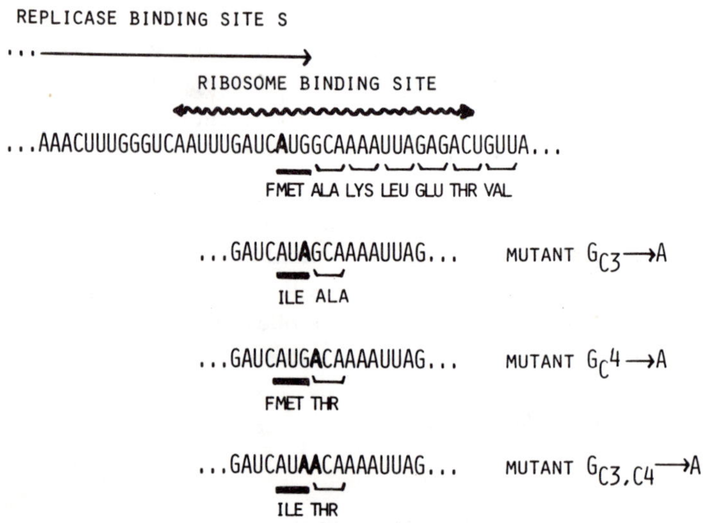

FIG. 3. Nucleotide sequence around the ribosome binding site of the coat cistron of wild type Qβ RNA, and mutants generated by site-directed mutagenesis at the 3rd and 4th position of the coat cistron (C_3 and C_4) (Taniguchi & Weissmann 1978).

site (Kolakofsky *et al.* 1973). The 70S Qβ RNA ribosome complex was used as template for Qβ replicase. Synthesis proceeded up to the position corresponding to the 16th nucleotide of the coat cistron (Kolakofsky *et al.* 1973). The ribosome was then dislodged by treatment with EDTA and stepwise synthesis was carried out, leading to insertion of $\overline{\text{HOC}}$MP into the positions complementary to the fourth and third nucleotides of the coat cistron (see Taniguchi & Weissmann 1978 for details).

Qβ plus strands were synthesized on the substituted minus strands and replicated *in vitro*. The ratio of wild type : mutant ($G_{C3} \rightarrow A$) : mutant ($G_{C4} \rightarrow A$) : mutant ($G_{C3,C4} \rightarrow A$) was found to be 1 : 1.8 : 1.6 : 4.5. We have not yet determined whether any of the mutant RNAs are infectious.

The ribosome binding capacity of Qβ RNA with G→A transitions in positions C_3 and/or C_4. A preparation of ^{32}P-labelled Qβ RNA consisting of the species described above was bound to ribosomes under initiation conditions and the 70S complex was treated with RNAse A. The [^{32}P]RNA retained by the ribosomes was isolated and the C_3/C_4 region was analysed. The relative binding efficiencies of wild type, mutant ($G_{C3} \rightarrow A$), mutant ($G_{C4} \rightarrow A$), and mutant ($G_{C3,C4} \rightarrow A$) were estimated to be 1 : < 0.1 : 2.8 : 0.33. Thus, both mutant RNAs lacking the A-U-G triplet had a reduced ribosome-binding capacity. This suggests that the $tRNA_F^{met}$-AUG interaction contributes substantially to the stabilization of the 70S ribosome complex.

It is striking that ribosomes were bound more efficiently to mutant ($G_{C4} \rightarrow A$) RNA than to wild-type RNA, and to mutant ($G_{C3,C4} \rightarrow A$) more efficiently than to mutant ($G_{C3} \rightarrow A$) RNA. Perhaps the nucleotides flanking the codon and the anticodon contribute to the stability of the interaction with fMet-tRNA. The mutants RNAs ($G_{C4} \rightarrow A$) and ($G_{C3,C4} \rightarrow A$) have the possibility of forming an additional A-U base pair as compared to wild-type RNA and mutant ($G_{C3} \rightarrow A$) RNA, respectively (Taniguchi & Weissmann 1978).

SITE-DIRECTED MUTAGENESIS IN CLONED β GLOBIN COMPLEMENTARY DNA (Müller *et al.* 1978)

We have used the plasmid PβG (Efstratiadis *et al.* 1977) to determine whether the principle of site-directed mutagenesis can be applied to DNA.

Introduction of N^4-hydroxydCMP into the positions corresponding to amino acids 121 and 122 of the rabbit β globin gene

Plasmid PβG, which contains an almost complete DNA copy of rabbit β globin mRNA, has a single EcoRI site within the globin gene region corresponding to amino acids 121 and 122 (cf. Fig. 4). Single-strand nicks were introduced by partial digestion with EcoRI and the nicked plasmid was incubated with *E.coli* DNA polymerase I and dATP, d$\overline{\text{HOC}}$TP and [α-^{32}P]dCTP, to incorporate the nucleotide analogue into the immediate vicinity of the nick. A control with dTTP instead of d$\overline{\text{HOC}}$TP was run in parallel.

Isolation and characterization of EcoRI-resistant PβG DNA

Four of the nine possible mutations resulting from the replication of d$\overline{\text{HOC}}$MP-substituted PβG should impart EcoRI resistance to PβG. To quantitate and isolate such mutants, d$\overline{\text{HOC}}$MP-substituted PβG as well as the

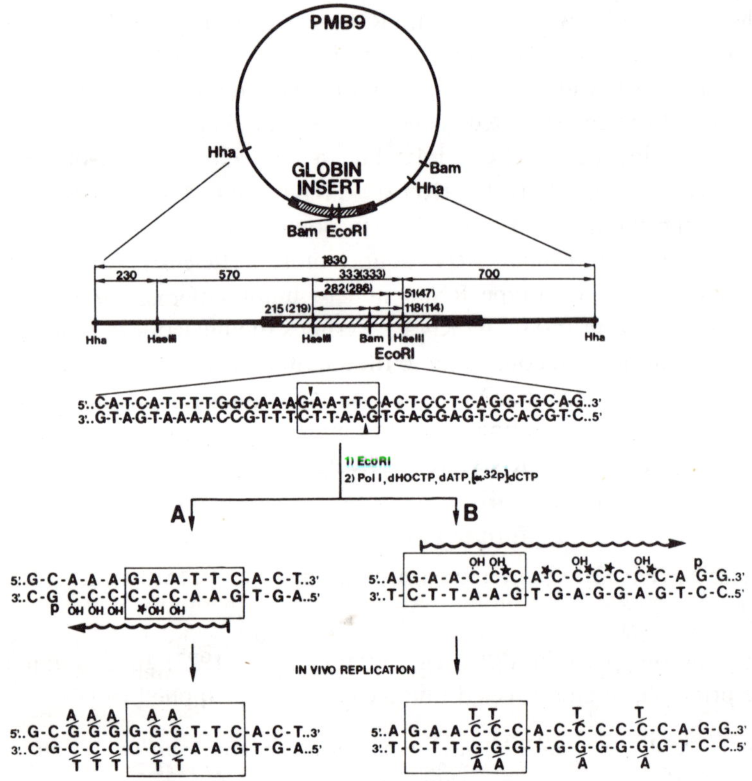

FIG. 4. Partial map of PβG (data from Efstratiadis *et al.* 1977) and scheme of procedure used in site-directed mutagenesis. (From Müller *et al.* 1978.)

'dTMP-substituted' control and untreated PβG (form I) were transfected into *E.coli* HB101 and tetracycline-resistant (plasmid-containing) colonies were isolated. Samples of plasmid DNA derived from the three preparations, as well as some of the original PβG DNA, were treated with EcoRI. The residual infectivity of all control preparations was about 0.07–0.08% of the undigested samples, while for the DNA derived from the d$\overline{\text{HOC}}$MP-substituted sample it was 1.9%.

To ascertain whether the colonies resulting from transfection with EcoRI-treated DNA were indeed due to EcoRI resistance of the plasmids, clones from each experiment were cultured and plasmid DNA was isolated. All 24 cloned plasmids derived from the d$\overline{\text{HOC}}$MP-substituted preparation, but only one of 24 from the 'dTMP-substituted' sample and eight of 98 plasmids from the control experiments with PβG were resistant to EcoRI. Most probably, a small fraction of the EcoRI-cleaved control DNA had been recircularized *in vivo,* restoring EcoRI-sensitive PβG. The average EcoRI resistance of control DNA was calculated to be 0.005%, and that of the mutagenized sample 1.9%, a 380-fold higher value.

Analysis of EcoRI-resistant PβG DNA

Restriction analysis showed that EcoRI resistance of the plasmids from the control experiments was in all cases due to extensive deletions which eliminated the EcoRI site (Meyer *et al.* 1978), probably as a consequence of the EcoRI-cleavage carried out during the selection procedure (cf. Lai & Nathans 1974).

Seven EcoRI-resistant plasmids derived from d$\overline{\text{HOC}}$MP-substituted preparations were chosen at random and their nucleotide sequence around the erstwhile EcoRI-site was determined (Fig. 5). As shown in Fig. 6, three plasmids had one, three had two and one had three AT→GC changes, all located within the mutagenized region (cf. Fig. 4).

Having established that site-directed mutagenesis is in principle feasible, it is necessary to extend the approach so as to make any region of the DNA susceptible to nucleotide substitution. An appropriate method which makes use of a set of cloned globin-specific DNAs of different lengths, to be used as primers for substrate-limited elongation on an intact circular DNA, is currently being developed. Our immediate intent is to introduce modifications into the putative ribosome binding site of PβG and to study their effect on protein initiation.

It is clear that site-directed mutagenesis will eventually be of use not only to elucidate the function of DNA sequences in transcription, translation and control, but also to generate hormones and enzymes with modified properties.

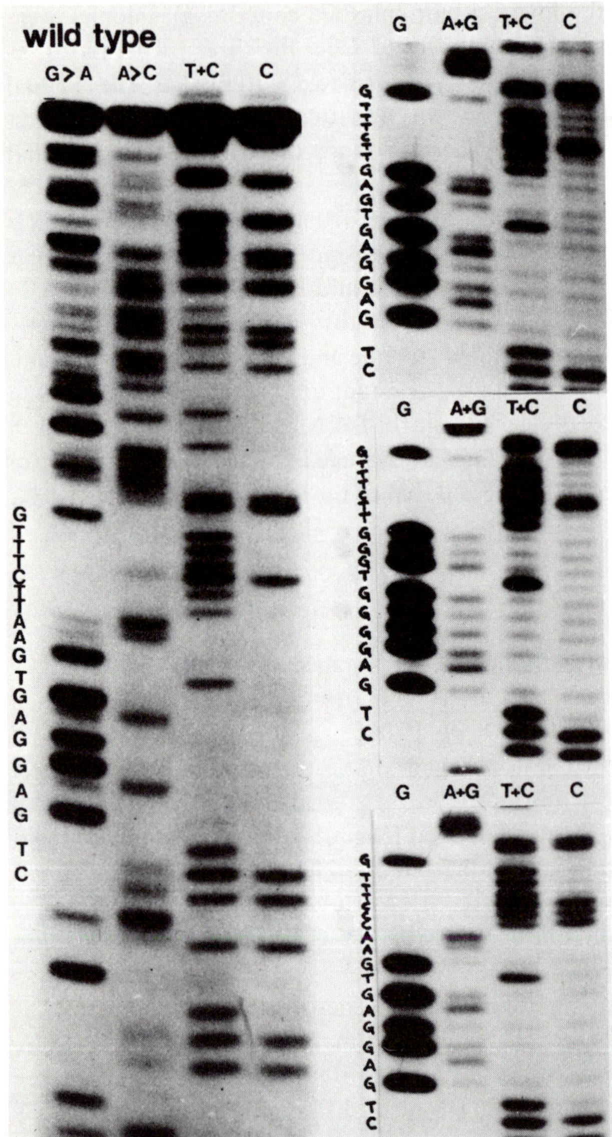

FIG. 5. Maxam-Gilbert sequence analysis of the region around the (modified) EcoRI site. (a) Wild type; (b) mutant 22.3; (c) mutant 22.14; (d) mutant 22.15. (From Müller *et al.* 1978.)

```
                      119    120    121    122    123    124    125
                      gly    lys    glu    phe    thr    pro    gln
wild type PβG  +  5' G-G-C-A-A-A|G-A-A-T-T-C|A-C-T-C-C-T-C-A-G
               -     C-C-G-T-T-T|C-T-T-A-A-G|T-G-A-G-G-A-G-T-C  5'

                      gly    lys    glu    LEU    thr    pro    gln
PβG 22.1       +  5' G-G-C-A-A-A|G-A-A-C-T-C|A-C-C-C-C-T-C-A-G
                                        *        *
               -     C-C-G-T-T-T|C-T-T-G-A-G|T-G-G-G-G-A-G-T-C  5'

                      gly    lys    glu    PRO    thr    pro    gln
PβG 22.2       +  5' G-G-C-A-A-A|G-A-A-C-C-C|A-C-T-C-C-T-C-A-G
                                        * *
               -     C-C-G-T-T-T|C-T-T-G-G-G|T-G-A-G-G-A-G-T-C  5'

                      gly    lys    glu    LEU    thr    pro    gln
PβG 22.3       +  5' G-G-C-A-A-A|G-A-A-C-T-C|A-C-T-C-C-T-C-A-G
                                        *
               -     C-C-G-T-T-T|C-T-T-G-A-G|T-G-A-G-G-A-G-T-C  5'

                      gly    lys    glu    SER    thr    pro    gln
PβG 22.13      +  5' G-G-C-A-A-A|G-A-A-T-C-C|A-C-T-C-C-T-C-A-G
                                          *
               -     C-C-G-T-T-T|C-T-T-A-G-G|T-G-A-G-G-A-G-T-C  5'

                      gly    lys    glu    PRO    thr    pro    gln
PβG 22.14      +  5' G-G-C-A-A-A|G-A-A-C-C-C|A-C-C-C-C-T-C-A-G
                                        * *      *
               -     C-C-G-T-T-T|C-T-T-G-G-G|T-G-G-G-G-A-G-T-C  5'

                      gly    lys    GLY    phe    thr    pro    gln
PβG 22.15      +  5' G-G-C-A-A-A|G-G-G-T-T-C|A-C-T-C-C-T-C-A-G
                                  * *
               -     C-C-G-T-T-T|C-C-C-A-A-G|T-G-A-G-G-A-G-T-C  5'

                      gly    lys    glu    phe    thr    pro    gln
PβG 12.6       +  5' G-G-C-A-A-A|G-A-G-T-T-C|A-C-T-C-C-T-C-A-G
                                    *
               -     C-C-G-T-T-T|C-T-C-A-A-G|T-G-A-G-G-A-G-T-C  5'
```

FIG. 6. Nucleotide sequence around the EcoRI site of wild-type PβG and EcoRI-resistant PβG DNA generated by site-directed mutagenesis. (From Müller *et al.* 1978.)

References

BATSCHELET, E., DOMINGO, E. & WEISSMANN, C. (1976) The proportion of revertant and mutant phage in a growing population, as a function of mutation and growth rate. *Gene 1,* 27-32

DOMINGO, E., FLAVELL, R. A. & WEISSMANN, C. (1976) In vitro site-directed mutagenesis: generation and properties of an infectious extracistronic mutant of bacteriophage Qβ. *Gene 1,* 3-25

EFSTRATIADIS, A., KAFATOS, F. C. & MANIATIS, T. (1977) The primary structure of rabbit β-globin mRNA as determined from cloned DNA. *Cell 10,* 571-585

FEIX, G., POLLET, R. & WEISSMANN, C. (1968) Replication of viral RNA, XVI. Enzymatic synthesis of infectious viral RNA with non-infectious Qβ minus strands as template. *Proc. Natl. Acad. Sci. U.S.A. 59,* 145-152

FLAVELL, R. A., SABO, D. L., BANDLE, E. F. & WEISSMANN, C. (1974) Site-directed mutagenesis: generation of an extracistronic mutation in bacteriophage Qβ RNA. *J. Mol. Biol. 89,* 225-272

FLAVELL, R. A., SABO, D. L., BANDLE, E. F. & WEISSMANN, C. (1975) Site-directed mutagenesis: effect of an extracistronic mutation on the in vitro propagation of bacteriophage Qβ RNA. *Proc. Natl. Acad. Sci. U.S.A. 72,* 367-371

FRANZE DE FERNANDEZ, M. T., HAYWARD, W. S. & AUGUST, J. T. (1972) Bacterial proteins required for replication of phage Qβ ribonucleic acid. *J. Biol. Chem. 247,* 824-831

GOELZ, S. & STEITZ, J. A. (1977) *Escherichia coli* ribosomal protein S1 recognizes two sites in bacteriophage Qβ RNA. *J. Biol. Chem. 252,* 5177-5179

KAMEN, R. I. (1970) Characterization of the subunits of Qβ replicase. *Nature (Lond.) 228,* 527-533

KOLAKOFSKY, D., BILLETER, M. A., WEBER, H. & WEISSMANN, C. (1973) Resynchronization of RNA synthesis by coliphage Qβ replicase at an internal site of the RNA template. *J. Mol. Biol. 76,* 271-284

KONDO, M., GALLERANI, R. & WEISSMANN, C. (1970) Subunit structure of Qβ replicase. *Nature (Lond.) 228,* 525-527

LAI, C.-J. & NATHANS, D. (1974) Deletion mutants of Simian virus 40 generated by enzymatic excision of DNA segments from the viral genome. *J. Mol. Biol. 89,* 179-193

MEYER, F., MÜLLER, W., PALMIERI, M. & WEBER, H. (1978) Occurrence of deletions among the progeny of plasmids treated with restriction enzymes. *Experientia (Basel), 34,* 948

MÜLLER, W., WEBER, H., MEYER, F. & WEISSMANN, C. (1978) Site-directed mutagenesis in DNA: generation of point mutations in cloned β globin cDNA at the positions corresponding to amino acids 121 to 123. *J. Mol. Biol., 124,* 343-358

SABO, D. L., DOMINGO, E., BANDLE, E. F., FLAVELL, R. A. & WEISSMANN, C. (1977) A guanosine to adenosine transition in the 3′ terminal extracistronic region of bacteriophage Qβ RNA leading to loss of infectivity. *J. Mol. Biol. 112,* 235-252

SENEAR, A. W. & STEITZ, J. A. (1976) Site-specific interaction of Qβ host factor and ribosomal protein S_1 with Qβ and R17 bacteriophage RNAs. *J. Biol. Chem. 251,* 1902-1912

SPIEGELMAN, S., HARUNA, I., HOLLAND, I. B., BEAUDREAU, G. & MILLS, D. (1965) The synthesis of a self-propagating and infectious nucleic acid with a purified enzyme. *Proc. Natl. Acad. Sci. U.S.A. 54,* 919-927

TANIGUCHI, T. & WEISSMANN, C. (1978) Site-directed mutations in the initiator region of the bacteriophage Qβ coat cistron and their effect on ribosome binding. *J. Mol. Biol. 118,* 533-565

WEISSMANN, C. (1974) The making of a phage. *FEBS (Fed. Eur. Biochem. Soc.) Lett. 40,* S10-S18

WEISSMANN, C. (1978) Reversed genetics. *Trends Biochem. Sci. 3,* N109-N111

Discussion

Crick: We badly need to know which the promoter sequences are in eukaryotes. I would suggest that you focus on that by the method you described of altering a sequence. You may have to worry about the different states of the various polymerases but if you could read along and say that that is a promoter sequence of a certain strength in eukaryotes, you would be in a very strong position.

Weissmann: We certainly plan to do that, but it is very difficult.

Pontecorvo: At present the real bottleneck lies in *in vitro* translation, does it not? The situation is like that of the Etruscan language: we know the alphabet, we know a few words, but we don't know the language. The essential step is to go ahead with *in vitro* translation.

Weissmann: I am extremely confident. First of all, as we shall hear from John Gurdon, one can introduce DNA into the nuclei of *Xenopus* oocytes and obtain translation products. The situation is a bit complex now as we don't have strictly homologous systems. The ideal thing would be to put *Xenopus* genes into *Xenopus* eggs, rabbit globin DNA into rabbit cells or, as is in fact our plan, put mouse chromosomal DNA back in the mouse cell. A very potent system is available: the Friend cell system. In this, haemopoietic cells are arrested at a stage of development where they do not yet make globin; dimethylsulphoxide (DMSO) and a variety of other compounds induce the cells to express haemoglobin. We want to put a genetic marker into the mouse β-globin chromosomal gene so that we can recognize its translation product and differentiate it from the resident β-globin genes. For example one might introduce a mutation which changes a neutral amino acid into an acidic one, or something like that, which would allow easy separation by biochemical techniques. We would introduce this cloned gene in an appropriate vector into the Friend cell and induce the Friend cell to differentiate by adding DMSO. Then we hope that both the resident gene and the inserted gene will be induced to produce their product in concert, and we will be able to distinguish them because of the genetic marker.

Ruddle: Why not use a human gene?

Weissmann: Because that is not a homologous situation.

Ruddle: Is that important?

Weissmann: If it fails it will be important; if it works it is fine.

Bodmer: But unless you get the vector back into the chromosomes, isn't the vector-mediated system with polyoma unlikely to respond to the same control systems with DMSO induction?

Weissmann: I don't think that this is necessarily so, but it is a possibility.

Brenner: It also depends on whether you think that deep down in every higher organism there is an *E.coli* struggling to get out!

Ruddle: It has been shown by Deisseroth & Hendrick (1978) that the human α-globin gene on human chromosome 16 introduced by cell hybridization is expressed in Friend cells.

Bodmer: Are they expressed with DMSO?

Ruddle: Yes.

Brenner: The question is whether, in order to get sensible information, it will be sufficient to add the gene anywhere, in any vector, or whether you will have to make transductants of the cells—that is, recombinants of the sequences so that they go in the correct place.

Jacob: But this is complementary DNA (cDNA), isn't it?

Weissmann: We are talking of chromosomal DNA. There is a whole set of experiments in which mutations can be put into the intervening sequences which will prevent their processing. One might speculate that in some diseases like thalassaemia RNA precursors will be synthesized but no β-globin will be formed, maybe due to mutations which affect the processing.

Gurdon: Professor Pontecorvo, do you prefer the use of a true *in vitro* system (test tubes) to the alternative *in vivo* system (injected living cells)?

Pontecorvo: Yes. It is a lot simpler.

Crick: He is a reductionist! The thing is surely clear: one tries the simplest system first; it may be slightly artifactual so one checks it against different stages of *in vivo* or *in vitro* systems. There is no real conflict here.

Brenner: This is simple molecular biology, but there is also the problem of assessing the value of genes that may have to work in entire organisms to produce their effects. Could you comment on the very long-term idea of putting the genes back into organisms?

Weissmann: I think that John Gurdon's system will eventually lend itself to this approach. It should be possible to do the injections in such a way that one ultimately gets development of the embryo and takes the inserted DNA through the complete cycle. The other approach that we are discussing with Dr Koprowski is the injection of cloned DNA into mouse blastocysts. In constructing vectors for this approach we are using polyoma DNA linked to a bacterial plasmid, say pBR 322. This entity, which we might call a double vector since it can grow in both animal cells and *E.coli*, has great advantages. One can clone DNA and make biochemical amounts of these entities in *E.coli*, then put them in the animal cell where they will replicate, express or be integrated. In addition, we would like to have a selective marker on the vector because when we put the hybrids into Friend cells the frequency with which they establish themselves may not be very high. We are planning to use thymidine kinase-negative Friend cells and a thymidine kinase gene on the vector so that we can select for those cells which have been transformed. To this system one can add the genes and their surrounding regions that one is interested in. By introducing appropriate mutations at predetermined sites the functions of control and other regions of the DNA can be explored. It may be possible to get mice which are thymidine kinase-minus, add some selective pressure by dampening the *in vivo* synthesis of thymidine, and give the cells which carry the vector certain advantages. Thus, one might get most of the embryo populated with the DNA hybrid. Those are dreams at present.

Henry Harris: I don't really think they are dreams. People have got quite a long way with this kind of thing. David Martin has used mutant teratocarcinomas that are deficient in the gene for hypoxanthine phosphoribosyltransferase (HPRT) (Dewey *et al.* 1977). The animal can be populated with the mutant $HPRT^-$ cells. It is not at all difficult to put a chromosome fragment bearing a good HPRT gene into such cells by cell fusion techniques, and I am sure that the corrected cells would populate an animal. Sometimes a selective advantage is generated in this way, because the $HPRT^+$ cells in many situations will overgrow the $HPRT^-$ cells.

Brenner: Has anyone ever seen a case of recombination at the original site or is everything an addition to the genome?

Ruddle: All instances of herpes simplex thymidine kinase or mammalian HPRT insertion have occurred at sites unrelated to any homologous loci in recipient cells.

Brenner: That may prove to be a technical barrier which we find we have to cross. It all depends on whether a larger region of the chromosome is needed in order to express the regulation.

Weissmann: We can always fall back on yeast. Hinnen *et al.* (1978) have found that some genes taken out of yeast can be put back and integrated in the original site along with whatever they had been linked to.

Henry Harris: I think such experiments will work with mammalian cells. I don't think we shall have to fall back on yeast.

References

DEISSEROTH, A. & HENDRICK, D. (1978) Human α-globin gene expression following dependent gene transfer into mouse erythroleukemia cells. *Cell 15,* 55-64

DEWEY, M. J., MARTIN, D. W. Jr., MARTIN, G. R. & MINTZ, B. (1977) Mosaic mice with teratocarcinoma-derived mutant cells deficient in hypoxanthine phosphoribosyltransferase. *Proc. Natl. Acad. Sci. U.S.A. 74,* 5564-5568

HINNEN, A., HICKS, J. B. & FINK, G. R. (1978) Transformation of yeast. *Proc. Natl. Acad. Sci. U.S.A. 75,* 1929-1933

Genetics in an oocyte

J. B. GURDON, D. A. MELTON and E. M. DE ROBERTIS

MRC Laboratory of Molecular Biology, Cambridge

Abstract The combination of DNA cloning in bacterial plasmids and DNA injection into frog oocyte nuclei permits a novel type of genetic analysis in which the function of defined sequences of DNA may be more readily and precisely investigated than before. Purified segments of chromosomal DNA containing 5S genes of *Xenopus* have been injected into the nuclei of *Xenopus* oocytes and the products of these genes recognized as labelled 5S RNA. Since short linear molecules of DNA are degraded in oocytes, the function of particular regions of DNA near a 5S gene is best determined by excising these regions and joining them to other kinds of DNA. This is achieved most simply by inserting regions of 5S DNA into a plasmid. Our results with α-amanitin show that the activity of a 5S DNA promoter inserted into a plasmid can be detected independently of the plasmid's own promoters, using DNA injection into oocytes. The observation that plasmids containing variant 5S genes result in the synthesis of abnormal 5S RNA molecules illustrates the possibility of mutating recombinant plasmids and then testing the biological effect of these mutations by DNA injection.

The prospect of achieving a detailed understanding of the composition and function of genetic material has been substantially improved by the recent spectacular advances in methods of isolating, propagating and sequencing individual genes or segments of DNA. Such procedures do not, however, provide any direct test of the function or significance of particular sequences of DNA. In conventional genetic analysis, the function of genes and adjacent segments of DNA is determined by the collection and mapping of mutations which affect a particular kind of genetic expression. We discuss here an alternative to conventional genetic analysis; this takes advantage of our recent finding that purified genes or cloned segments of DNA can be correctly transcribed when injected into the nucleus of a frog oocyte.

GENETIC ANALYSIS BY GENE INJECTION

The usual form of genetic analysis entails the collection of mutants, usually selected in eukaryotes by their possession of a morphological or biochemical abnormality. Each mutation can be mapped by standard genetic methods. After collecting and mapping many such mutants, it is possible to define a region of DNA which determines a particular type of genetic expression. It is hard to take this type of analysis to a more profound level unless the particular segments of DNA which carry the mutations can be isolated and cloned.

The use of gene injection coupled with gene cloning can in principle make possible a more precise type of genetic analysis by a simpler procedure. It furthermore bypasses the usual limitation of the need to use an organism with a short reproductive cycle. The principle of this procedure is as follows. Segments of DNA are inserted into plasmids and grown in bacteria. The coding capacity of each clone of DNA is then determined either by screening for hybridization to a known gene product (e.g. globin complementary DNA), or by injecting DNAs into oocytes and looking for the synthesis of new RNAs or proteins. Plasmids carrying the inserted segments of DNA are then grown again in bacteria under conditions of mutagenesis or their DNA may be mutated *in vitro*. Mutated clones are then screened by oocyte injection to find ones in which the gene product is altered. DNA from mutant clones thus collected may be sequenced to determine the precise genetic changes (e.g. substituted bases) which have taken place. In this way individual base changes or deletions may be related to known defects in the production or composition of known gene products. According to this procedure the precise biochemical basis of a mutation may be determined first, and its biological effects subsequently investigated.

The immediate aim of this type of analysis is to determine the biological activity of known nucleotide sequences in and around a gene. This information may lead on to an investigation of how such genes are controlled, as explained at the end of this article. We now discuss the extent to which injected oocytes satisfy the requirements of this type of analysis.

BACKGROUND TO DNA INJECTION EXPERIMENTS WITH OOCYTES

Living oocytes of *Xenopus* were used to assay the activity of purified macromolecules in message-injection experiments originating some years ago (Gurdon *et al.* 1971). The first success in transcribing purified genes in living oocytes depended on the discovery that they must be injected into the nucleus (germinal vesicle) of an oocyte (Mertz & Gurdon 1977). Clear evidence for the accurate transcription of injected DNA resulted from the use of *Xenopus* 5S DNA (Brown & Gurdon 1977); the existence of detailed information about the normal product of 5S genes made it possible to recognize, by oligonucleotide analysis, the transcription of *X.borealis* 5S DNA in *X.laevis* oocytes, thereby demonstrating the correct transcription of virtually homologous genes. Transcripts synthesized by some kinds of injected DNA are correctly translated in frog oocytes, as shown by the synthesis of SV40 virion proteins 1 and 3 (De Robertis & Mertz 1977; Gurdon *et al.* 1978). The early stages in this work and its relationship to other injection experiments with amphibia have been reviewed (Gurdon 1977).

GENERAL CHARACTERISTICS OF DNA TRANSCRIPTION IN INJECTED OOCYTES

In most experiments, 5 ng of DNA is injected (or $\sim 10^9$ copies of small genes such as SV40, plasmids, or 5S genes). This is more than 100 times the total DNA content of an oocyte nucleus (35 pg) and about 400 times its chromosomal DNA content (see Gurdon 1974, for background explanation). When much smaller amounts of 5S DNA are injected, the efficiency of transcription may reach 20% of that with which the oocyte's own endogenous 5S genes are transcribed (Gurdon & Brown 1978). With larger injection amounts, only a few per cent of the injected genes are thought to be transcribed, but there is no incorrect transcription from the excess genes injected, and there is no transcription from genes injected into oocyte cytoplasm (Mertz & Gurdon 1977; Brown & Gurdon 1977). Therefore, at least for those genes so far tested, it seems that some of the DNA deposited in the nucleus is transcribed entirely correctly, while the rest is not transcribed at all. Transcription starts within an hour of DNA injection, and takes place accurately as soon as it starts (Gurdon & Brown 1978). Therefore it is convenient in most experiments to inject DNA already mixed with a suitable label (usually [^{3}H]GTP), and to freeze oocytes about 6 hours after injection. Experiments involving the injection of DNA containing transfer RNA genes have shown that the final product, tRNA, is correctly processed from a precursor and secondarily modified when transcribed in oocytes (Kressmann *et al.* 1978*a*; Cortese *et al.* 1978). The nucleus of an oocyte is extraordinarily large ($\sim 400\ \mu$m diameter) and it does not require special skill to deposit some DNA in the nucleus of about half of all oocytes injected. If, in certain experiments, it is desirable to succeed in a higher percentage of cases, it may be helpful to centrifuge oocytes bringing the nucleus to the surface (Kressmann *et al.* 1978*b*). However centrifugation can be harmful to oocytes and tends to reduce their transcriptional activity.

Accurate transcription of injected DNA has been best documented for genes which code for 5S RNA and transfer RNA, and is less certain for other kinds of DNA. From Table 1, it can be seen that genes normally transcribed by polymerase III work very well in this system. Some genes normally transcribed by polymerase II are also transcribed by this enzyme in living oocytes, but not necessarily with the same accuracy as in normal cells. It is also established that some genes (ribosomal DNA) normally transcribed by polymerase I are correctly transcribed in injected oocytes though it is not known if this is done by the correct enzyme. We conclude that, for some kinds of gene at least, the transcription of injected DNA is sufficiently accurate for this system to be used for genetic analysis.

An important advantage of gene injection into oocytes is the information it can give about immediate gene products and about the process of transcription. In normal cells the direct products of a gene can be seen only after

TABLE 1

The transcription of different types of purified genes after injection into oocyte nuclei (Arabic numerals in the last four columns refer to references below)

Type of gene[a]	*Normal polymerase requirement*	*α-Amanitin sensitivity*[b]	*Synthesis of correct product*	*Correct sites of initiation and termination*[a]	*Strand and spacer gene selectivity*
Dytiscus rDNA	I	—	—	Yes (EM)[1]	—
Xenopus rDNA (in plasmid)	I	—	—	Yes (EM)[2]	—
SV40	II	II[3,10]	Yes[3,4,5,6]	—	—
Plasmid (pMB9)	—	II[7]	—	—	—
Drosophila histone (in plasmid)	II	—	Yes[5]	—	—
Sea urchin histone (in plasmid)	II	—	Yes[8]	—	—
Xenopus 5S DNA	III	III[10]	Yes[9]	Yes[9]	Yes[9]
Xenopus tDNA	III	—	Yes[11,12]	—	—
Nematode tDNA	III	III[13]	Yes[13]	—	—

[a]rDNA, DNA containing 28S and 18S ribosomal genes. tDNA, DNA containing transfer RNA genes. EM, evidence from electron microscopy of nuclear spreads.
[b]The Roman number indicates the type of polymerase used, as judged by sensitivity to α-amanitin (for a review, see Roeder *et al.* 1976).

References
(1) Trendelenburg *et al.* 1978; (2) Trendelenburg & Gurdon 1978; (3) Mertz & Gurdon 1977; (4) Laskey *et al.* 1978; (5) De Robertis & Mertz 1977; (6) Gurdon *et al.* 1978; (7) Table 2 of this paper; (8) Etkin 1978, personal communication; (9) Brown & Gurdon 1977; (10) Gurdon & Brown 1978; (11) Kressmann *et al.* 1978*a*; (12) Kressmann *et al.* 1978*b*; (13) Cortese *et al.* 1978.

several hours of labelling, by which time the initial product may have been extensively processed. Because of the enormous number of copies of *one kind* of gene injected into an oocyte, up to half of all transcripts synthesized by that oocyte can be the products of that type of gene. Thus transcripts of injected genes can be seen within 10 min of label being provided (Gurdon & Brown 1978). Precise information may be gained about the transcription of known genes by using electron microscopy to examine spreads of nuclei recovered from DNA-injected oocytes. The methods required were previously worked out for amphibian oocytes (Miller & Bakken 1972; Scheer *et al.* 1977). Though technically difficult, this procedure has been successfully applied to oocytes injected with *Dytiscus* ribosomal DNA (Trendelenburg *et al.* 1978), and with a plasmid bearing *Xenopus* ribosomal genes (Trendelenburg & Gurdon 1978). By this procedure, it should be possible to identify the position at which transcription is initiated and terminated in cloned segments of DNA; this is a type of information to which there is no direct access by other current procedures.

THE ANALYSIS OF SMALL SEGMENTS OF DNA IN INJECTED OOCYTES

The DNA molecules so far discussed are relatively large — several thousand base pairs long, that is, much too long to be useful for determining the function of lengths of DNA which might be expected to serve as a promoter, repressor-binding sequence, etc. A survey of the effect of differing DNA configuration on transcriptional activity has yielded the following conclusions (Wyllie *et al.* 1978). All DNA molecules deposited in the cytoplasm of an oocyte are degraded, apparently by an endonuclease activity. On the other hand, DNA molecules deposited in the nucleus are entirely stable so long as they are circular, but linear molecules are progressively degraded at their ends. The stabilization of DNA molecules in an oocyte nucleus is related to the presence of a component normally restricted to the nucleus and not to the exclusion from the nucleus of endonuclease activity (Wyllie *et al.* 1977). It is known that DNA molecules stabilized in the nucleus have become associated with an equal amount of protein, and contain nucleosomes (Wyllie *et al.* 1978). Such molecules have probably been converted into mini-chromosome-like structures. Linear DNA molecules injected into an oocyte nucleus are not degraded extremely rapidly, and it has been possible to demonstrate that a single 5S gene repeat of *Xenopus*, 700 base pairs long, contains its own promoter (Brown & Gurdon 1978). For the analysis of much shorter lengths of DNA, it is preferable to use other procedures.

The simplest procedure is to insert pieces of DNA into plasmids, and to inject circular recombinant DNA. For this to be a valid approach, we must be

TABLE 2

The α-amanitin sensitivity of plasmid and insert transcription in oocytes

DNA injected	*α-Amanitin injected*	*pMB9 transcripts*	*5S transcripts*
pMB9	—	6.5	—
	1 μg/ml	0.7	—
pMB9-5S	—	3.9	18.1
	1 μg/ml	0.2	31.5
None	—	0.1	0.5

pMB9 is a plasmid. pMB9-5S is a plasmid containing four adjacent repeats of *Xenopus laevis* 5S DNA inserted into the Hind III site (pXlo.8 kindly supplied by D. D. Brown). About 5 ng of DNA, with or without α-amanitin as indicated, was injected into oocytes which were also supplied with [^{3}H]GTP. After 6 h, oocytes were frozen; RNA was extracted and either hybridized to pMB9 DNA on a filter or electrophoresed to determine the amount of [^{3}H]5S RNA. The numbers in the last two columns represent the percentage of total [^{3}H]RNA which hybridized to pMB9 DNA, or which migrated on a gel as 5S RNA.

sure that the transcription of DNA inserted into a plasmid is dependent on *its own* promoter and not on a promoter contained in the plasmid. Table 2 shows that α-amanitin, at the concentration expected to suppress polymerase II activity, suppresses transcription from plasmid DNA, but does not at all decrease that of the inserted 5S genes (Fig. 1). Table 2 also shows that the termination of transcription of 5S genes in plasmid DNA takes place normally, since there is no read-through of transcription into the plasmid region in the presence of α-amanitin. Even in the absence of α-amanitin, the efficiency of transcription of 5S genes is about 50 times greater than that of the surrounding plasmid DNA (Table 3). Both these observations show that the 5S DNA promoters and terminators are used for transcription when these genes are inserted into a plasmid.

It is a curious but reproducible observation that α-amanitin substantially enhances 5S RNA synthesis from plasmids containing inserted 5S genes (Table 2). A similar observation has been made with plasmids containing

→

FIG. 1. Gel analysis of the transcription of the recombinant plasmid pXlo.8 (see legend to Table 2) in *Xenopus* oocytes. Oocytes were injected with a mixture of DNA and [^{3}H]GTP and frozen after 5 h. RNA was extracted and analysed by polyacrylamide gel electrophoresis and fluorography. The arrow indicates the position of 5S RNA.

A, no DNA injected. B, pXlo.8 DNA (5 ng/oocyte) injected. C, pXlo.8 DNA (5 ng/oocyte) and α-amanitin (1 μg/ml in injection solution of 50 nl) injected.

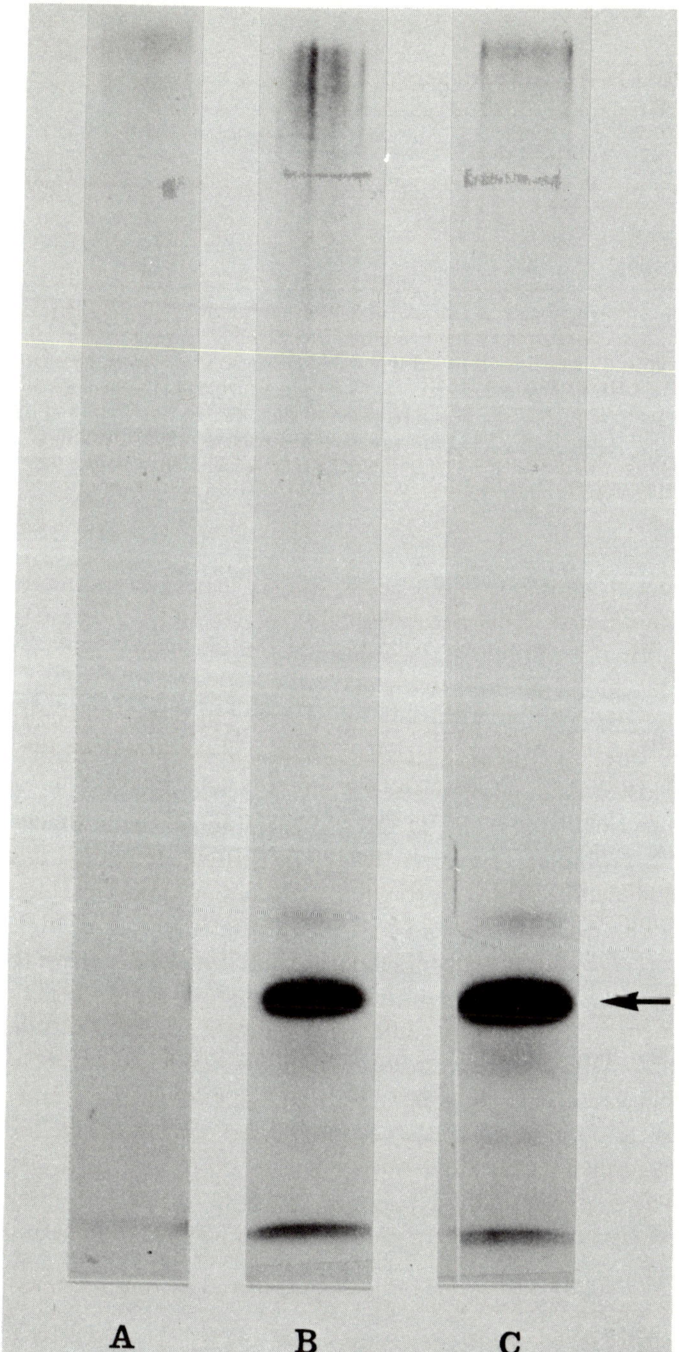
A
B
C

TABLE 3

The transcription of plasmid and insert DNA in oocytes

	pMB9	*5S genes*	*Ratio* $\frac{5S\ genes}{pMB9}$
DNA (Kb)	5.2	0.5	0.09
Transcripts (% of total)	3.9	18.1	4.6
Relative rate of transcription	—	—	50.1

DNA from the recombinant plasmid pX1o.8 (see legend to Table 2) was injected with [^{3}H]GTP into oocytes. After 6 h, RNA was extracted and hybridized to pMB9 DNA or analysed on gels; the results came from one of the experiments included in Table 2. The top row shows the length (Kb = 1000 base pairs) of DNA in the plasmid region or 5S gene insert region of the recombinant molecules. The second row shows the rate of transcription of these two regions (pMB9 of 5S transcripts as a % of total transcripts). The third row gives the relative rate of transcription (second row divided by top row). It is assumed that the efficiency of recognition of transcripts of hybridization and by gel analysis is similar.

tRNA genes (Cortese *et al.* 1978); tRNA genes, like 5S genes, are normally transcribed by polymerase III. This effect could be explained if polymerase II molecules which normally initiate in the plasmid region sometimes read through into the 5S region, not always on the correct strand. α-Amanitin might eliminate this read-through, permitting more efficient transcription of the 5S region by polymerase III.

The use of plasmids containing an inserted DNA sequence should serve to determine whether an isolated segment of DNA containing a structural gene also contains its own promoter. However, it may be difficult to extend this type of analysis to the identification of promoters, regulatory sequences, etc., when these are not immediately followed by a DNA sequence coding for a recognizable product. This difficulty could be overcome if it were possible to ligate, by enzymic means, a small segment of DNA containing a promoter to another DNA segment known *not* to contain a promoter, such as certain untranscribed satellite DNAs. The addition of a promoter sequence to satellite DNA should cause it to be transcribed, and this would be readily detectable by hybridization of labelled RNA from injected oocytes to satellite DNA. Preliminary results suggest that this design of experiment can give meaningful information. Eventually it might be possible to develop along these lines a general assay procedure for the recognition of promoters and regulatory sequences.

GENETICS IN AN OOCYTE — REALITIES AND POSSIBILITIES

The reality of being able to carry out a genetic analysis as outlined above has been demonstrated for 5S genes of *Xenopus* cloned in bacteria. Of 15 independent clones containing a single 5S gene insert, two have resulted in the synthesis of 5S RNA of atypical length, when tested in an oocyte (Brown & Gurdon 1978). In one case the synthesis of shorter than normal molecules implies the existence of premature termination; in the other, the production of extra-long molecules suggests a genetic change which causes a weakening of the normal termination signal. It is not known whether these changes represent mutations which arose during genetic manipulation and cloning, or whether some 5S genes of this type are present in normal cells. Sequencing of the 5S genes in these aberrant clones will be able to show the exact molecular basis of the functional changes.

It is clear that oocyte injection works very well as an assay procedure for genes normally transcribed by polymerase III. It has yet to be shown that polymerase I and polymerase II genes are handled correctly, i.e. that their transcription in injected oocytes depends on the use of their own promoters. However if we assume that such classes of genes will in time be shown to work correctly in oocytes, it is appropriate to consider some of the uses to which oocyte injection may eventually be put. We must make one other assumption, namely that the length of DNA which can be readily cloned in plasmids or in λ phage (about 15 000 base pairs) is sufficient to include a promoter as well as other regions of DNA presumed to control the promoter's activity. In this case, it should be possible by oocyte injection to determine the presence and location of promoter regions in cloned segments of chromosomal DNA. It would then be possible to isolate proteins from different cell types which have a very high binding affinity for certain types of cloned DNA, as has been done by Weideli and collaborators (1978). The biological effect of these DNA-binding proteins can then be tested by using oocytes as a neutral transcription assay system. This procedure would include the screening of random DNA clones so as to identify some which contain genes expressed in somatic cells but not in oocytes, and other clones which are expressed in oocytes but not in somatic cells. In such a way, it may be possible to use oocytes to help identify molecules which control the activity of genes or their promoters.

Though frogs do not impinge significantly on human biology, we hope to have shown here that their oocytes may have uniquely favourable properties which permit them to be used for an unconventional, but potentially very precise, type of genetic analysis. The success with which DNAs of several unrelated species are transcribed in frog oocytes could mean that this experimental system may have a wide general application.

ACKNOWLEDGEMENTS

We thank D. D. Brown for collaboration in much of the work reviewed, and for supplying DNA of the recombinant plasmid pXlo.8.

References

BROWN, D. D. & GURDON, J. B. (1977) High fidelity transcription of 5S DNA injected into *Xenopus* oocytes. *Proc. Natl. Acad. Sci. U.S.A. 74,* 2064-2068

BROWN, D. D. & GURDON, J. B. (1978) Cloned single repeating units of 5S DNA direct accurate transcription of 5S DNA when injected into *Xenopus* oocytes. *Proc. Natl. Acad. Sci. U.S.A. 75,* 2849-2853

CORTESE, R., MELTON, D. A., TRANQUILLA, T. & SMITH, J. D. (1978) Cloning of nematode tRNA genes and their expression in the frog oocyte. *Nucl. Acids Res. 5,* 4593-4611

DE ROBERTIS, E. M. & MERTZ, J. E. (1977) Coupled transcription-translation in DNA-injected *Xenopus* oocytes. *Cell 12,* 175-182

GURDON, J. B. (1974) *Control of Gene Expression in Animal Development,* Harvard University Press, Cambridge, Massachusetts

GURDON, J. B. (1977) Egg cytoplasm and gene control in development. *Proc. R. Soc. (Lond.) B Biol. Sci. 198,* 211-247

GURDON, J. B. & BROWN, D. D. (1978) The transcription of 5S DNA injected into *Xenopus* oocytes. *Dev. Biol. 67,* 346-356

GURDON, J. B., LANE, C. D., WOODLAND, H. R. & MARBAIX, G. (1971) The use of frog eggs and oocytes for the study of messenger RNA and its translation in living cells. *Nature (Lond.) 233,* 177-182

GURDON, J. B., WYLLIE, A. H. & DE ROBERTIS, E. M. (1978) The transcription and translation of DNA injected into oocytes. *Phil. Trans. R. Soc. Lond. B Biol. Sci. 283,* 367-372

KRESSMANN, A., CLARKSON, S. G., PIROTTA, V. & BIRNSTIEL, M. L. (1978*a*) Transcription of cloned tRNA gene fragments and subfragments injected into the oocyte nucleus of *Xenopus laevis. Proc. Natl. Acad. Sci. U.S.A. 75,* 1176-1180

KRESSMANN, A., CLARKSON, S. G., TELFORD, J. L. & BIRNSTIEL, M. L. (1978*b*) Transcription of *Xenopus* $tDNA_1^{met}$ and sea urchin histone DNA injected into the *Xenopus* oocyte nucleus. *Cold Spring Harbor Symp. Quant. Biol., 42,* 1077-1082

LASKEY, R. A., HONDA, B. M., MILLS, A. D., MORRIS, N. R., WYLLIE, A. H., MERTZ, J. E., DE ROBERTIS, E. M. & GURDON, J. B. (1978) Chromatin assembly and transcription in eggs and oocytes of *Xenopus laevis. Cold Spring Harbor Symp. Quant. Biol. 42,* 171-177

MERTZ, J. E. & GURDON, J. B. (1977) Purified DNAs are transcribed after microinjection into *Xenopus* oocytes. *Proc. Natl. Acad. Sci. U.S.A. 74,* 1502-1506

MILLER, O. L. & BAKKEN, A. H. (1972) Morphological studies of transcription. *Acta Endocrinol. Suppl. 168,* 155-177

ROEDER, R. G., SCHWARTZ, L. B. & SKLAR, V. E. F. (1976) Function, structure and regulation of eukaryotic nuclear RNA polymerases, in *Molecular Biology of Hormone Action (34th Symp. Soc. Dev. Biol.),* pp. 29-52, Academic Press, New York

SCHEER, U., TRENDELENBURG, M. F., KROHNE, G. & FRANKE, W. W. (1977) Lengths and patterns of transcriptional units in the amplified nucleoli of oocytes of *Xenopus laevis. Chromosoma (Berl.) 60,* 147-167

TRENDELENBURG, M. & GURDON, J. B. (1978) Transcription of cloned *Xenopus* ribosomal genes visualized after injection into oocyte nuclei. *Nature (Lond.) 276,* 292-294

TRENDELENBURG, M., ZENTGRAF, H., FRANKE, W. W. & GURDON, J. B. (1978) Transcription patterns of amplified *Dytiscus* ribosomal DNA after injection into *Xenopus* oocyte nuclei. *Proc. Natl. Acad. Sci. U.S.A. 75,* 3791-3795

WEIDELI, G., SCHEDL, P., ARTAVANIS-TSAKONAS, S., STEWARD, R., YUAN, R. & GEHRING, R. (1978) Purification of a protein from unfertilized eggs of *Drosophila* with specific affinity for a defined DNA sequence and the cloning of this DNA sequence in bacterial plasmids. *Cold Spring Harbor Symp. Quant. Biol. 42,* 693-700

WYLLIE, A. H., GURDON, J. B. & PRICE, J. (1977) Nuclear localization of an oocyte component required for the stability of injected DNA. *Nature (Lond.) 268,* 150-152

WYLLIE, A. H., LASKEY, R. A., FINCH, J. & GURDON, J. B. (1978) Selective DNA conservation and chromatin assembly after injection of SV40 DNA into *Xenopus* oocytes. *Dev. Biol. 64,* 178-188

Discussion

Henry Harris: I accept your argument but how long does it take to methylate a strip of DNA?

Gurdon: In general methylation works very efficiently if one of the two strands of DNA is already methylated. In normal circumstances where there is, say, replicating DNA, it would probably be a very rapid process. Where neither of the strands is methylated it might take longer, but I can't guess how long. Another piece of information is that oocytes have no detectable methylating activity as far as we know. So I think it is unlikely, but not formally disproved, that the DNA is remethylated before being used. We would like to be able to test that. We ought to recover the injected DNA and ask whether it has been methylated. The difficulty is that many of the injected genes are not in fact transcribed, as seen in the electron microscope pictures. Therefore to ask that question we would have to isolate the injected genes which are being transcribed. That is technically difficult but not impossible. We haven't done that yet.

Henry Harris: Have you any ideas about how you might look for the presence of a gene which is located some distance away from the structural gene and which might be affecting the expression of the structural gene?

Gurdon: The approach which I have described depends on having the control regions within the kind of size of DNA which can be cloned. If the control region was so far away from the gene that you would never get it into one cloned piece of DNA, this approach would not work.

Henry Harris: Can you think of ways of putting in two defined pieces of cloned DNA, where one does not contain a known structural gene?

Gurdon: What we can do is not bother about a structural gene at all. All we are really interested in is a promoter. We hope that the control sequences that govern the activity of a promoter are near it. If there is a large gap between the promoter and the structural gene, in principle that won't matter because we would not necessarily use that structural gene anyway. What we are really interested in is the regulation of the activities of a promoter.

Brenner: The argument says, roughly speaking, that if eukaryotic DNA is randomly cloned, one out of 10 or one out of 20 genes of average size might contain the promoter. So one might as well look for promoters by assaying 20 random clones by your method and asking later, with the same techniques, what that is connected to, by just mapping the clones.

Gurdon: Henry Harris is worried that there might be a promoter with its control sequence far away.

Henry Harris: I am not worried about it. I am just wondering whether you have some ideas on how to get at that problem.

Brenner: We can probably guess the maximum distance from the human haemoglobins, where the structure of the region is known.

Weissmann: The problem is that the first transcript that we know of, which is 1500 nucleotides long and from which the messenger is carved out, has not been proved to be the primary transcript.

Brenner: But aren't human haemoglobin γ, β and δ all on a piece of about 10 kilobases? Certainly γ is expressed differently from β, although δ might pose some questions.

Henry Harris: In the sort of system I work with one would select for a structural gene and use γ-irradiation to strip off everything on either side of it. As the radiation dose is increased, less and less adventitious material is retained with the selected structural gene, and you can see whether a great increase or decrease in the amount of the structural gene product occurs at a given point.

Gurdon: That is part of the general approach discussed earlier, in which one moves segments of DNA around and sees whether the control sequence is still connected with the gene. One removes bits of it until it isn't connected. That can clearly be done and it may turn out that there is no problem to be discussed.

Bodmer: You said that there was some RNA polymerase in areas that weren't being transcribed. Does that mean that the polymerase moves round the non-transcribed bits and doesn't fall off?

Gurdon: The sizes of the polymerase and the nucleosome are rather similar. If an expert like Michael Trendelenburg points it out to me, I believe I see a larger blob at the base of a transcript than where nucleosomes are believed to be present. One can certainly not conclude that the blobs in between the transcribed regions are necessarily polymerase.

Bodmer: What about using an antibody to the RNA polymerase to see whereabouts it is?

Gurdon: The resolution of labelled antibodies seen by autoradiography would be too low.

Gardner: Is the difference between your satellite–satellite and satellite–28S transcription, namely a 12-fold improvement in the latter, to be interpreted as showing that there are good and bad promoters, or is it that there is contamination?

Gurdon: I mentioned in my talk some preliminary results of an experiment carried out in collaboration with Douglas Melton. Calf satellite DNA is not naturally transcribed and is also transcribed to only a minimal extent after injection into *Xenopus* oocytes. Douglas Melton ligated satellite DNA (a) to satellite DNA (containing no promoter), (b) to a 28S ribosomal DNA fragment (believed not to have a promoter), or (c) to an 18S ribosomal DNA fragment (believed to contain a promoter). After injection of these samples independently into oocytes, substantial transcription of the satellite DNA was found only in the 18S ribosomal DNA–satellite DNA complex. The 28S ribosomal DNA–satellite DNA complex gave a little more satellite transcription than the satellite–satellite background level. This is not due to contamination. One possible interpretation is that there is a very weak promotional effect within the 28S ribosomal DNA segment. There have been other suggestions that this could be true. On the other hand, it may be that as we increase the total length of DNA, this itself slightly raises the background of the satellite transcription. The satellite insert is 1400 base pairs long and two of these make 2800; whereas a 28S segment adds another 5000 base pairs. Transcription in injected oocytes is favoured by increased DNA length; the ideal control would have been about five satellites joined together. Technically that is very difficult and Mr Melton has had to work very hard to make this work at a high efficiency. What most people do is to ligate and then have a very strong selection system for the successfully ligated molecules.

Brenner: That is called cloning!

Gurdon: To show a result one has to get sufficiently efficient ligation to produce *enough* of the right combinations in the injected DNA.

Bodmer: Why don't you clone them again?

Gurdon: That would require biohazard facilities not available to us, and would therefore be prohibited.

Ruddle: Your experimental system might be well suited to studying the integration of foreign DNA into chromosomes and the propagation of integrated sequences through the developing organism. There is also the possibility of recovering integrated sequences in the germ plasm. Are you looking into this?

Gurdon: We are starting experiments on that but most of these things have technical difficulties.

Ruddle: Can you inject large amounts of DNA without damaging the oocyte?

Gurdon: No. The fertilized eggs that are used for this are extraordinarily sensitive to the injection of large amounts of DNA (Gurdon 1974). It is not clear why that is so, but it doesn't matter too much. One is still able to inject into eggs about 100 000 genes of SV40 size instead of 1000 million, as can be done for oocytes.

Ruddle: So it really depends on the efficiency of integration?

Gurdon: Yes.

Nossal: Could you say a bit more about why oocytes allow transcription of artificially introduced genes to take place readily? If you could get enough DNA into a somatic cell, what would happen?

Gurdon: An oocyte is transformed rather rapidly by hormones into the cell which we call an egg, which then goes through a large number of divisions very quickly. Probably in preparation for this, an oocyte has a vast excess of all the components that are essential for very early development—DNA and RNA polymerases, histones, proteins, ribosomes, mitochondria and so on. These molecules are sufficient for at least 10 000 cells that are formed very rapidly after fertilization. In this respect there is nothing qualitatively special about oocytes. I think it is mainly a quantitative advantage that oocytes have.

Klein: In your SV40 experiment, why is virion protein, a late product, expressed? How is the early/late switch controlled? Is it technically possible to see whether the DNA of an inactivated X chromosome would be transcribed?

Gurdon: We believe that there is no replication of SV40 DNA in oocytes, which makes it surprising if there is late gene expression, such as one gets with VP3, which is expressed readily. An oocyte may have in it, among these other things, a reserve of the component that is normally provided by SV40 replication. It is also just conceivable that when a few SV40 molecules are injected in a semi-replicated state they are somehow able to complete replication. We think that SV40 replication can be completed but not initiated in oocytes; this could be sufficient to provide a replication-dependent component needed for late gene expression. Incidentally, predominantly late message is accumulated in injected oocytes (Laskey *et al.* 1978).

On the second question, I should say that we would like to transplant nuclei with inactive genes in the X chromosome and follow their reactivation. We feel that is likely to be technically difficult because we may not get enough product with the limited number of genes that can be introduced when whole nuclei are injected. Purified heterochromatin from an X chromosome would be a more favourable material for this experiment—if it exists.

Jacob: What is the ratio of transcribed to non-transcribed circular plasmids in your conditions?

Gurdon: That is a very difficult question to answer. With electron microscope work, as you know, it is questionable whether one can ever make a valid quantitative estimate. Most of the injected DNA is in a big clump, not evenly spread out. One has to look for just those DNA molecules which happen to be isolated. A very rough guess is that 1 or 2% of injected DNA molecules are transcribed. This fits with other biochemical data we have, when large doses of DNA are injected. When we put in limited amounts of DNA the figure of 20% efficiency seems to be right. The electron microscope work seems to show that genes are either fully transcribed or not at all. Injected oocytes provide an unusual situation where the genes are in gross excess of the normal polymerase capacity of the cells. Something may happen to the DNA in front of the gene, maybe near a promoter region, which essentially makes the gene open. RNA polymerases can then freely attach and the open gene becomes fully packed. There could have been another situation where all RNA polymerases were just spread out evenly over all the DNA, in which case occasional transcripts would be seen on most of the genes. Probably the answer is that under our standard conditions, with a very large DNA input, a minority of injected genes are used, though we can have a higher efficiency by putting in smaller numbers of genes.

Bodmer: You said that the oocyte was the same as the somatic cell but quantitatively different. In the somatic cell, we know by hybridization that the inactive X is not reactivated. So if one did get it by nuclear transplantation, it seems to me that that would be as one might expect—the oocyte is very different from purified DNA.

Gurdon: I didn't mean to imply that there are no qualitative differences between oocytes and somatic cells. Clearly there are some because when we inject nuclei we find oocyte-specific genes being turned on (De Robertis & Gurdon 1977). We know that there are some molecules in an oocyte which have an oocyte-specific effect. I was dealing specifically with the question of why the oocyte is so good at transcribing injected DNA.

Weissmann: Did you check the size of the transcript in the satellite insert? It should be pretty long. Is the satellite sequence transcribed all the way through or are there fortuitous terminations somewhere?

Gurdon: We have not checked the size. We used a 1400 base pair length of satellite DNA. I would be rather surprised if it isn't transcribed all the way through. We would have some difficulty in seeing it if it weren't.

Brenner: Are the injected DNAs all linear molecules?

Gurdon: Yes. That also limits the efficiency with which the experiment works.

Koprowski: How sure are you that SV40 DNA is not replicated?

Gurdon: We injected SV40 DNA with a density label. This is a very effective way of recognizing complete SV40 replication when eggs are used (Laskey & Gurdon 1973). Eggs are the product of oocytes and they are extremely active in inducing DNA synthesis. The same experiment done with oocytes shows no evidence of replication. Of course one can never be sure that there aren't a small minority of genes which are replicated. There is always a limit to the sensitivity of any experiment. If you like to say that just one molecule being replicated is enough to make the necessary components, we can't exclude that.

Koprowski: We have found similar problems after infection of preimplantation mouse embryos with SV40. In contrast to mouse somatic cells which are non-permissive for SV40 infection, embryonic cells were damaged by SV40, which inhibited formation of blastocysts after infection of two-cell embryos. Although immunofluorescence assays (with anti-SV40 antibody) indicated the presence of viral capsid protein, we were not certain whether replication of the virus actually took place. When SV40-infected embryos were exposed to α-amanitin, however, formation of viral capsid protein was inhibited without deleterious effect on the embryonic cells (Abramczuk *et al.* 1978). α-Amanitin blocks synthesis of SV40 capsid protein in susceptible (human or primate) somatic cells through inhibition of polymerase II (Jackson & Sugden 1972). Therefore, we came to the conclusion that SV40 is capable of synthesizing capsid protein after infection of preimplantation embryos.

Although susceptible to infection with SV40, preimplantation mouse embryos are resistant to infection with polyoma virus, for which mouse somatic cells are permissive. Even microinjection of the virus (Abramczuk *et al.* 1978) into two-cell embryos will not result in infection of the embryos. Only at the blastocyst stage does resistance to polyoma infection decrease and resistance to SV40 infection increase.

Gurdon: Oocytes are very favourable for biochemical work. One can inject mouse eggs and other sorts of eggs, and indeed somatic cells, but the oocytes are so large that the analysis of 5S RNA is normally conducted, in our hands, on material from one-fifth of one cell.

Peto: Do oocytes also differ from ordinary cells in not letting methylation continue, or in deliberately demethylating DNA? It has been suggested that cellular differentiation involves specific methylation as a switching mechanism, and that oocytes might deliberately demethylate their DNA in order to reset all such switches to zero.

Gurdon: That could happen if you do not accept our observation (Brown & Gurdon 1978) that methylated DNA from red blood cells is immediately transcribed. We assume that the DNA that we inject, which is methylated, is not demethylated before it is transcribed. If we could formally disprove demethylation, we would eliminate demethylation as a derepressing mechanism.

Henry Harris: Is there any enzyme that takes methyl groups off nucleic acid?

Weissmann: There could be repair.

Fincham: You were talking of the possibility of looking for regulatory molecules that you might inject, which might interact with your promoter sequences. Supposing you can find such molecules, have you thought of looking for the DNA sequences which code for them and reconstructing some of the networks and cascades that have been postulated from work with eukaryotes?

Gurdon: That is looking a certain distance ahead. In principle I think it ultimately comes down to how much biochemical work one can do, and whether one has enough material to find very small amounts of message which make small amounts of protein. One would like to use DNA-binding studies as a way of fishing out small amounts of one type of protein. That doesn't necessarily give the message or the gene but it seems to be a considerable help towards it. This is a problem we haven't thought of in detail.

Brenner: This is like cloning the genes for the processing enzymes in *E.coli* and hoping they haven't got inserts themselves!

Milstein: Will the egg be able to process the intervening sequences?

Gurdon: One would expect the SV40 transcripts to be processed to make VP1 and VP3. My assumption is that oocytes will process out intervening sequences, and that they may even be a useful source of processing enzymes. In the case of microinjected yeast tyrosine tRNA genes, which have an intervening sequence of 14 nucleotides, it has recently been shown that the oocyte can splice the transcripts correctly (De Robertis & Olson 1979).

References

Abramczuk, J., Vorbrodt, A., Solter, D. & Koprowski, H. (1978) Infection of mouse preimplantation embryos with simian virus 40 and polyoma virus. *Proc. Natl. Acad. Sci. U.S.A. 75,* 999-1003

Brown, D. D. & Gurdon, J. B. (1978) Cloned repeating units of 5S DNA direct accurate transcription of 5S RNA when injected into *Xenopus* oocytes. *Proc. Natl. Acad. Sci. U.S.A. 75,* 2849-2853

De Robertis, E.M. & Olson, M.V. (1979) Transcription and processing of yeast tyrosine tRNA genes microinjected into frog oocytes. *Nature,* in press

Gurdon, J. B. (1974) Molecular biology in a living cell. *Nature (Lond.) 248,* 772-776

Jackson, A. H. & Sugden, B. (1972) Inhibition of α-amanitin of simian virus 40-specific ribonucleic acid synthesis in nuclei of infected monkey cells. *J. Virol. 10,* 1086-1089

LASKEY, R. A. & GURDON, J. B. (1973) Induction of polyoma DNA synthesis by injection into frog egg cytoplasm. *Eur. J. Biochem. 37,* 467-471

LASKEY, R. A., HONDA, B. M., MILLS, A. D., MORRIS, N. R., WYLLIE, A. H., MERTZ, J. E., DE ROBERTIS, E. M. & GURDON, J. B. (1978) Chromatin assembly and transcription in eggs and oocytes of *Xenopus laevis. Cold Spring Harbor Symp. Quant. Biol. 42,* 171-177

DNA repair defects and chromosome instability disorders

PAUL E. POLANI

The Prince Philip Research Laboratories, Guy's Hospital Medical School, London

'The motto ... should be, Seek simplicity and distrust it'
(Alfred North Whitehead, The Tarner Lectures, 1919)

Abstract Xeroderma pigmentosum (XP), Fanconi anaemia (FA), ataxia telangiectasia (AT) and Bloom disease (BS) are four rare autosomal recessive disorders in which there is defective DNA repair and/or chromosome instability and proneness to malignancy. Between 80 and 90% of patients with XP have a defect, demonstrable at cell level, of excision of DNA lesions induced by ultraviolet rays, while the remainder have a cellular error of post-replication repair. XP cells are also deficient in repairing DNA damage caused by a variety of chemical mutagens. There are at least five different complementation groups of the first, or classical, type of XP (A to D, etc.). Apparently group C patients, as well as those with defective post-replication repair, do not show the progressive neurological illness found in a proportion of the other patients. AT is heterogeneous clinically and genetically. Clinically it presents with a progressive neurological illness, progressive telangiectases and a developmental disorder of the thymus. AT is characterized by sensitivity to X-rays and AT cells are unable to repair γ-ray-induced damage to bases in the DNA. It appears that in many cases of the disorder a chromosomally marked cellular clone is found. In BS the main defect, which results in growth retardation, sun-induced lesions of the face and susceptibility to infection, appears to be a slow DNA chain maturation during DNA synthesis. An increase of sister chromatid exchanges is characteristically seen in the chromosomes of cultured BS cells. In FA, in which there is progressive pancytopenia with eventual bone marrow exhaustion and a tendency to haemorrhage and infection, the cellular defect seems to consist of faulty removal or repair of cross-links in the DNA. In this condition, as in BS and AT, various structural chromosome changes are detected in cultured cells.

Patients with XP develop skin cancers in early life and often malignant melanomas. In the other three disorders, in which an immune deficiency is often present, leukaemia and related proliferative disorders are a frequent cause of death while other malignancies also occur. There is some evidence that points to an increased risk of malignancy in heterozygotes who carry the FA and AT genes.

There are many aspects of the human chromosome complement that could usefully be reviewed at this time and the embarrassment is one of choice. Perhaps the clearest realization of the long strides made by human cytogenetics, indeed by human genetics, comes from considering the paper by L. S. Penrose, in the 1959 Ciba Foundation Symposium on *Biochemistry of Human Genetics,* other papers in that Symposium, and the papers in the *British Medical Bulletin* of 1961, written in the early days of modern human cytogenetics. The chain explosion which had then just started was due to four major technical advances: first, in the beginning, the methods of culturing, blocking cell division at metaphase and spreading of bone marrow and fibroblast chromosomes; second the introduction by Moorhead *et al.* (1960) of a technique for culturing transformed lymphocytes; thirdly the introduction by Caspersson *et al.* (1969) of fluorescence chromosome banding and the extension by him and his collaborators (Caspersson *et al.* 1970), and by many others, of fluorescence banding and many other banding methods to the study of human chromosome morphology—normal, abnormal and variant; finally, and with far-reaching consequences for the study and mapping of the human genome (and many other problems of cell biology, malignancy, etc.), the introduction of relatively easy and efficient techniques of cell fusion (Harris & Watkins 1965), which, aided by cell selection procedures (Littlefield 1964), opened up new ways of studying cell hybridization (Barski *et al.* 1960; Ephrussi *et al.* 1964). We are at the threshold of major advances from the combined application of knowledge accruing from chemical and physical studies on the organization of DNA and of chromosomes, on detailed structuring of the gene, and from the applications of recombinant DNA technology and DNA sequencing to a variety of molecular problems. These advances augur well for a deeper understanding of the organization of the human chromosomes and genes, for the mapping of the human genes, and for an understanding of the way genes of eukaryotes function, the way they are controlled, especially during differentiation and development, and the deviations from normal control in abnormal situations.

It is clear that within this vast field subjects for review abound. In the light of the topics selected for discussion at this meeting it would seem appropriate to consider the chromosome instability syndromes and their relation to malignancy. However, I shall preface my review of the better-defined disorders of this group with a consideration of the behaviour of the larger human acrocentric chromosomes, which may turn out to be more generally relevant to the malignancy aspects of these syndromes.

ACROCENTRIC CHROMOSOMES AND CENTRIC FUSIONS

Centric fusions (Robertsonian translocations; Table 1) are practically half of all human translocations, both balanced and unbalanced (see also Cohen 1971). They arise in a manner which appears to be independent of parental or at least maternal age (Polani *et al.* 1960; Matsunaga & Tonomura 1972); a paternal age effect has been suggested for a small proportion of specific

TABLE 1

Proportion, per thousand births, of chromosome anomalies in population cytogenetic studies of about 57 000 newborn infants (see text) (modified from Hook & Hamerton 1977)

		Males		*Females*		*Both sexes*[a]			
Numerical	Sex chromosomes								
	47,XYY	0.93				0.46			
	47,XXY	0.93	2.6			0.46	1.30		
	Mosaics	0.37				0.19			
	Others	0.37				0.19		2.05	
	45,X			0.10		0.052			
	47,XXX			1.04	1.51	0.520	0.75		
	Mosaics			0.37		0.18			3.49
	Autosomes								
	+ D					0.05			
	+ E					0.12		1.44	
	+ G					1.25			
	Other					0.02			
Structural	Balanced Robertsonian translocations								
	D/D					0.72	0.91		
	D/G					0.19		1.95	
	translocations & inversions						1.04		
	Unbalanced Robertsonian translocations					0.07			2.55
							0.19		
	Unbalanced translocations					0.12			
	Inversions & deletions						0.11	0.60	
	Unclassified						0.30		
	Total abnormalities (balanced & unbalanced)								6.04 (1:166)

[a]Assuming that as many female as male newborn infants had been studied

centric fusions between G autosomes (Penrose 1962). Centric fusions are rare between homologues (see e.g. Cohen 1971; Hsu *et al.* 1973) and balanced carriers of these can only arise through post-fertilization errors and can only produce unbalanced offspring. Isochromosomes of acrocentrics can arise as germinal structural chromosome mutations through misdivision of the centromere and their behaviour is complex (Polani *et al.* 1965). In this case they can be either unbalanced or, if balanced, probably associated with mosaicism. They tend to produce unbalanced zygotes. Non-homologous centric fusions on the other hand are the commoner types and, though balanced carriers for these anomalies produce a proportion of unbalanced zygotes, they also produce carriers like themselves as well as chromosomally normal children. They are of three different major types: between two D group or two G group chromosomes, or between a D and a G. The frequencies with which different types are produced have only become better known recently, following banding studies (but see Cohen 1971 for the results of autoradiographic acrocentric chromosome identification). It can be estimated that somewhat less than half the centric fusions are between two D group chromosomes. By contrast centric fusions between a chromosome 21 and a 22 seem to be very rare, probably less than 10% of balanced centric fusions. The remainder of just over half the centric fusions are between a D and a G autosome. In the latter case they seem very rarely to involve a chromosome 22, so that they are mostly related to the production of interchange trisomy 21. One should remember that this may well be the result of a biased ascertainment of the frequency of D/G fusions. D/D centric fusions seem more commonly to involve a number 13 and a 14, while chromosome number 15 is seldom involved in fusion with another acrocentric (Cohen 1971). The D/21 centric fusions involve most commonly a number 14 while fusion of 21 with a number 13 is unusual and with a number 15 exceptional. Thus there appears to be some peculiarity of behaviour, particularly of chromosome number 15, in respect to centric fusion. Interestingly, the exceptional balanced carriers of centric fusions between two chromosomes 13 or two 14 have a normal somatic phenotype. By contrast it would appear that centric fusions, apparently balanced chromosomally between two chromosomes 15, may often result in developmental abnormality, specifically the Prader-Willi syndrome (Fraccaro *et al.* 1977; A. C. Berry & P. E. Polani, unpublished work; see also Hawkey & Smethies 1976; Bartsch-Sandhoff 1977; other chromosomal abnormalities have been detected in this condition, for example centric fusions between two G autosomes and possibly four out of five such patients may have a normal chromosome complement).

What could dictate this peculiarity of number 15 is unknown. However

Angell & Jacobs (1975; see also Latt *et al.* 1974) after labelling human chromosomes with bromodeoxyuridine (BUdR) for one cycle, noticed asymmetrical labelling of the constitutive heterochromatin of the paracentromeric region of chromosomes 1, 9, 15 and 16 and of the distal long arm of the Y chromosome, which are the human chromosomes with the largest blocks of paracentromeric and other constitutive heterochromatins. The situation is like that of the paracentromeric regions of the mouse chromosomes, where the abundant murine satellite DNA is mostly situated. The asymmetrical labelling suggests an asymmetrical distribution of the thymine–adenine base pairs between the two DNA strands with predominance of thymine in one and of adenine in the other of the two DNA strands (Lin *et al.* 1974*a*). Asymmetry has in fact been demonstrated chemically, after the isolation of mouse satellite DNA and study of base composition of the two separate strands (Flamm *et al.* 1967). It is suggested that this arrangement of DNA may favour on the one hand, and limit on the other, the formation of centric fusions in this species (Lin & Davidson 1974). It is thus possible that similar constraints pertain to some human chromosomes. However not all rodents display the asymmetrical labelling of paracentromeric heterochromatin (Lin *et al.* 1974*b*), nor, as we have seen, is it apparent on all human chromosomes. Possibly, whether the arrangement is visible with the relatively crude BUdR labelling method depends on the size of the paracentromeric heterochromatin in these regions of the individual chromosomes. Given some assumptions, in the mouse in one of the two DNA strands of a paracentromeric region the average number of thymine residues could be as many as a few million and the adenines about half the value, while the other strand would show a reciprocal complementary arrangement. Speculating further, one might consider this DNA difference as a method for 'flagging' a special chromatid, at the time of DNA synthesis, near the centromere. It could be important particularly to mark this region, which is the last to synthesize DNA at the end of the S phase of the cell cycle and could be the key region to chromosome movement and segregation at mitosis. One might see the importance of marking a DNA strand if selection of strand (and thus of chromatid) were an essential part of the dynamics of cell proliferation and were a requirement for stem cell selection as postulated by Cairns (1975). One could visualize that strand selection could be part of the mechanism for the conservation of histone assembly during semiconservative DNA replication. If indeed the distribution of old histones is selectively maintained in one strand of the replicating chromosomal DNA (Leffak *et al.* 1977), labelling the centromere could be the basic requirement for a mechanism essential for the preservation of 'the identity and sequence of nucleosomes in the chromatin of a stem cell line' as visualized by Cairns 'while allowing diversity in daughter DNA duplexes' (Felsenfeld 1978). Cairns (1975), being of the opinion that most mutations probably arose at DNA replication (hence the risk to stem cells), had hypothesized strand selection in stem cells as a mechanism which could segregate one DNA strand perennially to stem cells, thus safeguarding their DNA message from mutation of somatic origin as an immortal and pure blueprint. In this way, the mortal daughter cells would always collect the DNA molecules with the younger parental strand, while maintaining in the stem mother cells an immortal strand through successive replications of the stem cell DNA. Cairns felt that the immortal strands must be 'marked in some way so that sister chromatids can be distinguished at the centromere.' Non-random segregation of chromatids at mitosis generally, but with no special reference to stem cells, had earlier been observed in mouse embryonic primary cell cultures and, to a lesser extent, in an established Chinese hamster cell line (Lark *et al.* 1966) in keeping with a bacterial model of chromosome replication (Lark 1966). The observations of Herreros & Giannelli (1967), which pointed to segregation of parental strands at successive chromosome replications to alternate planes of cell division, fit the idea that chromatid segregation is non-random, as do the results of Rosenberger & Kessel (1968) in *Aspergillus* that suggest that chromatids with coeval DNA strands segregate at mitosis as a unit. Also Odartchenko & Keneklis (1973), having labelled the sperm chromosomes, had findings in early cleavage of fertilized mouse ova which suggested that the chromatin of blastomeres segregated according to parental origin, again without reference to stem cell properties. Recently Barlow (1978), using autoradiography, has obtained in *Zea* a pattern of autoradiographic grain segregation over the stem-cell-containing quiescent centre which seems consistent with Cairns' idea. Experiments by Potten *et al.* (1978) on mouse intestinal crypts and lingual papillae suggest results that although far from conclusive are nevertheless 'compatible with there being a small number of cells (1–5 in the stem cell region of each crypt and 1 at the base of each column of cells in the tongue), that selectively segregate their old and new DNA strands.' So it looks as if the immortal strand hypothesis and its relevance to some malignancies, at least, is worthy of serious consideration.

GENE AND CHROMOSOME REPAIR DEFECTS AND PRONENESS TO MALIGNANCY

The conditions that I shall describe in this paper are heterogeneous, and the implication that they are all 'repair' defects is not necessarily supported by the available evidence. They certainly result in DNA or in overt chromosome damage (or both), but in most the defective mechanism is known only in outline, or by inference. Nevertheless their similarity in respect to proneness to malignancy, and possibly in terms of their general molecular relationship, justifies their being considered together. As for the differentiation between gene and chromosome defects, some of the distinction is pragmatic, in the same way that a differentiation is made between mutagens and clastogens. In what follows, while outlining the clinical aspects, I shall concentrate on the chromosome and the general molecular aspects of four disorders: xeroderma pigmentosum (XP), Fanconi anaemia (FA), ataxia telangiectasia (AT) and Bloom disease (BS). I shall review the first more extensively than the others, since more is known about the molecular defect in XP, and I shall only summarize the other three and update our knowledge of them: I reviewed these three for an earlier Ciba Foundation symposium (Polani 1976) and only a few key references from that paper will be cited again.

Xeroderma pigmentosum

(a) Clinical aspects. XP is a relatively rare autosomal recessive disorder and affected individuals are not very unusually the issues of consanguineous parents. It is variously estimated to affect 1 in 65 000 (Dorn 1959) to 1 in 250 000 (Robbins *et al.* 1974) persons and is found in all races, though there is an apparent clustering of XP in the Middle East. The condition is genetically heterogeneous even more so than clinically. Apparently an autosomal dominant mild variety has also been described in a Scottish family (Anderson & Begg 1950).

XP is primarily a disease of the sun-exposed skin, but the biochemical disorder affects all cells of the body that have been tested and other systems than the skin may show abnormalities. The skin lesions were originally described by Kaposi in 1874. Usually from early infancy, sometimes later, there is sun sensitivity even on minimal exposure, shown by prolonged erythema with oedema and sooner or later blistering—truly for the XP patient the sun is the enemy. The sensitivity is for rays in the u.v. region of the solar spectrum, especially between 250 and 320 nm. Fairly quickly, by about two years of life, freckles develop in the exposed areas, which become dry and scaly. After the freckling, telangiectasias appear with depigmentation, skin atrophy and then proliferative lesions. These are actinic keratoses and verrucose hyperkeratotic papules, and the skin of the face acquires a thin, pinched appearance. Histologically the skin shows disorganization of the epidermis with atrophy of the Malpighian layer and focal proliferation of the basal layer. The rete pegs proliferate and are stuffed with melanin, while the melanocytes proliferate and look polymorphic. Keratoacanthomas, which seem to follow a spontaneous cycle of verrucose growth and regression, often develop. Basal or squamous cell carcinomas arise, generally in early childhood though sometimes only in adolescence, and are often multiple, while malignant melanomas, which may also be multifocal, are also prevalent (for a clinical description see Giannelli 1976). By 20 years of age every Caucasian XP patient has developed skin cancers so that the incidence of this type of malignancy is several orders of magnitude greater than the population incidence (Scott & Straf 1977). Twenty-two of 50 XP patients in Japan had skin cancers, the exceptions being young children and five older patients with nearly normal repair of u.v. damage (see below, Takebe *et al.* 1977). The figures for melanoma are similarly depressing. If malignant melanomas occur in 1 in 10 000 persons in the general population, the prevalence in XP patients has been found by some to be as high as 50%, or 5000 times higher, and the melanomas occur early. However, much lower figures of around 3% (300 times the average prevalence) have also been quoted in older, retrospective studies (Robbins *et al.* 1974). Internal tumours are not very common, but neurinomas, fibromas, angiomas and sarcomas are seen. One of our patients, aged 15, has developed a cerebellar medulloblastoma (S. A. Pawsey *et al.*, unpublished work). Short stature is sometimes described but is not a prominent feature of the disorder except in patients with major neurological manifestations. Especially common are scars and contractures of the eyelids with their attendant complications, or involvement of the conjunctiva or the cornea, at times with malignancy.

An important feature of XP is the frequent and progressive neurological involvement related to an anomaly of the CNS starting prenatally. It cannot be exclusively considered a postnatal progressive abiotrophy. The neurological syndrome, first described by De Sanctis & Cacchione in 1932, may affect perhaps one in six or more patients with XP. It presents with microcephaly with progressive intellectual deterioration, hearing loss, choreoathetosis or ataxia and spastic quadriplegia. Quite in contrast with this neurological picture of brain disease is the presence of lower motor neuron involvement with absent tendon reflexes, with neuropathic electromyogram and muscle biopsy, but with normal nerve conduction velocity. Neuropathological findings, which could well represent a primary lesion, are essentially neuronal loss affecting pyramidal and Purkinje cells, as well as cells of the basal ganglia and dentate nuclei (Robbins *et al.* 1974; Yano 1950; Reed *et al.* 1965).

(b) Basic defect. A dramatic change in our understanding of XP was brought about in 1968 when Cleaver showed that XP cells have a grossly

reduced ability to repair u.v.-induced damage to their DNA. Normal cells insert new bases in lieu of the damaged ones, but XP cells, including those from patients with the De Sanctis–Cacchione disorder, were inefficient in this. Cleaver also found (1969) that single-strand breaks in DNA were repaired efficiently by the abnormal cells and suggested that the XP defect must involve an early step of 'excision repair', a form of dark repair. Later it was shown that there are seemingly different excision repair defects in XP (complementation groups) and also that some XP cells have normal excision repair but a disabled post-replication repair system (XP variant: see below).

Cell fusion has been used to investigate the genetic heterogeneity of XP in experiments involving fusion of XP cells of different family sources. The first experiments were done by Bootsma *et al.* (1970) and expanded by de Weerd-Kastelein (1974; see also Kraemer *et al.* 1975; S. A. Pawsey *et al.*, unpublished work). At present five different complementation groups of 'excision XP' (A–E) have been described (Table 2) and it seems probable that others exist. Table 2, which includes 12 propositi studied at Guy's (S. A. Pawsey *et al.*, unpublished work) as well as the results of de Weerd-Kastelein (1974) and of Kraemer and his collaborators (1975), shows that A, C and D are the commonest groups among Caucasians and it would seem that neurological complications are not seen in XP subjects of group C. This group is not seen among the Japanese, in whom D also appears to be under-represented (Takabe *et al.* 1977).

Systems of repair of DNA damaged by u.v. light were well known from the study of bacteria, and especially repair-defective mutants, after the discovery by Kelner (1949) of a system for repairing u.v. damage, based on revival by

TABLE 2

Distribution of complementation groups of 'excision XP' determined by cell fusion experiments in three series of unrelated patients studied by de Weerd-Kastelein, by Kraemer and by Pawsey, and their colleagues (see text; Table I of Giannelli 1978)

Complementation group	*No. of patients*	
	Total	*No. with neurological disorders*
A	11	8
B	1	1
C	13	Nil
D	6	2
E	1	Nil
All	32	11

visible light. This system, if allowed to act after u.v. exposure, permitted survival rates of irradiated microorganisms by a factor in excess of 10^5. Thus microorganisms have developed three effective repair systems for the harassing damage which follows u.v. irradiation: photoreactivation, excision and recombination repair, in addition to repair used for other forms of DNA damage (reviewed by Howard-Flanders 1973).

Setlow (1968) showed that photons absorbed by DNA cause a variety of photoproducts, of which pyrimidine dimers are the best known and could be the biologically most important. It is the essence of the XP defect that it is these dimers which the abnormal cells are largely incapable of excising. Normal cells, by contrast, can perform what is a complex series of coordinated operations to excise the main photo-lesions in DNA. After incision under endonuclease control, the damaged part of DNA is excised, a new patch of DNA is polymerized, using the integer strand as a template for repair synthesis, and finally a ligase seals the newly synthesized to the old part of the DNA molecule (Cleaver 1974*a*). The presence of a dimer distorts the DNA helix considerably: some hydrogen bonds are opened and the helix is unwound a few degrees. As a consequence both transcription and replication of DNA before cell division are impaired.

A number of techniques, mostly indirect, can be used to demonstrate or infer the excision and the subsequent DNA repair defect: for example, impaired cell survival after irradiation, retention of dimers in irradiated DNA, the persistence of sites sensitive to a specific endonuclease isolated from *Micrococcus luteus,* the failure to incorporate BUdR in the course of repair synthesis and the consequent failure of photolytic DNA sedimentation shift, the inability to reactivate irradiated viruses, the failure to demonstrate—autoradiographically, visually or photometrically—unscheduled DNA synthesis after u.v. irradiation using a labelled nucleoside (Setlow 1975). Incidentally this latter technique proved sufficiently simple and reliable for prenatal diagnosis of XP as carried out by us on two occasions (Ramsay *et al.* 1974; S. A. Pawsey *et al.,* unpublished work). Regan *et al.* (1971) has suggested the use of the BUdR incorporation technique for the same purpose.

Some of the techniques give specific information that an early step in the complex excision repair process is faulty and, specifically, the incision and removal of dimers, as we know when these are measured directly. Assessment of single-strand breaks (which would correspond to the incision step of excision repair) by measurements of DNA sedimentation, when related to the number of dimers present, could pinpoint the step at fault, if the method of measuring were sufficiently sensitive. Using such a method Cleaver (1974*b*) has claimed that in some XP cell lines there is a block after incision whereas

Fornace *et al.* (1976) have claimed in most cases an incision defect. Recently another method of detecting incision has been successfully used and has shown that the XP cells (at least those from three complementation groups, see below) do not perform this step in the way that normal cells do (Cook *et al.* 1978). The method (Cook & Brazell 1975, 1976*a,b*) consists of mildly lysing cells to release 'nucleoids'. These nucleoids consist of DNA depleted of nuclear protein which remains supercoiled, circular and compact so that nucleoids sediment fast. An uncoiling of nucleoids, and thus slower sedimentation, can be achieved by the addition of ethidium bromide in increasing concentration; but at a certain concentration the original negative coiling reverses to positive and again speeds up sedimentation of the nucleoid DNA. Breakage of a DNA strand by γ-radiation, for example, decreases the speed of sedimentation of nucleoids and eliminates the biphasic response to ethidium bromide. When the nucleoids test was applied to u.v.-irradiated XP fibroblasts these behaved like unirradiated normal cells and gave no hint of DNA nicking as seen in irradiated normal cells. Thus, by this criterion, in complementation groups A, C and D the defect is one of incision. Whether this is due to a defective endonuclease is unknown because, unlike the situation in bacteria, the specific enzyme(s) has not been isolated (see below). One knows, however, that not more than four-fifths of dimers induced by low doses of u.v. irradiation are normally removed, that the repair patch is about 100 nucleotides long (this is taken to reflect, possibly, the organization of the chromosomal DNA), and that the repair is normally error-free (Cleaver 1968; Setlow *et al.* 1969) so that efficient repair prevents the accumulation of mutational changes. By contrast, when repair is defective, as in XP cells, both chemical agents and u.v. irradiation introduce mutations whose frequency is proportional to the extent of the defect (Maher & McCormick 1976), but, at least in the excision-defective cells, the frequency is not disproportionate when normal and XP cells receive equitoxic irradiation. Because of the existing uncertainty about the precise nature of the defect in 'excision XP', experiments probing the enzymic efficiency of XP cells, or assessing the activity of purified 'incision' enzymes on XP chromosomes, are especially interesting. The various enzymes involved in repair and the methods that can be used in their study are discussed by Grossman (1974). Tanaka *et al.* (1977) have shown the ability of endonuclease V of T4 phage to restore unscheduled DNA synthesis when the phage is introduced into u.v.-irradiated XP cells. Extracts of XP fibroblasts of the commoner complementation groups are capable of excising dimers once the chromosomal DNA (of *E. coli*) has been incised (Cook *et al.* 1975) or when purified u.v.-irradiated preparations of DNA from phage, *E. coli* or man are used; however, extracts of cells of XP group A cannot excise dimers on u.v.-irradiated XP chromatin, although similar extracts of normal cells do this well (Mortelmans *et al.* 1976). It follows logically that incision, but not excision, of damaged DNA, by complementation group A at least, is abnormal, and that the abnormality may make it difficult for the incision enzyme to gain access to the chromosomal DNA in its normally organized state. Thus it could be that the enzyme itself is normal and that the primary defect might reside elsewhere, either in a factor which allows the enzyme access to the damaged DNA or even in some feature required for the recognition of the damaged site. A specific protein capable of binding to damaged DNA, but not to dimers, is normally present in XP cells (Feldberg & Grossman 1976). Doniger & Grossman (1976) have isolated a human exonuclease (correxonuclease) which can initiate hydrolysis at single-strand breaks and can remove pyrimidine dimers in its path. Doniger & Grossman found (1977) that this nuclease was present in XP cells (groups A and C, and variant). As for more direct attempts at looking at endonucleases active on u.v.-irradiated DNA, Duker & Teebor (1975) found nothing unusual in extracts of XP cells, while Kuhnlein *et al.* (1976) found a marked deficiency of an endonuclease for apurinic sites in cells from complementation group D, but only a slight defect in those of other complementation groups and in the XP variant. The enzyme had a higher than normal K_m in the D group cells and also in those of the A group. So, while possible defective enzymes seem identifiable in XP cells, the situation is not sufficiently clear, although the last-named workers suggest that the defective endonuclease might be responsible for the neurological dysfunction of group A and D patients. On the other hand a number of other enzymes concerned in DNA repair would appear to be normal (Pedrini *et al.* 1974).

It is clear that whatever defect(s) produces 'excision XP', its nature must be very complex. Giannelli & Pawsey (1974) investigated this complexity, using virus-mediated cell fusion and a statistical nuclear dosimetry approach in experiments in which protein/enzyme synthesis was inhibited and unscheduled DNA synthesis was assessed over the XP nuclei. The complementation kinetics in these experiments showed that normal nuclei can provide enough correcting 'enzyme' to normalize unscheduled DNA synthesis of two or three XP nuclei, without the necessity of synthesizing any anew during the time of the experiment. In Giannelli & Pawsey's system, cells from groups A and E were complemented rapidly while those of group C and even more those of group E were corrected very slowly. Their results also suggested that the defective 'enzyme' in XP was polymeric, might be assembled in a repair organelle (see also Cleaver *et al.* 1972), and might involve acceptor molecules; thus the whole system might be susceptible to a variety of different derangements. Indeed the markedly different complementation kinetics suggested that the mutations resulting in the 'excision XP' groups A, C and D might involve different loci (S. A. Pawsey *et al.,* unpublished work). Because cells from groups A and C were complemented better by normal cells fused to them when protein synthesis was blocked, S. A. Pawsey *et al.* (unpublished work) think that the products of the wild alleles from the presumptive loci A and C normally combine to form a stable compound. There are, in addition, many other findings which have been taken to suggest that the XP mutations might involve changes of controlling rather than of structural genetic elements (Setlow 1978).

The severity of the repair defect in 'excision XP' has been commented on already and it is clear that the degree of abnormality in cells as measured by unscheduled DNA synthesis may vary from very grave, i.e. less than 2% of normal (complementation group A), to around 3–7% (group B), to 10–25% (group C) and up to 55% for group D (Robbins *et al.* 1974; Cleaver & Bootsma 1975; Bootsma *et al.* 1975), or possibly even higher (Takebe *et al.* 1977). However, with the exception of the 'XP variant' (see below), the problem of clinically 'mild' 'excision XP' with high levels of repair is a difficult one. It has been claimed that u.v. sensitivity of fibroblasts correlates with the tendency to neurological involvement (Andrews *et al.* 1976). Preliminary findings at Guy's Hospital (F. Giannelli, personal communication, 1978) do not lend support to this suggestion, and indeed the question of correlation between degrees of repair efficiency *in vitro* and clinical severity is in need of further study, as is that of the levels of repair efficiency of the different complementation groups (Giannelli 1976). Some correlation is claimed between the severity of the cellular defect and the clinical disorder (Bootsma 1978) but the problem is made complex by the vagaries of exposure of XP patients to sunlight and to other damaging agents, balanced by the degree of protection that affected persons will wish and be able to use.

Following the observation that, in otherwise characteristic examples of XP, cells could show a normal excision faculty (see Cleaver & Carter 1973), Lehmann *et al.* (1975) showed that these cells have a defect of DNA *synthesis* after u.v. irradiation, an error of post-replication repair (Lehmann *et al.* 1977). The condition was called XP variant. Some cases of variant may have a later onset than classical XP (Bootsma *et al.* 1975) and neurological complications are not seen in patients with this genetic defect. That XP variant cells are competent to incise damaged DNA is also shown by the nucleoid experiments because the nucleoids made from them behave like those of normal cells after u.v. irradiation (Cook *et al.* 1978). It seems that about one patient with XP in every five has this variant defect. Normal cells cope with DNA lesions (including dimers) which are left unrepaired by somehow by-passing these lesions at their next replicative DNA synthesis and thus arriving, albeit with some difficulty, at synthesizing DNA of high molecular weight. Abnormal, but relatively slight, difficulty in this process is experienced by 'excision XP' cells; but cells from patients with the XP variant are seriously delayed in producing fully mature DNA chains. Their disability is enhanced if caffeine or theophylline are added to the medium in which u.v.-irradiated cells are cultured. Caffeine seems to exercise a specific effect on DNA repair (Kihlman 1977) of normal cells too, but these cells can rely largely on excision repair for the removal of dimers: thus the effect is small. When cells are defective in excision repair a slight additional detrimental effect on post-replication repair is manifest (Cleaver & Bootsma 1975). By contrast, in the XP variant this repair mechanism is seriously impaired and caffeine sensitivity is marked. This suggests that there may be two steps in normal post-replication repair, one of which is caffeine-sensitive and the other, and major step, caffeine-insensitive. It is this major step that would be affected by the mutation of the XP variant. If this is what happens, we should anticipate that a mutant of the other step could be found whose cells would not be affected by caffeine. Another suggestion is that there is a single step and that the mutant post-replication repair enzyme is unsuccessful in competing with caffeine for the single-stranded DNA, the object of repair (Lehmann *et al.* 1975). However the interpretation of caffeine effects is not easy as caffeine may act indirectly and, because it is quickly handled by cells, its activity in DNA repair may be far from primary (Cleaver & Bootsma 1975). Whatever the influence of caffeine on post-replication repair (see also Kihlman 1977), it was thought that it leads to excessively long patches of newly synthesized DNA, possibly around 1000 nucleotides; however, recent evidence suggests that the infilling may be more modest (Setlow 1978). There is also evidence from cellular mutation rates that the cells in XP variant are particularly more error-prone than the excision-repair-defective XP cell (Maher *et al.* 1976) but also more resistant to the killing effect of u.v. (Maher *et al.* 1975).

Repair in XP cells is defective not only for u.v. radiation but also for a variety of chemical substances. Thus, axiomatically, the wild-type products which are abnormal in XP can be taken to have a broad spectrum of activities on different, but presumably similar, types of DNA damage. It is assumed that those agents, which cause damage usually repaired by the excision system, are especially damaging to 'excision XP' cells, whereas those which ultimately cause strand breaks are handled efficiently. It is seen that the former often lead to damage that not only is ineffectively repaired but also, and in consequence, leads to mutation and/or cell death. There is especial difficulty in classifying the chemical agents: most agents can cause a variety of lesions to the DNA, and some of these are efficiently repaired by XP cells, others are repaired but in an abnormal way, and yet other lesions cannot be repaired; the evidence is that this applies both to cells from 'excision XP' and those from the variant. As an example of the difficulties encountered by XP cells, when the repair patches produced in response to a damaging agent have been found to be long in *normal* cells, then XP cells have been destroyed by that chemical agent (Regan & Setlow 1974). Equally some lesions caused by alkylating agents, for example at the N^7 position of guanine, are handled efficiently by XP cells, while those at the O^6 position are not (Goth-Goldstein 1977). Setlow (1978) has provided a list of proficiencies and deficiencies experienced by XP cells. Among the chemical agents which are particularly injurious, 4-nitroquinoline-1-oxide is one which produces effects most similar to those induced by u.v. Other especially damaging agents are amino-fluorene derivatives, aflatoxin and benzanthracene derivatives. Conversely XP cells seem proficient in their handling of methyl- and ethyl-methane-sulphate and mitomycin C. Giannelli (1978) has made a useful and extensive compilation of the existing data on agents abnormally toxic to XP cells, distinguishing the various complementation groups of 'excision XP' from the XP variant cells, and has listed those agents which do not induce normal levels of repair in these types of cells. As well as being of theoretical interest these studies on the action of chemical and physical agents, other than u.v., on XP cells have a practical value. For example, it seems wise that when malignancies in XP patients are being treated the more mutagenic and damaging agents should be avoided, to minimize the risk that, while one malignancy is being attacked, the formation of another is encouraged. In this context it is of interest that in XP, malignant cells, just as other body cells that have been tested, seem to retain their basic repair defect, again a fact that may be relevant to the treatment of malignancy in XP (Robbins *et al.* 1974). Incidentally it does not seem that transformation of XP cells by SV40 reveals any special differences in these cells compared with normal cells (reviewed in Cleaver & Bootsma 1975).

Another aspect of the effect of the primary lesion in XP that has received attention is that of sister chromatid exchanges (SCEs). While a slight excess of spontaneous SCEs has been reported this has not been consistently found (Alhadeff & Cohen 1976; de Weerd-Kastelein *et al.* 1977). Exposure of normal fibroblasts to u.v. light results in an increase of SCEs (see Kato 1977) which is apparently more marked in XP cells of group A to D but not in E or XP variant cells (de Weerd-Kastelein *et al.* 1977). Also 4-nitroquinoline-1-oxide seems to induce an excess of SCEs in XP-virus-transformed fibroblasts on which the agent was tested compared with control cells (Wolff *et al.* 1977), but the effect depends on the duration of treatment (Stitch *et al.* 1976). Another point of interest is that spontaneous chromosome aberrations of the types seen in the chromosome instability disorders to be discussed later are not seen in XP. Nonetheless there is some evidence that u.v. irradiation finds XP cells more prone to chromosome breaks than normal cells (Parrington *et al.* 1971; Marshall & Scott 1976).

Perhaps this is the moment to ask a question on the origin of neurological complications in a sunlight sensitivity syndrome. We can only make hypotheses but there is the important fact that some chemicals cause damage to the chromosomal DNA which is repaired by involving at least parts of the same repair system as incises pyrimidine dimers. Whether some endogenous chemical products exist that could affect the nervous system and other body cells in XP—and we know that they all express the defective phenotype—in a way similar to that of the u.v.-mimetic drugs is not known. Clearly, however, and perhaps additionally to internal chemicals, some external chemical pollutants could be responsible for untoward effects on somatic tissues, including the nervous system. In this respect it is useful to recall the persistence of O^6-alkylated guanine in the rat brain after administration of ethylnitrosourea, which induces tissue-specific tumours (Goth & Rajewsky 1974). Persistence of lesions in the DNA could be responsible on the one hand for changes associated with malignant transformation and on the other hand for other errors in the read-out of the chromosomal DNA as well as for additional profound effects which may lead to cell death. It could be hypothesized that normal individuals are able to deal adequately with DNA lesions in the brain (in the example with O^6-alkyl guanine), so avoiding mishaps. An excess of alkylation, or a deficiency of repair, would bring about regressive cellular changes. An inability to deal with apurinic sites has, we have seen, been suggested as the mechanism underlying neurological damage (Kuhnlein *et al.* 1976).

The mechanism whereby cancers are induced in the cells of XP patients is fundamentally unknown. Clearly, as mutagenic and carcinogenic activities of many chemicals are interrelated it is easy to speculate that unrepaired damage to the chromosomal DNA, probably appearing at critical sites and possibly acting cumulatively, could eventuate in malignancy. In addition, cells in XP patients subjected to u.v. radiation are continuously generated with a changeable genotype which may well adapt them to proliferation and to resistance in an environment in which inflammatory and other lesions occur constantly. However the most interesting experiments in this respect are those that have probed the influence of a single unrepaired lesion, pyrimidine dimers, on the origin of malignancy. A revealing experiment was conducted in the micro-cyprine fish *Poecilia*. These vertebrates, like the amphibia, rely heavily on enzymic photoreactivation for repair of u.v. damage to their DNA. The relevant enzyme specifically monomerizes the dimers—and these photo-lesions only—produced by light at 254 nm, by using the energy supplied by photons in the 320–420 nm range. Hart & Setlow (1975) subjected tissue homogenates of the fish to u.v. irradiation at high fluence, injected the cells into gynogenetically naturally cloned fish and, with a dose of about 2×10^5 thyroid cells, noted that thyroid carcinomata arose in 100% of the injected cells. Conversely, treatment of the cell homogenates, after irradiation at 254 nm, with light at 360 nm (to activate the photoreactivating enzyme) reduced the incidence of thyroid cancers by at least 10-fold. The u.v.-mimetic drug, acetoxy-acetyl-amino-fluorene, produced similar effects when allowed to act on the cellular homogenate. The conclusion is that DNA lesions, specifically dimers (with u.v.), are at the origin of cancers when repair of these lesions is deficient. Indeed, in the cited experiments, the probability that a 'random dimer' might produce neoplastic transformation was of the order of 10^{-10}. The ultimate relationship between DNA lesion and cancer is, however, not automatically explained by the demonstration of damage which leads to somatic mutation. Clearly the mutational aspects are paramount to cancer induction and to the subsequent establishment of a malignancy, but activation of latent viruses by mutation and the effects of mutations on the efficiency with which cells are surveyed are, minimally, some of the other synergies that must be taken into consideration. Here lie some of the problems of the relevance of cellular and animal experiments to the natural, *in vivo* situation in man. Setlow (1978) has looked at the dosage aspects of the problem and has considered not only the less heavy exposure to which human XP cells would be subjected *in vivo* but also other possibilities of compensation and repair, including enzymic photoreactivation. Nevertheless one must also consider in man the length of exposure over months and years, the possibly synergistic action of other agents, and the possibility of damage to cells endowed with stem proliferative characteristics.

There are suggestions that heterozygotes for 'excision XP' may have, overall, a slightly impaired efficiency of excision repair (S. A. Pawsey *et al.*, unpublished work). That cells of such heterozygotes have less enzyme available for repair than normal cells has been found by a method of cell-fusion complementation dosimetry and it has been shown that heterozygote nuclei are less able to correct XP nuclei in heterokaryons than are normal nuclei (Giannelli & Pawsey 1974). However the nucleoid technique (Cook *et al.* 1978) had not distinguished, with the limits of the experiments done, between heterozygote and wild-type cells. The matter is of importance because of the knowledge which is accruing that heterozygotes for some recessive chromosome instability disorders (see below) may be more malignancy-prone than the population at large. When cellular effects can be shown in heterozygotes that mimic, even if on a minor scale and after suitable 'stresses', the effects seen in the homozygote mutant cells, it is perhaps axiomatic to consider that the carriers may be more prone to mutational and/or carcinogenic influences than those who are wild-type homozygotes at the loci under scrutiny.

Ataxia telangiectasia

(a) Clinical aspects. This protean disease was first described by Syllaba & Henner in 1926 as a progressive choreoathetosis with ocular telangiectasia in three sibs. Redescribed in 1941 by Louis-Bar in a case at autopsy, it became recognized only in 1957/1958 when another element of the syndrome, proneness to sinopulmonary infection, was added, and thymic absence or hypoplasia together with absence of ovaries were noted at necropsy. When the disease was reviewed by Boder in 1975, 484 cases had been reported, with 42 autopsies, and Boder estimated a prevalence of between 1 in 50 000 and 1 in 30 000 children. In 2 of 45 families the parents were first cousins and from this one could arrive at a much lower estimate of the incidence of AT with very wide confidence limits. Such an estimate is extremely crude, depends on the basic level of first-cousin marriages in the populations from which the data are derived, and is influenced by genetic heterogeneity as well. The disease is said to have a frequency of about 1 in 8000 among Moroccan Jews. Neurologically the disorder is progressive, with an ataxia of early onset, choreoathetosis, oculomotor signs, dysarthria, drooling and postural anomalies. Also the telangiectasia is progressive and may involve the perinasal region and, more rarely, the antecubital and popliteal fossae and the dorsa of the hands. The sinopulmonary infection is severe and progressive and is the most common cause of death. Somatic growth is retarded, and mental backwardness of mild degree is apparent in later childhood in one-third of cases, but is mild and not progressive. Progeric changes of hair and skin are common, together with freckling. Senile keratosis and even basal cell carcinomas of the face have occasionally been noted in young patients. Insulin-resistant diabetes seems common. Pigmentary changes like those seen in advanced radiodermatitis, as well as vitiligo, café-au-lait spots and hyperpigmented macules are not unusual. Twelve of the 41 reported autopsies (Boder 1975), which were all in young subjects (mean age 11.6 years), had leukaemia or other lymphoreticular

malignancy, including one IgM monoclonal gammopathy (mean age 9.0 years). In addition one patient had died at 19 years with a gastric adenocarcinoma and one other at 13 years with a medulloblastoma. Of 46 collected malignancies 65% were lymphoreticular, 16% were leukaemia, and 19% involved other organs (Kersey *et al.* 1975). The neuropathological findings are complex but essentially there is cerebellar cortical atrophy which involves granular and Purkinje cells and appears to be primary, together with cell loss in the dentate nuclei and olives. Multinucleated neurons, or neurons and pituitary cells with large nuclei, are an interesting feature of the disease. Similarly, cells with large nuclei have been observed in practically all organs of the body. Histological skin changes suggest a venular affection and skin atrophy resembling that associated with age and with actinic damage. At necropsy the ovaries have been noticed to be absent or hypoplastic. The thymus is probably always present but seriously maldeveloped and abnormal: it consists entirely of epithelial stroma and is similar to the fetal organ before its lymphoid differentiation (Biggar & Good 1975). Lymphocytes are scanty and Hassal's corpuscles absent. The peripheral lymph nodes and the tonsils are depleted of germinal centres and plasma cells. Antibody production, especially IgA and IgE, often seems inadequate and serum IgM is of low molecular weight. Lymphocytes are not easily transformed by phytohaemagglutinin, and the patients fail to show delayed hypersensitivity and tolerate homografts excessively well. So there is evidence of T but also B cell dysfunction.

Another puzzling feature of the disease is that levels of serum α-fetoprotein range from 40 to 2500 ng/ml compared with the normal of about 20 ng/ml. It appears that liver function tests may be abnormal. This sketch of AT disregards the clinical heterogeneity of the disease, with a tendency of the different clinical patterns to run true in individual families (Hecht 1977).

(b) Basic defect. In 1966 it was reported that 20–30% of cultured lymphocytes showed chromosomal breaks. Subsequent studies confirmed this and stressed also the high frequency of endoreduplicated and polyploid cells. More recently Cohen *et al.* (1975) have also found an increase in structural chromosome changes (breaks and gaps, dicentrics and rings, and interchanges and interchange configurations) but have observed these changes not only in cultured lymphocytes but also in cultured fibroblasts. Furthermore the latter cells displayed more chromosome damage than the former, an unusual behaviour compared with that of the other chromosome instability disorders. Collateral evidence of fibroblast involvement in AT, and thus of the more generalized nature of the AT dysfunction, comes from the observed prolonga-

tion of culture doubling time (Siegel *et al.* 1975). However the most interesting development in the cytogenetics of AT was the observation that in patients with this disease a special cell clone seems to develop which gradually and slowly proliferates. It is not clear what the relationship of the specific clone is to the lymphoreticular malignancy (Oxford *et al.* 1975), often a lymphosarcoma, which AT patients develop (Kersey *et al.* 1973). The characteristic of this clone, which by 1976 had been described in 15 patients (Francke 1977), but possibly in more than 40 altogether (Hecht 1977; Cohen *et al.* 1975; Oxford *et al.* 1975), is that a D chromosome is invariably involved. Whenever identified it has proved to be a number 14 with a break-point at band 12 of the long arm. It is becoming evident that in addition to the structural changes already discussed, stable rearrangements are found also in the cultured fibroblasts (Oxford *et al.* 1975), and it has been observed that the cells containing the marker D chromosome may have, on occasions, few other chromosomal changes (see also Nelson *et al.* 1975) and that the emergence of the 'D clone' is age-related.

Important to the molecular biology of the defect was the realization that patients with AT were unduly sensitive to X-radiation (Morgan *et al.* 1968; Cunliffe *et al.* 1975). Their lymphocytes were shown to be more sensitive *in vitro* than normal lymphocytes to X-irradiation, showing an excess of breaks and other evidence of chromosome damage (Higurashi & Conen 1973). This enhanced sensitivity to γ-radiation damage was recently confirmed on cultured fibroblasts (Taylor *et al.* 1975) and correlated (at least for one subject) with the reported *in vivo* sensitivity. However studies in alkaline sucrose gradients suggested that at least some AT cells are normally able to repair single-strand breaks. Also double-strand breaks seem to be efficiently repaired, judging from DNA sedimentation experiments conducted at various times after cell irradiation *in vitro* (see also Lehmann & Stevens 1977). Nevertheless examination of X-irradiated AT resting lymphocytes (irradiated in the G_0 phase) at the subsequent metaphase showed a large proportion of chromatid breaks and interchanges as well as deletions; together with the observed excess of acentric fragments, this would be an unexpected finding in normal cells, in which chromatid changes of any variety are seldom seen after irradiation at this stage in the cell cycle. This finding implies that there is a radiation repair defect with a high level of double-strand breaks or a failure to repair the broken segments and rejoin them (Taylor *et al.* 1976). It seems that the answer to these responses to X-irradiation may lie in the results of experiments with anoxic irradiation when the effect of X-ray is comparatively more marked in causing base damage than single-strand breaks (Paterson & Smith 1977). Two mutant cell lines were able to rejoin single-strand breaks normally but were clearly defective in repairing the X-irradiation damage. With an extract of *Micrococcus luteus* containing an endonuclease which makes an incision in the X-ray damaged DNA (at a nuclease-sensitive site) it was found that in some AT cells, compared with normal cells, this particular repair process of nuclease-sensitive sites proceeded very slowly (see also Paterson *et al.* 1976, 1977). This suggests that the radiosensitivity, at least in some cases of AT (three cell strains were tested), may depend on a defect of the excision of X-radiation-damaged bases (see also Lehmann 1978) by a mechanism different from that which deals with u.v. excision repair as shown by Paterson *et al.* (1976, 1977). Paterson & Smith (1977) find two different complementation groups in the endonucleolytic cleavage system and Arlett (1977*a*) defines at least three classes of AT cells, all γ-radiation sensitive: two with the endonuclease defect but capable of complementing each other, and one without a demonstrable defect of endonucleolytic activity (see also Paterson *et al.* 1977). Paterson tested the AT cells with other physical and with chemical agents, and compared them with other 'repair defects'. He concludes that there are specific repair systems for different lesions and thus that normal cells have a range of repair capacities. It is also clear that AT cells are unduly susceptible to damage by some chemicals but this has not yet been tested extensively (Hoar & Sargent 1976; Arlett 1977*b*). The detailed picture of the genetic heterogeneity of AT revealed by these studies is as yet unclear but it looks as if a minimum of three, but possibly more, different genetic defects may have been identified. What is clear is that AT is a genetic disorder of undue ionizing radiation sensitivity, that this sensitivity seems to be due characteristically to the difficulty of coping with radiation damage to DNA bases, and that the difficulty may be due in some cases to a faulty function of either of two endonuclease systems. How this molecular defect is related to the other features of the disease—the progressive neurological illness, the thymic and immunological disturbance, the α-fetoprotein abnormality and the seemingly common and specific chromosomal lesion—is completely obscure. It is to be noted (see also Hecht 1977) that the lesion of chromosome 14 at position 12 of the long arm may well be near the locus for nucleoside phosphorylase (NP) and it is possible that the immunodeficiency and involvement, especially of T cells, may be related to dysfunction of this enzyme. It is known that deficiency of NP can produce a combined immunodeficiency syndrome in a parallel way to that in which about half the children with the syndrome have adenosine deaminase (ADA) deficiency (Giblett *et al.* 1972; Chen *et al.* 1974; Hirschhorn 1977). In addition other metabolic defects of the purine pathways seem to be involved in these immunodeficiency disorders (Hirschhorn 1977). It had been thought that ADA deficiency exercised its effect through the accumulation of adenosine but Simmonds *et al.* (1978) suggest that it is deoxyadenosine which, accumulating as a result of the enzyme deficiency, blocks blastogenesis, as confirmed also by *in vitro* studies on lymphocytes. It is clear that NP deficiency, which would interfere with the direct catabolism of deoxyadenosine to adenine, and the indirect one via inosine (following the action of ADA), would have a similar effect to that produced by ADA deficiency.

The involvement of chromosome 14 and the lymphoproliferative malignancies taken together *are* of interest because this chromosome has been implicated in African Burkitt's lymphoma in which the presence of an additional terminal band on the long arm of this chromosome was noted by Manolov & Manolova (1972). Subsequent work has confirmed the presence of the marker chromosome 14 in Burkitt's lymphoma cells and in most established Burkitt lines (Jarvis *et al.* 1974), and has suggested that it is the result of a terminal translocation from the end of the long arm of chromosome 8 (Zech *et al.* 1976). This observation is of special interest in view of the association between the Epstein-Barr virus (EBV) and the African disease on the one hand and of 'non-developmental' trisomy of chromosome 8 and myeloproliferative disorders on the other (Philip 1975). However EBV-transformed normal cells or those from infectious mononucleosis did not show the specific translocation (Jarvis *et al.* 1974) although, conversely, it was found in North American Burkitt lymphoma (Kaiser-McCaw *et al.* 1977). Marker chromosomes 14 have been detected in other lymphoproliferative disorders, including Hodgkin's lymphoma (Fukuhara *et al.* 1976; see also Liang & Rowley 1978) and lymphosarcoma; in the latter, translocation of the long arm of 11 to that of 14 was reported (Fleischman & Prigogina 1977). It is of interest that specific band duplications of chromosome 1 correlate, instead, with myeloproliferative disorder (Gahrton *et al.* 1978; Rowley 1977, 1978; Kovacs 1978). Other large acrocentrics are involved in promyelocytic leukaemia, for example (15/17 translocation: Rowley 1977), and in retinoblastoma (deletion of the long arm of 13: Polani 1969; Wilson *et al.* 1969, 1977).

It is tempting to try to suggest a hypothesis for the relationship in AT between the site-specific change, the immunodeficiency and the malignant clone, as I have done before, and I am again led to think of an integrated (host-suppressed) virus. If the virus were on chromosome 14 near *NP* and were potentially capable of suppressing the activity of the *NP* locus, this might open the way to immunodeficiency. The virus itself, if unsuppressed by the AT mutant gene in double dose, would be responsible for the chromosome effect and presumably the subsequent clonal proliferation, as well as for the enzyme deficiency, and thus for the defect of maturation and development of some cells of the immune system and, perhaps, of the nervous system. How would the γ-radiation repair defect fit into this scheme? What is the connection, if any, between X-radiation sensitivity with normal single-strand break repair in cells from retinoblastoma patients with chromosome 13 deletion (Weichselbaum *et al.* 1977), and the X-ray sensitivity and chromosome 14 changes in AT? Naturally this attempt at a unitary view is a mere surmise and undoubtedly other hypotheses might fit the facts as well or better. It is worth repeating that the important facts to be brought together are (1) the specific chromosome break, (2) the possibly characteristic radiation sensitivity, but with good ability to perform strand repair, (3) the immunodeficiency, especially cell-mediated, (4) the neurological illness, and (5) the proneness to malignancy.

Bloom syndrome

Bloom syndrome (BS) was first recognized by Bloom in 1954 as a syndrome of stunted growth, sun sensitivity and telangiectatic erythema of the face of butterfly distribution. Later it became apparent that there was familial clustering of the condition (Szalay 1963; Wolf 1963). Together with a high frequency of consanguinity among the parents of—especially—the non-Jewish children with this condition, this suggested that BS was an autosomal recessive disorder. German (1969, 1973), who has worked extensively on BS, found that about half of the cases in his early studies were of Ashkenazi Jewish descent, and it seems that in these patients the Bloom gene derives from a Jewish population which had settled in south-east Poland and north-west Ukraine. We have seen four patients in this country, one in a family of British Jews, one in a Palestinian girl and two in unrelated families of Anglo-Saxon stock. There is an apparently high consanguinity rate among the parents of Gentile children with BS and from this a very approximate estimate could be made of the frequency of the condition, which is likely to be less than one in one million and could be considerably less. Among Ashkenazi Jewish families with BS children consanguineous marriages are much less frequent than among Gentiles, perhaps 5% compared with 50%, so that the frequency of the disorder in these populations would be correspondingly higher, affecting perhaps 1 in 100 000 neonates. German (1974) estimates an incidence among Ashkenazim of 1 in 160 000.

It has been suggested that some homozygous mutant fetuses may not survive to birth. Infants with BS are lighter at birth than infants with other point-mutational or chromosomal disorders, weighing on average only 1900 g at term. These infants grow up into small adults, reaching an average height of 145 cm. They have long and pointed faces and may show café-au-lait pigmentation of the skin. Mental retardation does not appear to be a frequent association. Deafness has been detected. BS patients are delicate and prone to respiratory and gastrointestinal infection, sometimes severe. They have small testes and seem to be infertile and possibly hypogonadal (Kauli *et al.* 1977). The telangiectatic erythema with scaliness is only exceptionally present at birth but develops with exposure to sun and can lead to skin atrophy and scarring. Indeed the response to u.v. light is abnormal. The affected subjects are immunologically impaired. They have low levels of all the immunoglobulin classes, IgG, IgA and IgM (e.g. Hütteroth *et al.* 1975). They respond weakly when challenged by a variety of antigens. The blood lymphocytes respond poorly to pokeweed mitogen, suggesting impairment of their B cells, while their T cells respond well to phytohaemagglutinin, but both B and T cells are present in normal proportions. In mixed lymphocyte cultures BS lymphocytes proved to react like poor responder but normal stimulator cells. Clearly some of these immunological defects must be responsible for the proneness to infection which these patients display, and the defects probably depend on anomalies of cellular proliferation in these patients. The other important reason for the reduced life expectancy of BS patients is their proneness to cancer. To the end of 1977 there were 74 well documented patients with this condition, 71 of whom were in the registry established by German (German *et al.* 1977), and 59 ranged in age from 3 to 43 years with a mean of 16.4 years. Of the 69 well-followed-up patients in this collection 13, or 19%, had developed a malignancy, six a leukaemia, three a lymphosarcoma or reticulum cell sarcoma, and four a gastrointestinal cancer. The average ages were 14.6 years for the leukaemias and related disorders and 36.2 years for the other malignancies.

The interest of BS centres around the inherent tendency for structural chromosome anomalies to occur and their difference from the changes observed in FA. The main cytogenetic findings are the presence in cultured cells of chromatid and isochromatid gaps and breaks, of acentric fragments, of dicentric and ring chromosomes and, especially, of cross-configurations, so-called quadriradials, resulting from chromatid interchange. By contrast with those in FA, in BS the crosses are symmetrical, are often close to the centromeres, which oppose each other, and very often seem to involve homologous chromosomes. It also seems that in addition to the characteristic quadriradials and dicentrics there is a peculiarity of distribution of changes in different chromosomes. Here much detail has come from banding studies

(Kuhn 1976) which also show that the breaks occur in the short dark bands of Q-banded chromosomes and in the interbands, and that the exchanges occur at homologous sites (Therman & Kuhn 1976). Also it seems that some chromosomes may be more often involved in exchanges with their partners than others. This has been claimed for chromosome number 1 but not number 2 (German 1974), and for number 17 but not number 18 (Therman & Kuhn 1976). The X chromosome (C. A. Bourgeois, personal communication 1976; Kuhn 1976) seldom takes part in quadriradial formation. The reason for non-participation of the X chromosome is obvious in males; in females it could be its out-of-phase DNA synthetic behaviour, and, if so, this could be an important clue more generally to the interchanges between homologues in this syndrome.

The other important observation on BS chromosomes stems from the original work of Zakharov & Egolina (1972) and of Latt (1973) (see Wolff 1977) on SCEs using BUdR incorporation and special staining techniques. There is evidence from the behaviour of ring chromosomes that SCEs occur spontaneously but some of them are determined by the incorporated BUdR in a dosage-dependent way (Latt & Juergens 1977). Normal lymphocytes with a standard BUdR dose contain an average of 5 to 10 SCEs per cell (or, say, 0.15 per chromosome), though there are variations which depend on culture conditions but also on subjects. Different chromosome segments seem to be differentially affected: for example the distal third of the long arm of chromosome 1, the middle of the short arm of 2 and of the long arms of 6 and 18, the middle and terminal parts of the long arm of the X, and the segments near the end of the long arm of 4. Normal cells treated with cross-linking agents such as mitomycin C can, depending on dose, show 10 times the number of exchanges of untreated cells (Latt 1974). Incidentally γ-radiation enhances SCE formation only moderately compared with the marked effect on chromosome structure, which suggests that only some X-ray-induced DNA lesions are related to, or rejoined by, SCEs. When BS cells were treated with BUdR it was found that instead of a base level of 1 to 14 SCEs per cell, 45 to 162 were observed (Chaganti *et al.* 1974). This, however, was not the case for Bloom heterozygotes whose cells behaved normally. Originally it was considered that BS cells might be unduly sensitive to BUdR but SCEs in excess were observed also with tritium labelling and, in any case, SCE formation in BS lymphocytes and bone marrow cells was found not to be BUdR dose-dependent (Shiraishi *et al.* 1976). Incidentally, bone marrow cells also show a characteristic excess of SCEs, as do BS fibroblasts in culture, but in bone marrow cells and fibroblasts the increment is less than that of cultured lymphocytes (Sperling *et al.* 1976).

A remarkable feature of BS has recently been reported in cells of 5 of 21 patients (German *et al.* 1977). Two discrete populations of cells were found, one of which had normal levels of SCEs and the other the characteristically high BS levels. The allegedly normal cells were between 2 and almost 50% of all cells. Chromosome aberrations were only found in the abnormal cells. It is tempting to ascribe the change to back mutation of the Bloom gene, or, more remotely, to a dosage effect resulting from chromosome errors; and there are other possible explanations for the phenomenon. This is not surprising as BS cells should mutate rather readily, even *in vivo*.

The basic defect in BS is unknown but there are peculiarities in the rate of replicative DNA chain growth in BS cells. Hand & German (1975), using an autoradiographic and measurement method on DNA fibres from cultured fibroblasts, find that the rate of growth of DNA chains during replication is significantly slower in BS than in control or FA cells. Giannelli *et al.* (1977) have shown that cultured BS fibroblasts tend to grow poorly in culture and spend more time than normal cells in a non-replicative (G_0) phase. Also, labelling experiments suggest that BS cells have a lower rate of DNA replication. In pulse chase experiments, sedimentation profiles of alkaline sucrose gradients showed that the newly synthesized DNA in BS has a consistently lower molecular weight than in control cultures. This extends the observation of Hand & German (1975) from a few replicons of a few chromosomes of replicating cells to a total population of cells, and suggests that the average rate of maturation of DNA chains attains a maximum size, under the culture conditions used by Giannelli *et al.* (1977), of about eight replicons of 30 nm each.

In order to explore further the defect in BS cells, Giannelli *et al.* (1977) exposed them to u.v. light, to which the fibroblasts proved three times more sensitive than controls in terms of survival. Experiments with u.v. light and caffeine, however, suggested a normal post-replication by-pass of u.v.-induced lesions in BS cells and Cleaver (1970) showed that at least the early incisional steps of excision repair are not affected in these cells. In view of the u.v. sensitivity one of the possibilities is that some interference with repair mechanisms nevertheless may exist and may also involve DNA polymerizations, possibly because repair and polymerization share an enzyme or enzyme system. The enzyme could be a DNA polymerase or some related polymerizing factor. These findings would explain the observed phenotype in BS of, on the one hand, poor growth and, on the other hand, light sensitivity.

If we accept that the essential defect in BS leads to a slowing up in DNA replication so that neighbouring replicating units remain unsealed or unwound for an excessively long time, it is easy to visualize that this may favour exchanges especially between synchronously replicating and spatially close chromosomal segments. This could explain (see also Hand & German 1975) the formation of SCEs and even the seemingly precise exchanges between homologous chromatids at homologous stretches. The argument might be pushed further by suggesting that the reason why the two female X chromosomes would seldom engage in homologous exchanges in BS cells (Kuhn 1976) is that they replicate asynchronously.

Shafer (1977) has produced an alternative view of the high frequency of SCEs in BS cells. His suggestion links up with another view of the formation of SCEs which he considers to be a method for by-passing spontaneous or induced DNA cross-links. This way Shafer takes into account the effects of mitomycin C in normal cells in relation both to sister chromatid (Latt 1974) and to homologous chromatid exchanges (Shaw & Cohen 1965; German & La Rock 1969) and the marked SCE-inducing effect of this and other bifunctional alkylating agents compared with the much weaker effect of the monofunctional ones (Perry & Evans 1975). According to Shafer, cross-links, if unrepaired because the process is defective or because the cross-links have occurred before DNA replication, are by-passed by SCEs whose function it is to do just this. However Shafer, like Kato (1977), does not exclude other mechanisms of SCE formation (for a review of this see Wolff 1977). In Shafer's view BS is an error of repair of cross-links with consequent accumulation of these and therefore of SCEs. Because of this, the SCE by-pass would become overloaded in BS and the presence of cross-links, having saturated the SCE system, would cause a spilling-over into homologous chromatid exchanges (see also FA).

I have already stressed how the u.v. cellular responses and the slow DNA replication may account for some features of the BS phenotype. Equally, mutational changes of chromosomes and of genes may be responsible for some of the phenotypic effects such as pigmentation defects or possible variation observed in BS cells with respect to SCEs. The sensitivity to certain chemicals and thereby the greater proneness to mutational changes may account for the proneness to malignancy which would be particularly evident for the more rapidly dividing cellular systems of the lymphoreticular and gut tissues. In addition we may wish to consider whether the excessive SCE formation in BS cells could have another detrimental effect, namely on stem cells, if we accept Cairns's hypothesis on the immortal DNA strand (see Cairns 1975; see also Conclusions).

Fanconi anaemia

In 1927 Fanconi first described a family with three children affected by a syndrome of pancytopenia, multiple malformations, hyperpigmentation and growth retardation. Detailed knowledge of the clinical features of the condition is derived mainly from the surveys of the literature by Gmyrek & Syllm-Rapoport (1964) and Beard and colleagues (1973), amounting to about 200 cases. The anaemia is hypoplastic after a hyperplastic phase with a mean age of onset of 6.5 years in boys and 8.5 years in girls. The hypoplasia and subsequent aplasia of the bone marrow are caused by a failure to form functioning marrow stem cells. Hyperpigmentation, both in the form of a fine reticular melanosis and of café-au-lait patches, is very common. In half the cases the skeleton is affected and the radial rays, in particular, are malformed. Renal anomalies are also common and general somatic development is impaired in at least half the cases both before and after birth. The average birth weight is only 2520 g. Microcephaly is present in almost half the cases, mental retardation in almost one-fifth, and deafness in about 7%. Hypogenitalism, at least in males, is also a feature of this disorder. FA is genetically determined as an autosomal recessive condition and this is both on the grounds of segregation analysis and of high parental consanguinity, the parents being seemingly unaffected (Schroeder *et al.* 1976). From the frequency of parental consanguinity we may derive an estimate of the frequency of FA. Assuming that FA is genetically homogeneous—which it may well not be—it could have a frequency of the order of one per million. However an estimate of 1 in 350 000 (Swift 1977) has been given and may be nearer the mark.

Chromosome instability in FA has attracted much attention since Schroeder *et al.* (1964) first showed a variety of structural chromosome changes in the cultured lymphocytes of two brothers with the disease. A high proportion of cells showed various anomalies such as gaps and breaks, mostly of chromatid type, and cross-configurations interpretable as a result of chromatid interchanges. In addition there were many cells with chromosome endoreduplication. Subsequent studies and reports have confirmed these findings and have added details on the position of gaps, breaks and translocations, and on the chromosomes involved. These studies have also clearly shown a great deal of variability in the chromosome changes, not only between different individuals but also in the same individual tested at different times. This variability is probably an important clue to what it means to have the FA molecular defect. One plausible explanation is that environmental stresses, and perhaps internal stresses as well, act on the cells and, through the molecular defect, cause changes in the chromosomal DNA, with consequent display of chromosomal instability. However, variations between different individuals can also be an expression of genetic heterogeneity and in addition some of the variation may be due to differences in cell culture protocols in different laboratories. Another important point is that the chromosome aberrations have been found, though more laboriously, in bone marrow cells, in both B and T lymphocytes (Bushkell *et al.* 1976) and in cultured fibroblasts. The fact that fibroblasts display the basic cytogenetic FA defect is especially important if we are to attribute the developmental anomalies and the neurological changes to proliferative anomalies of the relevant cells which, therefore, must also be assumed to manifest the molecular FA defect.

An important feature of the interchromosomal exchanges is that they usually involve non-homologous chromosomes, and the exchanges are therefore mostly asymmetrical. This is a feature which distinguishes the FA from the BS somatic interchanges. These interchanges are seen as quadriradiate, and at times triradiate, configurations. Interchange figures are much less frequent in FA than in BS, and more gaps and breaks are seen in FA (Therman & Kuhn 1976). After the introduction of banding techniques, chromosome studies pinpointed this characteristic even more clearly. Work by von Koskull & Aula (1973) had suggested that certain chromosome bands in one or other arm of numbers 1, 3, 6 and 13 were especially break-prone, that the sex chromosomes were rather break-free, and that the breaks were in the light chromosome bands. Work at Guy's Hospital, later confirmed by Dutrillaux *et al.* (1977), revealed neither autosome nor band preference. The breaks, too, were often located at the boundaries between bands. An additional important aspect of the interchromosomal effects was the relatively high frequency of exchanges which involved breaks near the centromere or, conversely, near the telomeres.

Many authors have investigated healthy relatives, parents and sibs of FA patients, generally with negative results. However a few examples of abnormalities similar to those in the patients but much less pronounced have been reported. A few examples of numerically unbalanced cells have also been reported.

Patients with FA have an increased risk of developing malignancies, particularly leukaemia, but also solid tumours of the gastrointestinal tract and the liver. Unfortunately, there are no good data on the size of the risk, although for childhood leukaemia, for example, the risk could be around one in 30. However there are other causes of mortality, from infection or from haemorrhage, which probably leads us to underestimate the true propensity to malignancy. On anecdotal evidence, carriers of the FA gene seem to be more at risk of malignant disease than average. Swift (1977) has attempted to obtain information on their propensity to malignancy. Using maximum likelihood estimates (Swift *et al.* 1974), he thinks that the overall risk of malignancy is three times the standard risk, but for leukaemia the figure is 12 times higher. Transformed into the proportion of persons with specific malignancies who are heterozygous for the FA gene, the figures suggest that 1% of all malignancies and 4% of all leukaemias occur in FA carriers whose overall population frequency is about 0.33% (see also Conclusions).

A finding of interest is the 12–24-fold higher transformability of FA cells compared with control cells by the hamster oncogenic virus SV40. This is a potential shared by cells that are irradiated, or aged in culture, or from a variety of abnormal conditions, for example with sex chromosome or autosomal imbalance, like trisomy 21 (see also Knudson *et al.* 1973). This transformability in FA is of interest in view of the suggested relationship between cell transformability *in vitro* and proneness to leukaemia *in vivo* (Miller & Todaro 1969: see also Whitehouse 1974). It is of interest that most observers have also found that cells of heterozygotes have a six to eight-fold greater transformability by SV40 than normal cells. It has been suggested that the high transformability of FA fibroblasts may reflect an error of DNA replication or repair in this disease. A related observation was made by Lubiniecki *et al.* (1977) on fibroblast cultures from FA patients. These cells expressed with high intensity the SV40 T antigen and, interestingly, this expression preceded the manifestations of the anaemia and pancytopenia in the homozygous mutants.

Meanwhile interest was beginning to focus on the molecular nature of the FA abnormality, after observations that FA cells were unduly susceptible to ionizing radiation and especially to the effects of bifunctional DNA alkylating agents. These physical and chemical agents are both lethal and the source of increased chromosome damage in FA cells. A clearer approach to an understanding of what may be the effect of the mutation in FA was obtained by the use of mitomycin C, an agent which induces cross-links between DNA strands, where it can act as a bifunctional alkylating agent. Mitomycin C was found to have two effects on the chromosomes of cultured lymphocytes (Latt *et al.* 1975). First it induced a marked excess of chromosome breaks and rearrangements in FA compared with control cells. Secondly an unusual result was obtained when BUdR was used to study the rate of formation of SCEs. Untreated FA cells yield normal numbers of SCEs by the BUdR method but, after mitomycin C, the increment of SCEs was less than half that observed in normal controls. Conversely, chromosome and chromatid damage was much increased by the agent. The findings suggested that the chromatid breaks may represent vicarious or overflow events positively correlated with the relative paucity of SCEs. In confirmation of this form of reasoning it was noted that many of the breaks occurred at sites of incomplete exchange. By contrast with these results from the use of a bifunctional agent, monofunctional alkylating substances produced more SCEs and, vicariously, fewer breaks, although SCE numbers were still below those induced in normal control cells. On the strength of these observations Latt *et al.* (1975) suggested that although FA cells competently repaired purine bases damaged by monofunctional alkylating agents and coped well with the secondary consequences of the damage, they had instead a specific disability in dealing with a special step in the repair of damage caused by cross-linking agents. It was interesting that cultured dermal FA fibroblasts under similar experimental conditions had near-normal rates of SCEs (Latt *et al.* 1975). Fujiwara & Tatsumi (1975) and Fujiwara *et al.* (1977) obtained results with mitomycin C which, at the chemical level, supported the idea of a faulty cross-link removal in FA cells compared with normal cells. Several hours after mitomycin treatment the DNA extracted from FA cells sedimented more rapidly than is usual in normal cells. This suggested that the unrepaired cross-links persisted in the chromosomal DNA and limited the possibility of it being denatured when extracted from the cell nuclei. As for the specific mutation that could interfere with the removal of cross-links—if such is, as many believe, the essential defect in FA—nothing precise can be said. It is known that in bacteria removal of cross-links is a function that falls to an endonuclease from genes *uvrA* and *uvrB* and to an exonuclease activity from the product of the *PolA* gene. Two cuts are made, one on either side of one of the attachments of the bifunctional agents. This results in the release of the agent from *one of its two DNA strands.* The cut strand is stabilized, pending its fairly rapid mending, by the product of the *uvrC* gene (Cole *et al.* 1976). Incidentally the same endonuclease is also responsible for the incision step in excision repair of pyrimidine dimers. In human cells a different endonuclease would appear to be involved, if one is to judge from the efficiency of XP cells in removing cross-links but not in incising their damaged DNA, and from the converse behaviour of FA cells (see below).

A radically different view of the defect in FA (and in BS) is taken by Shafer (1977). Suggesting that repair of cross-links happens mostly by a post-replicational process requiring the presence of both duplexes of DNA, Shafer hypothesizes that unrepaired cross-links, an impediment to DNA replication, can be by-passed by SCEs. In his view SCE formation, which segregates the cross-link to one of the daughter duplexes, is a two-step process. The first step is a strand displacement that takes place near the cross-link and allows the cross-linked strands to segregate together and be met by the nascent replicating DNA for which each strand is a template (the other two strands join in a similar manner to be met by their own nascent DNA strands). The second step is concerned with the ligation of the terminally adjoining strands. Shafer goes on to suggest that FA is a defect in this second step of the SCE by-pass mechanism, so that the cells display undue sensitivity to cross-links which, however, they are potentially able to repair normally. This fault would lead to the formation of fewer SCEs, many of which, furthermore, would be incomplete. Chromatid exchanges would tend to form at the sites of incomplete SCE formation. It should be noted that the rate of movement of replication forks in FA cells seems normal and this would militate against a defect in cross-link repair (Hand 1977; Iyer & Szybalski 1964).

So at the moment two views exist of the difficulty which FA cells seem to show in dealing with cross-links; there is either a defect in a step in cross-link removal or a defect in the mechanism of cross-link by-pass, namely a step in SCE formation when (and if) SCEs reflect this mechanism. The striking effect on normal cells of cross-linking alkylators compared with the effect, some three orders of magnitude less, of the monofunctional alkylators suggests that removal of bifunctional alkylators proceeds in some special way and that it is the *persistence of cross-links* that matters in FA rather than a single-step event possibly concerned with the removal of the 'half-detached' bifunctional agents. It is difficult to see why efficiency in removing the first half of such a cross-link, as apparently suggested by Latt *et al.* (1975), should not be matched by efficiency in the last step of removal, particularly as one would imagine that it is the first step that is somewhat more special or singularly specific to the handling of cross-links by mammalian cells. If it is correct to think that what matters in FA is the *persistence* of the cross-link then FA can be seen as a defect which allows this persistence; alternatively, as Shafer suggests, it is a defect of cross-link by-pass. By comparison, in BS the defect may be one which results in excessive SCE formation on the basis of persistently open DNA replication units (as in one of the models suggested by Kato 1977) (in keeping with Shafer's view it may then be a defect of excessive formation of cross-links which saturated the cross-link removal system and induced SCE as a by-pass phenomenon). Alternatively, as suggested by Shafer, the primary error might be one of faulty cross-link removal with consequent accumulation of cross-links at successive cell divisions and saturation of by-pass SCEs. At any rate in FA, whatever the disability, it would have the consequence of inducing chromosome breaks, gene mutation and impairment of cell division with consequent prolongation of the cell cycle, which is indeed observed (Dutrillaux & Fosse 1976). Finally it is important that FA cells are able to excise dimers—certainly those produced at low doses, but possibly generally (Setlow 1975; Regan *et al.* 1973)—and so rejoin single-strand breaks (Vincent *et al.* 1975; Taylor *et al.* 1975; Sheridan & Huang 1977) and exercise endonucleolytic activity on apurinic sites (Teebor & Duker 1975) induced by radiation and chemical mutagens, but not necessarily on other γ-induced lesions (Cerutti & Remsen 1976). Also, while the cell cycle is prolonged, the rate of DNA growth between replication points seems normal (Hand 1977). So this fact and the evidence from non-homologous chromosome exchanges indicate that a DNA replication defect is not the basis of FA.

CONCLUSIONS

Summarizing, XP is usually one of a series of defects of excision of u.v.-induced DNA lesions and similar lesions introduced by some mutagens, while in other cases it is a defect of the post-replication repair of these lesions. The former, but not the latter, may be complicated by progressive neurological illness. AT is a clinically and genetically protean disorder of chromosome instability, with sensitivity to γ-radiation which generally rests on an inability to remove DNA lesions rather than to mend strand breaks; it is characterized by malignancies and by the proliferation of a cell clone carrying a chromosome site-specific lesion whose relationship to the malignancies is not entirely clear. BS seems to have a defect consisting of slow DNA chain maturation during DNA synthesis with the formation of an excessive number of sister chromatid exchanges, while FA seems to depend on a defect in dealing with DNA cross-links. However, views differ on the basic molecular problem in these two disorders. All four disorders behave as rare autosomal recessives and in at least three of them there is evidence of genetic heterogeneity. All four are characterized by proneness to malignancy which in XP only is confined to solid tumours, mostly of the sun-exposed skin. In a way we are dealing with a situation analogous to, but contrasting with, the other inheritable malignancy-prone disorders that are due to single gene defects such as inherited retinoblastoma or polyposis coli. In the latter disorders there are gene changes which specifically, combining with other events, lead to malignant transformation of specific tissues and have been taken to be the first step on that road. In the 'repair' defects we are instead dealing with specific gene changes that lead to relatively non-specific mutagenesis at the DNA or chromosomal level from which, it would appear, malignancy subsequently evolves. So, basically, the mechanisms of carcinogenesis are different in the two sets of defects.

At any rate it seems that carcinogenesis in a broad sense can be related to faulty repair of genes and chromosomes. It is equally clear that, given the complexity of carcinogenesis, only a fraction of cancers can arise direct from this cause. Nevertheless repair defects, either primary or secondary to saturation of repair systems in different tissues and organs, must play an important role in the initiation of cancer. Thus the study of repair is relevant to an understanding of carcinogenesis.

Looking at these matters in greater detail, the first point to make is that these defects can be considered as having essentially two main cellular results: they affect cell proliferation and they foster or generate somatic mutations. In addition often there is serious impairment of immune functions, either humoral or cellular or both. Evidence on the effect on cell proliferation is clear when one considers, for example, the growth disturbances they cause, often prenatal; or the interference with normal brain and central sensory differentiation and function; or interference with developmental processes more generally; or the interference with stem cell function, whether the stem cells are those of the bone marrow, the immune system or the gonia. As for somatic mutations, these are of two broad types: chromosome and 'point' changes. It is clear that the cells generated with different chromosome complements carry different and variable mutations. These are ever-sporting and, by their nature, many will be cell-lethal, but some will turn out to be cell-viable and indeed may make the mutant cells highly adaptable to changing environmental circumstances. Somatic mutations, possibly chromosomal, may also underlie some of the changes we observe in these patients, for example patches of pigmentation or vitiligo. Or somatic mutations may underlie the patchiness of abnormal cells, as suggested in BS. As for somatic gene mutations (as opposed to visible chromosome mutations), these are particularly likely to emerge as an effect of the defects of DNA and chromosome repair. However, the real meaning of these phenomena hinges on whether we recognize the seeds of malignancy in these aspects of gene and chromosome mutation. I have already alluded to the chromosome changes in FA and BS. Clearly these can play a role in carcinogenesis, not only by making available a great variety of cells with different potentialities and a wide range of adaptability to environmental circumstance, but also by haphazardly manufacturing cells which may escape from the normal control of growth and proliferation. Chromosome translocations can result in position effects and, from them, unbalance in terms of duplications and deficiencies can derive by malsegregation at somatic cell division. Some of the deficiencies can result in the uncovering of recessive alleles concerned with malignant transformation. Chromosome breaks may conceivably allow oncogenic viruses access, or easier access, to the chromosomal DNA and act as sites for viral integration. There is experimental evidence for this view as the frequency of viral transformation of cells is increased by treatment with agents that cause DNA strand breaks, and some DNA-damaging chemicals favour viral transformation before their damage is fully repaired. On the other hand, experimentally, chromosome breaks and losses can favour cancers (Harris 1971), perhaps because at times loss of suppressor loci may release proviruses for action and for malignant transformation. Though viral agents are not looked upon as common causes of human cancers, one must forget neither the direct participation of viruses in cancer formation after virus infection nor the evidence which suggests that integrated viruses, at least in some laboratory species, may be common and their action might, conceivably, be released by chromosomal imbalance. But there are also other and more special aspects of carcinogenesis in the instability syndromes, in addition to the variable output of chromosomally unbalanced cells. In AT a specific clone may, at least at some times and in some ways, be a feature of the origin of malignant disease. In BS there might again be something special that might be related to the initiation of malignancy if we accept Cairns's working hypothesis (1975) of the immortal strand in stem cells, its relevance to the avoidance of carcinogenesis, and the effect of SCEs on the purity of the strand (see also Cairns 1977). However, perhaps the clearest example of a relation between a DNA lesion and malignancy is seen in XP and in the effect of pyrimidine dimers on malignant transformation, as shown by the elegant experiments by Hart & Setlow (1975) in *Poecilia formosa*. These point to the very presence of the mutational dimers as the key to malignancy. From this experiment we can extrapolate to try to understand the key to the origin of malignancy in the u.v.-exposed skin of XP patients. Exposure of cells to u.v.-mimetic drugs was capable, in the experiments, of bringing about malignancy. So, in repair errors like XP, malignant changes other than in the skin may result from the action of mutagens on organs to which such chemicals gained access or where the 'ultimate carcinogen' was made or became concentrated.

All in all these DNA repair and chromosome instability disorders lend support to mutational hypotheses of the origin of cancer, however the importance of mutation is conceived and whatever the meaning one attaches to the word mutation (see above). At the limit, and very broadly, this could be taken to mean point or frame-shift mutations, duplications, deficiencies, possibly position effects or even the activity of viral genomes, incorporated anew or released for malignant transformation. This leads us to the two (or more?) step theories of cancer (Ashley 1969; Armitage & Doll 1957; see also Knudson *et al.* 1973), where one at least of the steps (as in inheritable predispositions to malignancies) is mutational. Even an environmental origin for most cancers accepts the coinvolvement of DNA, and thus accepts mutational events while offering clear pragmatic prospects for prevention (Doll 1977). To the same two-step motif belongs the distinction between initiators and promoters, with the latter being thought of as agents which stimulate cells to proliferate excessively (Knudson *et al.* 1973) or as mechanisms that alter in some way, but also act at times through enhanced cell division, 'the frequency at which regulatory circuits are reset' in the cell (Cairns 1977).

Then, in some of the disorders considered here, namely FA, BS and AT, the role of an immune deficiency should be considered. What part faulty immune surveillance (Burnet 1970, 1974) plays in the initiation, but especially in the maintenance and spread of malignancy, can only be a matter for speculation. Present hypotheses, based mainly but not only on experimental evidence, and especially on work on solid tumours, are against the importance of immune surveillance in carcinogenesis (see e.g. Rygaard & Polvsen 1976: Möller & Möller 1976; Cairns 1977). Yet it is clear that faulty immunosurveillance and, for example, the role of the macrophages in defence against cancer are considered relevant (*Lancet* 1976; Eccles 1978; Woodruff & Speedy 1978). It is noteworthy that lymphoproliferative and related malignancies, in particular, seem to be linked with faulty immune functions in the syndromes considered here. It is thus possible that immune defences may be effective especially in the protection against these malignancies rather than against cancers arising in 'solid' tissues. Interestingly the immunological defects themselves can be seen as by-products of defective cell proliferation. It could further be speculated that not only in the disorders considered here but also more generally (but neither universally nor necessarily preponderantly), leukaemias and similar proliferative disorders may rely on weaker immune defences for their colonization of the host, and thus possibly may be at no invasive disadvantage from a fairly stable (as well as sometimes specific) karyotype. Whether leukaemias can pave their own way by producing in a direct manner an additional immune defect is yet another problem. Conversely (but neither universally nor necessarily preponderantly) solid tumours would perhaps 'adopt' a different strategy which did not have to by-pass inherently weak immune defences of the host. Such defences, even if strong, could be disregarded by these tumours because of the repertoire of high variability which neoplastic cells originating from solid tissues seem to be able to muster, display and expose to selection. Variability of this kind may on the one hand lead to the origin of the neoplastic cell (as in the chromosome instability syndromes) and on the other hand could make neoplastic cells invasive locally and generally; such variability is the result of a hypermutable chromosomal complement, from the power to undergo somatic recombination, from the power to haploidize, exposing recessives, for example, and perhaps even to regain vigour through a somatic alternative to sex.

The contribution to wider public health issues of the clear-cut diseases predisposing to malignancy and discussed here is small. However, in this context the behaviour of the asymptomatic heterozygous carriers of these recessive genes is of interest. It could be estimated that the carriers for AT could be at least between 1 in 150 and 1 in 100 and those for FA 1 in 300. Taking XP, BS, AT and FA together, 1 in 50 persons may be at the very least a carrier of one of the genes in question. For the carrier states of FA and AT Swift (1977) has provided evidence for an increased risk of malignancy which, depending on gene and allele, and type of malignancy, ranges from twice to twelve times the standard risk. Swift estimates that 5% of all persons dying from any malignancy before the age of 45 may carry the gene for AT and that 7% of all deaths from lymphoreticular malignancy below 45 occur in carriers of the AT gene. Expanding these views further, he estimates that to carry the AT and FA genes increases the risk of malignancy by a factor of two, taking all ages and malignancies together.

As is often the case, the study of the rare and unusual in medicine, the treasuring of exceptions, the attempt to unravel nature's experiments and the use of a genetic approach have opened unsuspected vistas. Although we have covered much ground in our approach to these rare conditions we are still only at the opening moves of a difficult but exciting stage in the study of the idiosyncrasies of human responses to environmental agents, in the study of human somatic mutation, and in the work on predisposition to cancers. Ultimately the paths to prevention lie in the lessons of epidemiology (Cairns 1977) being heeded through the avoidance of environmental carcinogens (Doll 1977), and in the detection and unravelling of individual idiosyncrasies towards chemical and physical agents which induce cancers.

ACKNOWLEDGEMENTS

It is a special pleasure to thank Dr F. Giannelli for much informative, interactive and critical discussion over the years, particularly on xeroderma pigmentosum. I am grateful to Miss Jane Lyttleton for questions, information and discussion on many problems of the chromosome instability syndromes. The financial support of the Spastics Society is gratefully acknowledged.

References

ALHADEFF, B. & COHEN, M. M. (1976) Frequency and distribution of sister chromatid exchanges in human peripheral lymphocytes. *Isr. J. Med. Sci. 12,* 1440-1447

ANDERSON, T. E. & BEGG, M. (1950) Xeroderma pigmentosum of mild type. *Br. J. Dermatol. 62,* 402-407

ANDREWS, A. D., BARRETT, S. F. & ROBBINS, J. H. (1976) Relation of DNA repair processes to pathological ageing of the nervous system in xeroderma pigmentosum. *Lancet 1,* 1318-1320

ANGELL, R. R. & JACOBS, P. A. (1975) Lateral asymmetry in human constitutive heterochromatin. *Chromosoma (Berl.) 51,* 301-310

ARLETT, C. F. (1977*a*) Lethal response to DNA-damaging agents in a variety of human fibroblast cell strains. *Br. J. Cancer 35,* 244

ARLETT, C. F. (1977*b*) Lethal response to DNA damaging agents in a variety of human fibroblast cell strains. *Mutat. Res. 46,* 106

ARMITAGE, P. & DOLL, R. (1957) A two-stage theory of carcinogenesis in relation to the age distribution of human cancer. *Br. J. Cancer 11,* 161-169

ASHLEY, D. J. B. (1969) The two 'hit' and multiple 'hit' theories of carcinogenesis. *Br. J. Cancer 23,* 313-328

BARLOW, P. W. (1978) The concept of the stem cell in the context of plant growth and development, in *Stem Cells and Tissue Homeostasis* (Lord, B. I. *et al.,* eds.), pp. 87-113, Cambridge University Press, Cambridge,

BARSKI, G., SORIEUL, S. & CORNEFERT, F. (1960) Biologie cellulaire: production dans des cultures *in vitro* de deux souches cellulaires en association, de cellulles de caractère 'hybride'. *C.R. Acad. Sci. (Paris) 251,* 1825-1827

BARTSCH-SANDHOFF, M. (1977) Fusion of homologous chromosomes (15q 15q) as cause of recurrent abortion. *Lancet 1,* 551

BEARD, M. E. J., YOUNG, D. E., BATEMAN, C. J. T., MCCARTHY, G. T., SMITH, M. E., SINCLAIR, L., FRANKLIN, A. W. & SCOTT, R. B. (1973) Fanconi's anaemia. *Q. J. Med. 42,* 403-422

BIGGAR, W. D. & GOOD, R. A. (1975) Immunodeficiency in ataxia telangiectasia, in *Immunodeficiency in Man and Animals* (Bergsma, D. *et al.,* eds.) *(Birth Defects: Original Article Ser. vol. 11, No. 1),*pp. 271-276, Sinauer Associates Inc., Sunderland, Massachusetts

BLOOM, D. (1954) Congenital telangiectatic erythema resembling lupus erythematosus in dwarfs. Probably a syndrome entity. *Am. J. Dis. Child. 88,* 754-758

BODER, E. (1975) Ataxia-telangiectasia: some historic, clinical and pathological observations, in *Immunodeficiency in Man and Animals* (Bergsma, D. *et al.,* eds.) *(Birth Defects: Original Article Ser. vol. 11, No. 1),* pp. 255-270, Sinauer Associates Inc., Sunderland, Massachusetts

BOOTSMA, D. (1978) Defective DNA repair and cancer, in *Research in Photobiology* (Castellani, A., ed.), pp. 455-468, Plenum Press, New York

BOOTSMA, D., MULDER, M. P., POT, F. & COHEN, J. A. (1970) Different inherited levels of DNA repair replication in xeroderma pigmentosum cell strains after exposure to ultraviolet irradiation. *Mutat. Res. 9,* 507-516

BOOTSMA, D., DE WEERD-KASTELEIN, E. A., KLEIJER, W. J. & KEYZEZ, W. (1975) Genetic complementation. Analysis of xeroderma pigmentosum, in *Molecular Mechanisms for Repair of DNA – Part B. Basic Life Sciences,* vol. 5 (Hanawalt, P. C. & Setlow, R. B., eds.), pp. 725-728, Plenum Press, New York

British Medical Bulletin (1971) Human genetics. *17,* 177-263

BURNET, M. (1970) *Immunological Surveillance,* Pergamon Press, Oxford

BURNET, M. (1974) *Intrinsic Mutagenesis: A Genetic Approach to Ageing,* pp. 158-187, MTP Press, Lancaster

BUSHKELL, L. L., KERSEY, J. H. & CERVENKA, J. (1976) Chromosomal breaks in T & B lymphocytes in Fanconi's anemia. *Clin. Genet. 9,* 583-587

CAIRNS, J. (1975) Mutation selection and the natural history of cancer. *Nature (Lond.) 255,* 197-200

CAIRNS, J. (1977) Some thoughts about cancer research in lieu of a summary, in *Origins of Human Cancer,* Book C: *Human Risk Assessment* (Hiatt, H. H. *et al.,* eds.) *(Cold Spring Harbor Conf. Cell Proliferation, vol. 4),*pp. 1813-1820, Cold Spring Harbor Laboratory, New York

CASPERSSON, T., ZECH, L., MODEST, E. J., FOLEY, G. E., WAGH, U. & SIMONSSON, E. (1969) Chemical differentiation with fluorescent alkylating agents in *Vicia faba* metaphase chromosomes. *Exp. Cell Res. 58,* 128-140

CASPERSSON, T., ZECH, L., JOHANNSON, C. & MODEST, E. J. (1970) Identification of human chromosomes by DNA-binding fluorescent agents. *Chromosoma (Berl.) 30,* 215-227

CERUTTI, P. A. & REMSEN, J. F. (1976) Gamma-ray excision repair in normal and diseased human cells, in *Biology of Radiation in Carcinogenesis* (Yuhas, J. M. *et al.*, eds.), pp. 93-101, Raven Press, New York

CHAGANTI, R. S. K., SCHONBERG, S. & GERMAN, J. (1974) The manyfold increase in sister chromatid exchanges in Bloom's syndrome lymphocytes. *Proc. Natl. Acad. Sci. U.S.A. 71,* 4508-4512

CHEN S.-H., SCOTT, C. R. & GIBLETT, E. R. (1974) Adenosine deaminase: demonstration of a 'silent' gene associated with combined immunodeficiency disease. *Am. J. Hum. Genet. 26,* 103-107

CLEAVER, J. E. (1968) Defective repair replication of DNA in xeroderma pigmentosum. *Nature (Lond.) 218,* 652-656

CLEAVER, J. E. (1969) Xeroderma pigmentosum: a human disease in which an initial stage of DNA repair is defective. *Proc. Natl. Acad. Sci. U.S.A. 63,* 428-435

CLEAVER, J. E. (1970) DNA damage and repair in light-sensitive human skin disease. *J. Invest. Dermatol. 54,* 181-195

CLEAVER, J. E. (1974*a*) Repair processes for photochemical damage in mammalian cells, in *Advances in Radiation Biology,* vol. 4 (Lett, J. T. *et al.,* eds.), pp. 1-75, Academic Press, New York

CLEAVER, J. E. (1974*b*) Sedimentation of DNA from human fibroblasts irradiated with ultraviolet light: possible detection of excision breaks in normal and repair-deficient xeroderma pigmentosum cells. *Radiat. Res. 57,* 207-227

CLEAVER, J. E. & BOOTSMA, D. (1975) Xeroderma pigmentosum: biochemical and genetic characteristics. *Annu. Rev. Genet. 9,* 19-38

CLEAVER, J. E. & CARTER, D. M. (1973) Xeroderma pigmentosum variants: influence of temperature on DNA repairs. *J. Invest. Dermatol. 60.* 29-32

CLEAVER, J. E., THOMAS, G. H., TROSKO, J. E. & LETT, J. T. (1972) Excision repair (dimer excision strand breakage & repair replication) in primary cultures of eukaryotic (bovine) cells. *Exp. Cell Res. 74,* 67-80

COHEN, M. M. (1971) The chromosomal constitution of 165 human translocations involving D group chromosomes identified by autoradiography. *Ann. Genet. 14,* 87-96

COHEN, M. M., SHAHAM, M., DAGAN, J., SHMUELI, E. & KOHN, G. (1975) Cytogenetic investigations in families with ataxia-telangiectasia. *Cytogenet. Cell Genet. 15,* 338-356

COLE, R. S., LEVITAN, D. & SINDEN, R. R. (1976) Removal of psoralen interstrand cross-links from DNA of *Escherichia coli:* mechanism and genetic control. *J. Mol. Biol. 103,* 39-59

COOK, P. R. & BRAZELL, I. A. (1975) Supercoils in human DNA. *J. Cell Sci. 22,* 261-279

COOK, P. R. & BRAZELL, I. A. (1976*a*) Conformational constraints in nuclear DNA. *J. Cell Sci. 22,* 287-302

COOK, P. R. & BRAZELL, I. A. (1976*b*) Detection and repair of single-strand breaks in nuclear DNA. *Nature (Lond.) 263,* 679-682

COOK, K., FRIEDBERG, E. C. & CLEAVER, J. E. (1975) Excision of thymine dimers from specifically incised DNA by extracts of xeroderma pigmentosum cells. *Nature (Lond.) 256,* 235-236

COOK, P. R., BRAZELL, I. A., PAWSEY, S. A. & GIANNELLI, F. (1978) Changes induced by ultraviolet light in the superhelical DNA of lymphocytes from subjects with xeroderma pigmentosum and normal controls. *J. Cell Sci. 29,* 117-127

CUNLIFFE, P. N., MANN, J. R., CAMERON, A. H. & ROBERTS, K. D. (1975) Radiosensitivity in ataxia-telangiectasia. *Br. J. Radiol. 48,* 374-376

DE SANCTIS, C. & CACCHIONE, A. (1932) L'idiozia xerodermica. *Riv. Sper. Freniatr. Med. Leg. Alienazioni Ment. 56,* 269-292

DE WEERD-KASTELEIN, E. A. (1974) *Genetic heterogeneity in the human skin disease xeroderma pigmentosum,* Doctoral thesis, Erasmus University, Rotterdam

DE WEERD-KASTELEIN, E., KEIJZER, W., RAINALDI, G. & BOOTSMA, D. (1977) Induction of sister chromatid exchanges in xeroderma pigmentosum cells after exposure to ultraviolet light. *Mutat. Res. 45,* 253-261

DOLL, R. (1977) Strategy for detection of cancer hazards to man. *Nature (Lond.) 265,* 589-596

DONIGER, J. & GROSSMAN, L. (1976) Human correxonuclease. Purification and properties of a DNA repair exonuclease from placenta. *J. Biol. Chem. 251,* 4579-4587

DONIGER, J. & GROSSMAN, L. (1977) Item 14: human correxonuclease. A repair exonuclease isolated from placenta. Papers presented at 2nd International Workshop, May 1976, Netherlands. Reports. *Mutat. Res. 46,* 114

DORN, H. (1959) Xeroderma pigmentosum. *Acta Genet. Med. Gemellol. 8,* 395-408

DUKER, N. J. & TEEBOR, G. W. (1975) Different ultraviolet DNA endonuclease activity in human cells. *Nature (Lond.) 255,* 82-84

DUTRILLAUX, B. & FOSSE, A.-M. (1976) Study of the cellular division cycle of normal and abnormal subjects using BrdU. *Ann. Genet. 19,* 95-102

DUTRILLAUX, B., COUTURIER, J., VIEGAS-PÉQUIGNOT, E. & SCHAISON, G. (1977) Localization of chromatid breaks in Fanconi's anemia, using three consecutive stains. *Hum. Genet. 37,* 64-71

ECCLES, S.A. (1978) Macrophage and cancer, in *Immunological Aspects of Cancer* (Castro, J.E., ed.), pp. 123-154, MTP Press, Lancaster

EPHRUSSI, B., SCALETTA, L.J., STENCHEVER, M.A. & YOSHIDA, M.C. (1964) Hybridization of somatic cells *in vitro,* in *Cytogenetics of Cells in Culture (Symp. Int. Soc. Cell Biol.)* vol. 3 (Harris, R.J.C., ed.), pp.13-25, Academic Press, New York

FANCONI, G. (1927) Familiäre infantile perniziosa-artige Anämie (perniziöses Blutbild und Konstitution). *Jb. Kinderheilk. 117,* 257-280

FELDBERG, R. & GROSSMAN, L. (1976) A DNA binding protein from human placenta specific for ultraviolet damaged DNA. *Biochemistry 15,* 2402-2408

FELSENFELD, G. (1978) Chromatin. *Nature (Lond.) 271,* 115-122

FLAMM, W. G., MCCALLUM, M. & WALKER, P. M. B. (1967) The isolation of complementary strands from a mouse DNA fraction. *Proc. Natl. Acad. Sci. U.S.A. 57,* 1729-1734

FLEISCHMAN, E. W. & PRIGOGINA, E. L. (1977) Karyotype peculiarities of malignant lymphomas. *Hum. Genet. 35,* 269-279

FORNACE, A. J. JR., KOHN, K. W. & KANN, H. E. JR. (1976) DNA single-strand breaks during repair of UV damage in human fibroblasts and abnormalities of repair in xeroderma pigmentosum. *Proc. Natl. Acad. Sci. U.S.A. 73,* 39-43

FRACCARO, M., ZUFFARDI, O., BUHLER, E. M. & JURIK, L. P. (1977) 15/15 translocation in Prader-Willi syndrome. *J. Med. Genet. 14,* 275-276

FRANCKE, U. (1977) Cytogenetics and somatic cell genetics: the impact of chromosome banding, in *Trends and Teaching in Clinical Genetics* (Bergsma, D. *et al.*, eds.) *(Birth Defects: Original Article Ser. vol. 13, No. 6),* pp. 79-103, Alan R. Liss, New York

FUJIWARA, Y. & TATSUMI, M. (1975) Repair of mitomycin C damage to DNA in mammalian cells and its impairment in Fanconi's anemia cells. *Biochem. Biophys. Res. Commun. 66,* 592-598

FUJIWARA, Y., TATSUMI, M. & SASAKI, M. S. (1977) Cross-link repair in human cells and its possible defect in Fanconi's anemia cells. *J. Mol. Biol. 113,* 635-650

FUKUHARA, S., SHIRAKAWA, S. & UCHINO, H. (1976) Specific marker chromosome 14 in malignant lymphomas. *Nature (Lond.) 259,* 210-211

GAHRTON, G., FRIBERG, K., ZECH, L. & LINDSTEN, J. (1978) Duplication of part of chromosome No. 1 in myeloproliferative diseases. *Lancet 1,* 96-97

GERMAN, J. (1969) Bloom's syndrome I. Genetical and clinical observations in the first twenty-seven patients. *Am. J. Hum. Genet. 21,* 196-227

GERMAN, J. (1973) Genetic disorders associated with chromosomal instability and cancer. *J. Invest. Dermatol. 60,* 427-434

GERMAN, J. (1974) Bloom's syndrome. II. The prototype of human genetic disorders predisposing to chromosome instability and cancer, in *Chromosomes and Cancer* (German, J., ed.), pp. 601-617, Wiley, New York

GERMAN, J. & LA ROCK, J. (1969) Chromosomal effects of mitomycin, a potential recombinogen in mammalian cell genetics. *Tex. Rep. Biol. Med. 27,* 409-418

GERMAN, J., SCHONBERG, S., LOUIE, E. & CHAGANTI, R. S. K. (1977) Bloom's syndrome IV. Sister-chromatid exchanges in lymphocytes. *Am. J. Hum. Genet. 29,* 248-255

GIANNELLI, F. (1976) Xeroderma pigmentosum, in *Scientific Foundations of Oncology* (Symington, T. & Carter, R. L., eds.), pp. 476-489, Heinemann Medical Books, London

GIANNELLI, F. (1978) Xeroderma pigmentosum and the role of DNA repair in oncogenesis. *Bull. Cancer (Paris) 65,* 323-334

GIANNELLI, F. & PAWSEY, S. A. (1974) DNA repair synthesis in human heterokaryons II. A test for heterozygosity in xeroderma pigmentosum and some insight into the structure of the defective enzyme. *J. Cell Sci. 15,* 163-176

GIANNELLI, F., BENSON, P. F., PAWSEY, S. A. & POLANI, P. E. (1977) Ultraviolet light sensitivity and delayed DNA-chain maturation in Bloom's syndrome fibroblasts. *Nature (Lond.) 265,* 466-469

GIBLETT, E. R., ANDERSON, J. E., COHEN, F., POLLARA, B. & MEUWISSEN, H. (1972) Adenosine-deaminase deficiency in two patients with severely impaired cellular immunity. *Lancet 2,* 1067-1069

GMYREK, D. & SYLLM-RAPOPORT, I. (1964) Fanconi's anemia. Analysis of 129 cases from the literature. *Z. Kinderheilk. 91,* 297-337

GOTH, R. & RAJEWSKY, M. F. (1974) Persistence of 06-ethylguanine in rat-brain DNA: correlation with nervous system-specific carcinogenesis by ethylnitrosourea. *Proc. Natl. Acad. Sci. U.S.A. 71,* 639-643

GOTH-GOLDSTEIN, R. (1977) Repair of DNA damaged by alkylating carcinogens is defective in xeroderma pigmentosum-derived fibroblasts. *Nature (Lond.) 267,* 81-82

GROSSMAN, L. (1974) Enzymes involved in the repair of DNA, in *Advances in Radiation Biology,* vol. 4 (Lett, J. T. *et al.,* eds.), pp. 77-129, Academic Press, New York

HAND, R. (1977) Human DNA replication: fiber autoradiographic analysis of diploid cells from normal adults and from Fanconi's anemia and ataxia telangiectasia. *Human. Genet. 37,* 55-64

HAND, R. & GERMAN, J. (1975) A retarded rate of DNA chain growth in Bloom's syndrome. *Proc. Natl. Acad. Sci. U.S.A. 72,* 758-762

HARRIS, H. (1971) The Croonian Lecture, 1971. Cell fusion and the analysis of malignancy. *Proc. R. Soc. Lond. B Biol. Sci. 179,* 1-20

HARRIS, H. & WATKINS, J. F. (1965) Hybrid cells derived from mouse and man: artificial heterokaryons of mammalian cells from different species. *Nature (Lond.) 205,* 640-646

HART, R. W. & SETLOW, R. B. (1975) Direct evidence that pyrimidine dimers in DNA result in neoplastic transformation, in *Molecular Mechanisms for the Repair of DNA,* Part B (Hanawalt, P. C. & Setlow, R. B., eds.), pp. 719-724, Plenum Press, New York

HAWKEY, C. J. & SMETHIES, A. (1976) The Prader-Willi syndrome with a 15/15 translocation. Case report and review of the literature. *J. Med. Genet. 13,* 152-157

HECHT, F. (1977) The nonrandomness of human chromosome abnormalities, in *Population Cytogenetics: Studies in Humans* (Hook, E. B. & Porter, I. H., eds.), pp. 237-250, Academic Press, New York

HERREROS, B. & GIANNELLI, F. (1967) Spatial distribution of old and new chromatid sub-units and frequency of chromatid exchanges in induced lymphocyte endoreduplications. *Nature (Lond.) 216,* 286-287

HIGURASHI, M. & CONEN, P. E. (1973) *In vitro* chromosomal radiosensitivity in 'chromosomal breakage syndromes'. *Cancer (Phila.) 32,* 380-383

HIRSCHHORN, R. (1977) Adenosine deaminase deficiency and immunodeficiencies. *Fed. Proc. 36,* 2166-2170

HOAR, D. I. & SARGENT, P. (1976) Chemical mutagen hypersensitivity in ataxia telangiectasia. *Nature (Lond.) 261,* 590-592

HOOK, E. B. & HAMERTON, J. L. (1977) The frequency of chromosome abnormalities detected in consecutive newborn studies—Differences between studies— Results by sex and by severity of phenotypic involvement, in *Population Cytogenetics: Studies in Humans* (Hook, E. B. & Porter, I. H., eds.), pp. 63-80, Academic Press, New York

HOWARD-FLANDERS, P. (1973) DNA repair and recombination. *Br. Med. Bull. 29,* 226-235

HSU, L.Y.F., KIM, G.J., SUJANSKY, E., KOUSSEFF, B. & HIRSCHHORN, K. (1973) Reciprocal translocations versus centric fusion between two No. 13 chromosomes. A case of 46,XX,−13+t (13;13) (p12;q13) and a case of 46, XY,−13,+t (13;13) (p12;p12). *Cytogenet. Cell Genet. 12,* 235-244

HÜTTEROTH, T.H., LITWIN, S.D. & GERMAN, J. (1975) Abnormal immune responses of Bloom's syndrome lymphocytes *in vitro. J. Clin. Invest. 56,* 1-7

IYER, V.N. & SZYBALSKY, W. (1964) Mitomycins and porfiromycin: chemical mechanism of activation and cross-linking of DNA. *Science (Wash. D.C.) 145,* 55-58

JARVIS, J.E., BALL, G., RICKINSON, A.B. & EPSTEIN, M.A. (1974) Cytogenetic studies on human lymphoblastoid cell lines from Burkitt's lymphomas and other sources. *Int. J. Cancer 14,* 716-721

KAISER-MCCAW, B., EPSTEIN, A.L., KAPLAN, H.S. & HECHT, F. (1977) Chromosome 14 translocation in African and North American Burkitt's lymphoma. *Int. J. Cancer 19,* 482-486

KAPOSI, M. (1874) Xeroderma, parchment skin, in *On Diseases of the Skin,* vol. 3 (Hebra, F. & Kaposi, M., eds.), pp. 252-258, New Sydenham Society, London

KATO, H. (1977) Mechanisms for sister chromatid exchange and their relation to the production of chromosomal aberrations. *Chromosoma (Berl.) 59,* 179-191

KAULI, R., PRAGER-LEWIN, R., KAUFMAN, H. & LARON, Z. (1977) Gonadal function in Bloom's syndrome. *Clin. Endocrinol.* 285-289

KELNER, H. (1949) Effect of visible light on the recovery of *Streptomyces griseus* conidia from ultraviolet irradiation injury. *Proc. Natl. Acad. Sci. U.S.A. 35,* 73-79

KERSEY, J.H., SPECTOR, B.D. & GOOD, R.A. (1973) Primary immunodeficiency diseases and cancer: the immunodeficiency-cancer registry. *Int. J. Cancer 12,* 333-347

KERSEY, J.H., SPECTOR, B.D. & GOOD, R.A. (1975) Primary immunodeficiency and malignancy, in *Immunodeficiency in Man and Animals* (Bergsma, D. *et al.,* eds.) *(Birth Defects; Original Article Ser. vol. 11 No. 1),* pp. 289-298, Sinauer Associates Inc., Sunderland, Massachusetts

KIHLMAN, B.A. (1977) *Caffeine and Chromosomes,* pp. 249-277, Elsevier Scientific Publishing Co., Amsterdam

KNUDSON, A.G., STRONG, L.C. & ANDERSON, D.E. (1973) Heredity and cancer in man. *Prog. Med. Genet. 9,* 113-158

KOVACS, G. (1978) Abnormalities of chromosome No. 1 in haematological malignancies. *Lancet 1,* 555

KRAEMER, K.H., COON, H.G., PETINGA, R.A., BARRETT, S.F., RAHE, A.E. & ROBBINS, J.H. (1975) Genetic heterogeneity in xeroderma pigmentosum: complementation groups and their relationship to DNA repair rates. *Proc. Natl. Acad. Sci. U.S.A. 72,* 59-63

KUHN, E. (1976) Localisation by Q banding of mitotic chiasmata in cases of Bloom's syndrome. *Chromosoma (Berl.) 57,* 1-11

KUHNLEIN, U., PENHOET, E.E. & LINN, S. (1976) An altered apurinic DNA endonuclease activity in group A and group D xeroderma pigmentosum fibroblasts. *Proc. Natl. Acad. Sci. U.S.A. 73,* 1169-1173

Lancet (1976) Macrophages v. cancer., *2,* 27-28

LARK, K.G. (1966) Regulation of chromosome replication and segregation in bacteria. *Bacteriol. Rev. 30,* 3-32

LARK, K.G., CONSIGLI, R.A. & MINOCHA, H.C. (1966) Segregation of sister chromatids in mammalian cells. *Science (Wash. D.C.) 154,* 1202-1205

LATT, S.A. (1973) Microfluorometric detection of deoxyribonucleic acid replication in human metaphase chromosomes. *Proc. Natl. Acad. Sci. U.S.A. 70,* 3395-3399

LATT, S.A. (1974) Sister chromatid exchanges, indices of human chromosome damage and repair: detection by fluorescence and induction by mitomycin C. *Proc. Natl. Acad. Sci. U.S.A. 71,* 3162-3166

LATT, S.A. & JUERGENS, L.A. (1977) Determinants of sister chromatid exchange frequencies in human chromosomes, in *Population Cytogenetics: Studies in Humans* (Hook, E.B. & Porter, I.H., eds.), pp. 217-236, Academic Press, New York

LATT, S.A., DAVIDSON, R.L., LIN, M.S. & GERALD, P.S. (1974) Lateral asymmetry in the fluorescence of human Y chromosomes stained with 33258 Hoechst. *Exp. Cell Res. 87,* 425-429

LATT, S.A., STETTEN, G., JUERGENS, L.A., BUCHANAN, G.R. & GERALD, P.S. (1975) Induction by alkylating agents of sister chromatid exchanges and chromatid breaks in Fanconi's anemia. *Proc. Natl. Acad. Sci. U.S.A. 72,* 4066-4070

LEFFAK, I.M., GRAINGER, R. & WEINTRAUB, H. (1977) Conservative assembly and segregation of nucleosomal histones. *Cell 12,* 837-845

LEHMANN, A.R. (1978) Repair processes for radiation-induced DNA damage, in *Effects of Ionizing Radiation on DNA. Physical, Chemical and Biological Aspects* (Hutterman, J., *et al.,* eds.) *(Molecular Biology, Biochemistry and Biophysics, vol. 27),* pp. 312-334, Springer, Berlin

LEHMANN, A.R. & STEVENS, S. (1977) The production and repair of double strand breaks in cells from normal humans and from patients with ataxia telangiectasia. *Biochim. Biophys. Acta 474,* 49-60

LEHMANN, A.R., KIRK-BELL, S., ARLETT, C.F., PATERSON, M.C., LOHMAN, P.M.H., DE WEERD-KASTELEIN, E.A. & BOOTSMA, D. (1975) Xeroderma pigmentosum cells with normal levels of excision repair have a defect in DNA synthesis after UV-irradiation. *Proc. Natl. Acad. Sci. U.S.A. 72,* 219-223

LEHMANN, A.R., KIRK-BELL, S., ARLETT, C.F., HARCOURT, S.A., DE WEERD-KASTELEIN, E.A., KEIJZER, W. & HALL-SMITH, P. (1977) Repair of ultraviolet light damage in a variety of human fibroblast cell strains. *Cancer Res. 37,* 904-910

LIANG, W. & ROWLEY, J.D. (1978) 14q + marker chromosomes in multiple myeloma and plasmacell leukaemia. *Lancet 1,* 96

LIN, M.S. & DAVIDSON, R.L. (1974) Centric fusion, satellite DNA, and DNA polarity in mouse chromosomes. *Science (Wash. D.C.) 185,* 1179-1181

LIN, M.S., LATT, S.A. & DAVIDSON, R.L. (1974*a*) Microfluorometric detection of asymmetry in the centromeric region of mouse chromosomes. *Exp. Cell Res. 86,* 392-395

LIN, M.S., LATT, S.A. & DAVIDSON, R.L. (1974*b*) Identification of human and mouse chromosomes in human-mouse hybrids by centromere fluorescence. *Exp. Cell Res. 87,* 429-433

LITTLEFIELD, J. (1964) Selection of hybrids from matings of fibroblasts *in vivo* and their presumed recombinants. *Science (Wash D.C.) 145,* 709-710

LOUIS-BAR, D. (1941) Quoted by Boder, E. (1975)

LUBINIECKI, A.S., BLATTNER, W.A., DOSIK, H., SUN, C. & FRAUMENI, J.F. Jr. (1977) SV40 T-antigen expression in skin fibroblasts from clinically normal individuals and from ten cases of Fanconi anemia. *Am. J. Hematol. 2,* 35-40

MAHER, V.M. & MCCORMICK, J.J. (1976) Effect of DNA repair on the cytotoxicity and mutagenicity of UV radiation and of chemical carcinogens in normal and xeroderma pigmentosum cells, in *Biology of Radiation Carcinogenesis* (Yuhas, J.M. *et al.,* eds.), pp. 129-145, Raven Press, New York

MAHER, V.M., BIRCH, N., OTTO, J.R. & MCCORMICK, J.J. (1975) Cytotoxicity of carcinogenic aromatic amides in normal and xeroderma pigmentosum fibroblasts with different DNA repair capabilities. *J. Natl. Cancer Inst. 54,* 1287-1294

MAHER, V.M., DUELLETTE, L.M., CURREN, R.D. & MCCORMICK, J.J. (1976) Caffeine enhancement of the cytotoxic and mutagenic effect of ultraviolet irradiation in a xeroderma pigmentosum variant strain of human cells. *Biochem. Biophys. Res. Commun. 71,* 228-234

MANOLOV, G. & MANOLOVA, Y. (1972) Marker band in one chromosome 14 from Burkitt lymphomas. *Nature (Lond.) 237,* 33-34

MARSHALL, R.R. & SCOTT, D. (1976) The relationship between chromosome damage and cell killing in UV-irradiated normal and xeroderma pigmentosum cells. *Mutat. Res. 36,* 397-400

MATSUNAGA, E. & TONOMURA, A. (1972) Parental age and birth weight in translocation Down's syndrome. *Ann. Hum. Genet. 36,* 209-219

MILLER, R.W. & TODARO, G.J. (1969) Viral transmission of cells from persons at high risk of cancer. *Lancet 1,* 81-82

MÖLLER, G. & MÖLLER, E. (1976) The concept of immunological surveillance against neoplasia. *Transplant. Rev. 28,* 3-15

MOORHEAD, P.S., NOWELL, P.C., MELLMAN, W.J., BATTIPS, D.M. & HUNGERFORD, D.A. (1960) Chromosome preparations of leukocytes cultured from human peripheral blood. *Exp. Cell Res. 20,* 613-616

MORGAN, J.L., HOLCOMB, T.M. & MORRISSEY, R.W. (1968) Radiation reaction in ataxia telangiectasia. *Am. J. Dis. Child. 116,* 557-558

MORTELMANS, K., FRIEDBERG, E.C., SLOR, H., THOMAS, G. & CLEAVER, J.E. (1976) Defective thymine dimer excision by cell-free extracts of xeroderma pigmentosum cells. *Proc. Natl. Acad. Sci. U.S.A. 73,* 2757-2761

NELSON, M.M., BLOM, A. & ARENS, L. (1975) Chromosomes in ataxia-telangiectasia. *Lancet 1,* 518-519

ODARTCHENKO, N. & KENEKLIS, T. (1973) Localization of paternal DNA in interphase nuclei of mouse eggs during early cleavage. *Nature (Lond.) 241,* 528-529

OXFORD, J.M., HARNDEN, D.G., PARRINGTON, J.M. & DELHANTY, J.D.A. (1975) Specific chromosome aberrations in ataxia telangiectasia. *J. Med. Genet. 12,* 251-262

PARRINGTON, J.M., DELHANTY, J.D.A. & BADEN, H.P. (1971) Unscheduled DNA synthesis, u.v.-induced chromosome aberrations and SV40 transformation in cultured cells from xeroderma pigmentosum. *Ann. Hum. Genet. 35,* 149-160

PATERSON, M.C. & SMITH, B.P. (1977) Ataxia telangiectasia. A γ-ray analogue of xeroderma pigmentosum. *Mutat. Res. 46,* 148

PATERSON, M.C., SMITH, B.P., LOHMAN, P.H.M., ANDERSON, A.K. & FISHMAN, L. (1976) Defective excision repair of γ-ray damaged DNA in human (ataxia telangiectasia) fibroblasts. *Nature (Lond.) 260,* 444-447

PATERSON, M.C., SMITH, B.P., KNIGHT, P.A. & ANDERSON, A.K. (1977) Ataxia telangiectasia – an inherited human disease involving radiosensitivity, malignancy and defective DNA repair, in *Research in Photobiology* (Castellani, A., ed.), pp. 207-218, Plenum Press, New York

PEDRINI, A.M., DALPRA, L., CIARROCCHI, G., PEDRALI NIY, G.C.F., SPADARI, S., NUZZO, F. & FALASCHI, A. (1974) Levels of some enzymes acting on DNA in xeroderma pigmentosum. *Nucleic Acids Res. 1,* 193-202

PENROSE, L.S. (1959) Genetical analysis in man, in *Biochemistry of Human Genetics (Ciba Found. Symp.),* pp. 9-22, Churchill, London

PENROSE, L.S. (1962) Paternal age and mongolism. *Lancet 1,* 1101

PERRY, P. & EVANS, H.J. (1975) Cytological detection of mutagen-carcinogen exposure by sister chromatid exchange. *Nature (Lond.) 258,* 121-125

PHILIP, P. (1975) Trisomy 8 in acute myeloid leukaemia. *Scand. J. Haematol. 14,* 140-147

POLANI, P.E. (1969) Autosomal imbalance and its syndromes, excluding Down's. *Br. Med. Bull. 25,* 81-93

POLANI, P.E. (1976) Cytogenetics of Fanconi anaemia and related chromosome disorders, in *Congenital Disorders of Erythropoiesis (Ciba Found. Symp. 37),* pp. 261-306, Elsevier/Excerpta Medica/North-Holland, Amsterdam

POLANI, P.E., BRIGGS, J.H., FORD, C.E. & CLARKE, C.M. (1960) A mongol girl with 46 chromosomes. *Lancet 1,* 721-724

POLANI, P.E., HAMERTON, J.L., GIANNELLI, F. & CARTER, C.O. (1965) Cytogenetics of Down's syndrome (mongolism) III. Frequency of interchange trisomics and mutation rate of chromosome interchanges. *Cytogenetics (Basel) 4,* 193-206

POTTEN, C.S., HUME, W.J., REID, P. & CAIRNS, J. (1978) The segregation of DNA in epithelial stem cells. *Cell 15,* 899-906

RAMSAY, C.A., COLTART, T.M., BLUNT, S., PAWSEY, S.A. & GIANNELLI, F. (1974) Prenatal diagnosis of xeroderma pigmentosum. Report of the first successful case. *Lancet 2,* 1109-1112

REED, W.B., MAY, S.B. & NICKEL, W.R. (1965) Xeroderma pigmentosum with neurological complications. *Arch. Dermatol. 91,* 224-226

REGAN, J.D. & SETLOW, R.B. (1974) Two forms of repair in the DNA of human cells damaged by chemical carcinogens and mutagens. *Cancer Res. 34,* 3318-3325

REGAN, J.D., SETLOW, R.B., KABACK, M.M., HOWELL, R.R., KLEIN, E. & BURGESS, G. (1971) Xeroderma pigmentosum: a rapid sensitive method for prenatal diagnosis. *Science (Wash. D.C.) 174,* 147-150

REGAN, J.D., SETLOW, R.B., CARRIER, W.L. & LEE, W.H. (1973) Molecular events following the ultraviolet irradiation of human cells from ultraviolet-sensitive individuals, in *Advances in Radiation Research: Biology and Medicine,* vol. 1 (Duplan, J. & Chapiro, A., eds.), pp. 119-125, Gordon & Breach, New York

ROBBINS, J.H., KRAEMER, K.H., LUTZNER, M.A., FESTOFF, B.W. & COON, H.G. (1974) Xeroderma pigmentosum: an inherited disease with sun sensitivity, multiple cutaneous neoplasms and abnormal DNA repair. *Ann. Intern. Med. 80,* 221-248

ROSENBERGER, R.F. & KESSEL, M. (1968) Nonrandom sister chromatid segregation and nuclear migration in hyphae of *Aspergillus nidulans. J. Bacteriol. 96,* 1208-1213

ROWLEY, J.D. (1977) Mapping of human chromosomal regions related to neoplasia: evidence from chromosomes 1 and 17. *Proc. Natl. Acad. Sci. U.S.A. 74,* 5729-5733

ROWLEY, J.D. (1978) Abnormalities of chromosome No. 1 in haematological malignancies. *Lancet 1,* 554-555

RYGAARD, J. & POVLSEN, C.O. (1976) The nude mouse *vs.* the hypothesis of immunological surveillance. *Transplant. Rev. 28,* 43-61

SCHROEDER, T.M., ANSCHUTZ, F. & KNOPP, A. (1964) Spontane Chromosomenaberrationen bei familiärer Panmyelopathia. *Humangenetik 1,* 194-196

SCHROEDER, T.M., TILGEN, D., KRÜGER, J. & YOGEL, F. (1976) Formal genetics of Fanconi's anemia. *Hum. Genet. 32,* 257-288

SCOTT, E.L. & STRAF, M.L. (1977) Ultraviolet radiation as a cause of cancer, in *Origins of Human Cancer,* Book A: *The Incidence of Human Cancer* (Hiatt, H.H. *et al.,* eds.) *(Cold Spring Harbor Conf. Cell Proliferation, vol. 4),* pp. 529-546, Cold Spring Harbor Laboratory, New York

SETLOW, R.B. (1968) The photochemistry, photobiology and repair of polynucleotides, in *Progress in Nucleic Acid Research and Molecular Biology,* vol. 8 (Davidson, J.N. & Cohn, W.E., eds.), pp. 257-295, Academic Press, New York

SETLOW, R.B. (1975) Relationships among repair, cancer and genetic deficiency: overview, in *Molecular Mechanisms for Repair of DNA,* Part B (Hanawalt, P.C. & Setlow, R.B., eds.), pp. 711-717, Plenum Press, New York

SETLOW, R.B. (1978) Repair deficient human disorders and cancer. *Nature (Lond.) 271,* 713-717

SETLOW, R.B., REGAN, J.D.,GERMAN, J. & CARRIER, W.L. (1969) Evidence that xeroderma pigmentosum cells do not perform the first step in the repair of ultraviolet damage to their DNA. *Proc. Natl. Acad. Sci. U.S.A. 64,* 1035-1041

SHAFER, D.A. (1977) Replication bypass model of sister chromatid exchanges and implications for Bloom's syndrome and Fanconi's anemia. *Hum. Genet. 39,* 177-190

SHAW, M.W. & COHEN, M.M. (1965) Chromosome exchanges in human leukocytes induced by mitomycin C. *Genetics 51,* 181-190

SHERIDAN, R.B. III. & HUANG, P.C. (1977) Single strand breakage and repair in eukaryotic DNA as assayed by S-1 nuclease. *Nucleic Acids Res. 4,* 299-318

SHIRAISHI, Y., FREEMAN, A.I. & SANDBERG, A.A. (1976) Increased sister chromatid exchange in bone marrow and blood cells from Bloom's syndrome. *Cytogenet. Cell Genet. 17,* 162-173

SIEGEL, E.B., ELMORE, E. & SWIFT, M. (1975) Growth and cell cycle abnormalities of cultured fibroblasts of ataxia-telangiectasia patients. (Annual meeting Am. Soc. Hum. Genet., Maryland, Oct. 9-11) *Am. J. Hum. Genet. 27,* 81A (abstr.)

SIMMONDS, H.A., PANAYI, G.S. & CORRIGALL, V. (1978) A role for purine metabolism in the immune response: adenosine deaminase activity and deoxyadenosine catabolism. *Lancet 1,* 60-63

SPERLING, K., GOLL, U., KUNZE, J., LÜDTKE, E.-K., TOLKSDORF, M. & OBE, G. (1976) Cytogenetic investigations in a new case of Bloom's syndrome. *Hum. Genet. 31,*47-52

STITCH, H.S., SAN, R.H.C., LAM, P.P.S., KOROPATNICK, D.J., LO, L.W. & LAISHES, B.A. (1976) DNA fragmentation and DNA repair as an *in vitro* and *in vivo* assay for chemical procarcinogens, carcinogens and carcinogenic nitrosation products, in *Screening Tests in Chemical Carcinogenesis* (Montesano, R. *et al.,* eds.) *(I.A.R.C. Scientific Publications, No. 12),* pp. 617-638, IARC, Lyon

SWIFT, M. (1977) Malignant neoplasms in heterozygous carriers of genes for certain autosomal recessive syndromes, in *Genetics of Human Cancer (Progress in Cancer Research and Therapy),* vol. 3 (Mulvihill, J.J. *et al.,* eds.), pp. 209-221, Raven Press, New York

SWIFT, M., COHEN, J. & PINKHAM, R. (1974) A maximum-likelihood method for estimating the disease predisposition of heterozygotes. *Am. J. Hum. Genet. 26,* 304-317

SYLLABA, L. & HENNER, K. (1926) Quoted by Boder, E. (1975)

SZALAY, G.C. (1963) Dwarfism with skin manifestations. *J. Pediatr. 62,* 686-695

TAKEBE, H., MIKI, Y., KOZUKA, T., FURUYAMA, J.-I., TANAKA, K., SASAKI, M.S., FUMIWARA, Y. & AKIBA, H. (1977) DNA repair characteristics and skin cancers of xeroderma pigmentosum patients in Japan. *Cancer Res. 37,* 490-495

TANAKA, K., HAYAKAWA, H., SEKIGUCHI, M. & OKADA, Y. (1977) Specific action of T4 endonuclease V on damaged DNA in xeroderma pigmentosum cells *in vivo. Proc. Natl. Acad. Sci. U.S.A. 74,* 2958-2962

TAYLOR, A.M.R., HARNDEN, D.G., ARLETT, C.F., HARCOURT, S.A., LEHMANN, A.R., STEVENS, S. & BRIDGES, B.A. (1975) Ataxia telangiectasia: a human mutation with abnormal radiation sensitivity. *Nature (Lond.) 258,* 427-429

TAYLOR, A.M.R., METCALFE, J.A., OXFORD, J.M. & HARNDEN, D.G. (1976) Is chromatid-type damage in ataxia telangiectasia after irradiation at G° a consequence of defective repair? *Nature (Lond.) 260,* 441-443

TEEBOR, G.W. & DUKER, N.J. (1975) Human endonuclease activity for DNA apurinic sites. *Nature (Lond.) 258,* 544-547

THERMAN, E. & KUHN, E.M. (1976) Cytological demonstration of mitotic crossing-over in man. *Cytogenet. Cell Genet. 17,* 254-267

VINCENT, R.A. Jr., SHERIDAN, R.B. III. & HUANG, P.C. (1975) DNA strand breakage repair in ataxia telangiectasia fibroblast-like cells. *Mutat. Res. 33,* 357-366

VON KOSKULL, H. & AULA, P. (1973) Nonrandom distribution of chromosome breaks in Fanconi's anaemia. *Cytogenet. Cell Genet. 12,* 423-434

WEICHSELBAUM, R.R., NOVE, J. & LITTLE, J.B. (1977) Skin fibroblasts from a D-deletion type retinoblastoma patient are abnormally X-ray sensitive. *Nature (Lond.) 266,* 726-727

WHITEHEAD, A.N. (1920) *The Concept of Nature,* The Tarner Lectures delivered in Trinity College, November 1919, pp. 143-163, Cambridge University Press, Cambridge

WHITEHOUSE, H.L.K. (1974) Chromosome integration of viral DNA: the open-replicon hypothesis carcinogenesis, in *Chromosomes and Cancer* (German, J., ed.), pp. 41-76, Wiley, New York

WILSON, M.G., MELNICK, J. & TOWNER, J.W. (1969) Retinoblastoma and deletion D(14) syndrome. *J. Med. Genet. 6,* 322-327

WILSON, M.G., EBBIN, A.J., TOWNER, J.W. & SPENCER, W.H. (1977) Chromosomal anomalies in patients with retinoblastoma. *Clin. Genet. 12,* 1-8

WOLFF, J. (1963) Congenital telangiectatic erythema resembling lupus erythematosus in dwarfs: Bloom's syndrome. *Arch. Dermatol. 87,* 764

WOLFF, S. (1977) Sister chromatid exchanges, in *Annual Review of Genetics,* vol. 11 (Roman, H.L. *et al.,* eds.), pp. 183-201, Annual Reviews, Palo Alto, California

WOLFF, S., RODIN, B. & CLEAVER, J. (1977) Sister chromatid exchanges induced by mutagenic carcinogens in normal and xeroderma pigmentosum cells. *Nature (Lond.) 265,* 347-349

WOODRUFF, M. & SPEEDY, G. (1978) Inhibition of chemical carcinogenesis by *Corynebacterium parvum. Proc. R. Soc. Lond. B Biol. Sci. 201,* 209-215

YANO, K. (1950) Xeroderma pigmentosum mit Störungen des Zentralnerven-systems: eine histopathologische Untersuchung. *Folia Psychiatr. Neurol. Jpn. 4,* 143-151

ZAKHAROV, A.F. & EGOLINA, N.A. (1972) Differential spiralization along mammalian mitotic chromosomes. I. BudR-revealed differentiation in Chinese hamster chromosomes. *Chromosoma (Berl.) 38,* 341-365

ZECH, L., HAGLUND, U., NILSSON, K. & KLEIN, G. (1976) Characteristic chromosomal abnormalities in biopsies and lymphoid-cell lines from patients with Burkitt and non-Burkitt lymphomas. *Int. J. Cancer 17,* 47-56

Discussion

Peto: Could you tell us more about the asymmetry which was seen in chromosome 15?

Polani: If one incorporates into the replicating chromosomes bromodeoxyuridine for one cycle and then does the necessary staining, one of the two chromatid strands will fluoresce brilliantly in the paracentromere region and the other one will not. The idea is that the fluorescent chromatid strand has incorporated bromodeoxyuridine in place of thymine and thus that there is asymmetry in the base pair composition. Professor Walker has shown that there is asymmetry of base type composition at the chemical level.

Walker: The two strands of satellite DNAs certainly have different sequences, but there is no evidence that all the AT-rich sequences in the mouse, for example, are on one strand in a given chromosome (Flamm *et al.* 1969). Indeed, there are several instances of inverted repeats in satellite DNA, so one strand must contain both kinds of sequence.

Polani: You had something like a 40/60 thymine/adenine difference between strands (Flamm *et al.* 1967).

Peto: So you are suggesting that this asymmetry exists centrally in chromosome 15 but not elsewhere in chromosome 15 and not on other chromosomes?

Polani: Yes, it exists only in the paracentromeric region. But it has been shown, as I said, with chromosomes 1, 9, 15 and 16 and these are the chromosomes which have rather a lot of heterochromatin in that region so they should be the ones in which it is easy to detect a difference. And it is easier to detect it in the mouse than in other species because of the large amount of paracentromeric heterochromatin.

Pontecorvo: Surely if the centromeric chromatin is a reiteration of a small number of bases, that chromatin is almost certainly asymmetrical and therefore it is bound to behave in that way.

Polani: Yes, the thing could be trivial. But it would be more interesting if the difference were real and had some functional meaning to it.

Bodmer: With the bi-arm chromosomes in the mouse, as Davidson showed, isn't there an asymmetry in the direction the replication goes across the two chromosomes (Lin & Davidson 1974)? That seems better evidence for an immortal strand than simple asymmetry and straightforward replication.

Polani: Yes, that is true of course. In the bi-arm chromosomes there is double asymmetry, north and south.

Peto: One side is all new and one side is all old, so there is not separate initiation of replication at different points with random joining of new and old. But one will see that anyway, in the absence of sister chromatid exchange.

There are suggestions that if stem cells which have been labelled for some time have just one round of division, which is not labelled, they become totally cold, which really means that all the old DNA consisted of a cold immortal strand paired with a hot transient strand, and the transient strand is then wholly expelled in a single division.

Polani: This seems true for one cell at the base of each column of cells of the papillae of the tongue of the mouse and for the intestinal crypts, where perhaps one to five cells behave in this way (see Potten *et al.* 1978, and my paper). So although the evidence isn't fully satisfactory it looks as if the immortal strand hypothesis may hold in these cases. There is some evidence in maize that goes the same way (Barlow 1978).

Brenner: The immortal strand hypothesis necessarily presupposes that in those cells no sister chromatid exchange ever happens. There is also a requirement for defining the polarity of the centromere segment.

Pontecorvo: Yes, but that is a matter of asymmetry, as I said before. Since there is centric chromatin based on reiteration of a short sequence, the same for all chromosomes, it is possible for the spindle to distinguish the two DNA strands.

Crick: It has to be the same way round for all chromosomes, which doesn't follow from the asymmetry of the composition.

Peto: The likelihood of all the old strands going into the committed daughter by chance alone is 1 in 2^{46}.

Polani: It is true that there must be no sister chromatid exchange to contaminate the purity of the strands.

Pontecorvo: If there is a little, it doesn't matter.

Brenner: The stem cells are going to go on doing this for a long time. Even a low incidence of sister chromatid exchange will result in segregation.

Henry Harris: Stem cell division does not go on all that long.

Brenner: How many stem cell divisions are there in the gut?

Henry Harris: Nobody can give you an exact answer.

Crick: It is quite low. It depends on whether you want total immortality or immortality in the region near the centromere, the second being a weaker form of the hypothesis.

Polani: In Bloom syndrome damage to the allegedly immortal strand of stem cells is just what would happen, through an excess of sister chromatid exchanges: if what we read in the cells *in vitro* is true *in vivo,* the strand purity is ruined.

Bodmer: So it is surprising that those patients are as normal as they are, with such fundamental defects.

Polani: Yes; apart from their small stature, these children develop perfectly normally.

Bodmer: It doesn't say much for the immortal strand idea, does it?

Peto: It all depends how many cells divide.

Bodmer: They have survived a long time by any normal criteria.

Polani: One would prefer to live to a ripe old age than to die of leukaemia around the age of 12.

Henry Harris: Many more cells might die in Bloom syndrome. You may be selecting from rather small numbers that survive.

Polani: One could argue of course that their stem cells are not all that good because, for example, their immune system doesn't function very well, they are tiny at birth, and their germinal stem cells, in their testes, may have impaired proliferative ability.

Bodmer: But 12 years is a long time for a fully functioning individual to survive with such a basic defect.

Crick: If we got cancer at the same rate as the mouse does we would all be dead by now. But that is another issue.

Henry Harris: Is there any information about relative rates of exchanges in the different chromosomes in Bloom syndrome? Do each of the chromosomes undergo exchanges at more or less the same rate?

Polani: Originally there was an idea that there was a difference in the different chromosomes and indeed in different chromosome arms. On the whole, the more people have looked at it, the less evidence there is of non-randomness in sister chromatid exchanges in Bloom disease. The only thing one can say is that within the chromosomes, the exchanges tend to be more in the interband regions than within the bands, but this doesn't answer your point.

Peto: You said there was some evidence from Barlow. Could you either describe it or give a reference?

Polani: Peter Barlow (1978) labelled the cells of the quiescent centre of the plant roots and obtained some evidence to support the idea of immortal strands.

Peto: The cells just go completely cold?

Polani: Yes, I think he went for that approach rather than the other.

Peto: A lot of theories about cancer *induction* suggest that initiation-type changes occur first and that these then have to be followed by epigenetic changes; so increasing the amount of mutagens that the *patients* are subjected to would be very dangerous indeed, while increasing the amount of promotional activity they suffer would have little effect because they are already being strongly promoted. Is there any experimental way of testing this?

Polani: Only *in vitro* on the cells.

Peto: Can you transform Bloom cells, for example? Human cells *in vitro* hardly ever transform.

Polani: I haven't seen any record of transformed Bloom cells but xeroderma pigmentosum cells have been transformed with SV40.

Neel: Were the heterozygote frequencies calculated before the knowledge of complementation classes was developed? In xeroderma pigmentosum, for instance, where some seven complementation groups are known, the heterozygote frequency for all the different alleles combined is probably at least as high as one would calculate from the assumption of a single locus.

Polani: Yes indeed, you are right. The heterozygote frequencies have not taken account of heterogeneity and could be out by a factor of 2 or 3.

Klein: J. German (personal communication) transformed lymphocytes from patients with Bloom's syndrome by Epstein-Barr virus, in order to obtain lines with a high frequency of sister chromatid exchange (SCE). Surprisingly, the lines had a very low rate of SCE. The virus must therefore select relatively normal cell variants, when it transforms, or introduce repair mechanisms of its own that prevent SCE.

What is known about the chromosomal anomalies that appear in the neoplasias of patients with ataxia telangiectasia? Kaiser-McCaw *et al.* (1975) have shown that anomalies of chromosome 14 are very common in normal cells in these patients. These anomalies are not of the same type, however, as seen in lymphomas and other malignancies of the haemopoietic system. Has anyone looked at the tumours that develop in the AT patients themselves?

Polani: It was you and Dr Laura Zech (1976) who showed that there was a translocation of terminal 8 to terminal 14 in Burkitt lymphoma. As for your second question, I am only aware of one or two malignancies in ataxia telangiectasia which have been studied sufficiently clearly to show that there was this clone involving chromosome 14 in the malignant cells. Oxford and colleagues (1975) are not very sure whether the clone is unequivocally relevant to the establishment of malignancy.

Koprowski: How many people with ataxia telangiectasia show no abnormality of chromosome 14?

Polani: This is very difficult to answer. Between 15 and 40 patients have been described with abnormalities of chromosome 14. This is probably the sum total of patients that have been looked at. Some time ago more than 480 patients with ataxia telangiectasia were reviewed, but this is the total of cases recorded, including those described very early, and very few of them have been studied chromosomally.

Koprowski: So, essentially, out of the total number of patients on record, a relatively small number showed abnormalities of chromosome 14?

Polani: That is what one would conclude. At least they have an abnormality of one of the D group chromosomes. Those that have been banded are few.

Henry Harris: When do the CNS abnormalities appear? Do they get progressively worse?

Polani: In both ataxia telangiectasia and xeroderma pigmentosum they get worse with time. The ataxia in ataxia telangiectasia is seen in early life but some changes in xeroderma pigmentosum are present at birth. These were described by De Sanctis & Cacchione in 1932 as xerodermic idiocy. These children are usually born mentally defective with small heads, but not invariably so, and then they develop progressive spasticity or choreoathetosis.

Henry Harris: What do the sections of the brains look like?

Polani: In xeroderma with neurological troubles they show simple atrophy, focal and patchy, without any reactive phenomena.

Henry Harris: So one thinks in terms of stochastic events, rather than field effects like pyramidal degeneration or something of that sort.

Polani: There are no signs of active degeneration, no ballooniform cells or anything of that sort.

Siniscalco: I wonder whether chromosome 14 has some peculiarity which makes it more likely to be involved in numerical and/or structural mutations. In the first two cases of X-autosomal translocations discovered, chromosome 14 was involved (Allderdice *et al.* 1972; Buckton *et al.* 1971). In view of this, shouldn't we be very conservative about attributing a specific meaning to the correlation found between trisomy 14 and certain types of malignancies?

Nossal: Could you round out the picture of these extraordinarily interesting DNA repair-defect diseases by bringing us up to date about the disease progeria, where the victims age prematurely, and where repair defects have been claimed by some?

Polani: There have been so many conflicting reports on progeria that I just don't understand it. Certainly the cells of progeric patients don't grow very well. They have a limited 'Hayflick' number.

Nossal: It is not clearly a DNA repair defect?

Polani: No, not clearly, and certainly not anything that could be independent of a population of ageing cells *in vitro* (see, for example, Cleaver 1978).

Scriver: You mentioned central nervous system findings in your patients with xeroderma pigmentosum. One can propose an embryological reason for that correlation of manifestations. Why don't heterozygotes have skin or CNS defects rather than malignancy of internal tissues?

Polani: I have no answer, though evidence on this is very shaky. At any rate, perhaps their body cells are susceptible to a variety of chemical mutagens which to other people are less damaging.

Scriver: But if they live 40 or 50 years they will have been exposed to ultraviolet radiation.

Polani: Yes. However, it doesn't seem to work like that. Probably repair of the specific u.v. dimers is quite good but repair of other damage, caused by chemical mutagens, may not be so good.

References

ALLDERDICE, P.W., MILLER, O.J., KLINGER, H.P., OPITZ, J.M. & PALLISTER, P.D. (1972) Demonstration of a spreading effect in an X-autosome translocation by combined autoradiographic and quinacrine-fluorescence studies, in *Human Genetics* (Proc. 4th Int. Congr. Paris, 6-11 September 1971) Abstracts, pp. 14-15

BARLOW, P.W. (1978) The concept of the stem cell in the context of plant growth and development, in *Stem Cells and Tissue Homeostasis* (Lord, B.I. *et al.*, eds.), pp. 87-113, Cambridge University Press, Cambridge,

BUCKTON, K.E., JACOBS, P.A., RAE, L.A., NEWTON, M.S. & SANGER, R. (1971) An inherited X-autosome translocation in Man. *Ann. Hum. Genet. 35,* 171-178

CLEAVER, J.E. (1978) Human inherited diseases with altered mechanisms for DNA repair and mutagenesis, in *Birth Defects* (Littlefield, J.W. *et al.,* eds.) *(Proc. 5th Int. Conf. Montreal, Aug. 1977), International Congress Series No. 432,* pp. 85-100, Excerpta Medica, Amsterdam

DE SANCTIS, C. & CACCHIONE, A. (1932) L'idiozia xerodermica. *Riv. Sper. Freniatr. Med. Leg. Alienazioni Ment. 56,* 269-292

FLAMM, W.G., MCCALLUM, M. & WALKER, P.M.B. (1967) The isolation of complementary strands from a mouse DNA fraction. *Proc. Natl. Acad. Sci. U.S.A. 57,* 1729-1734

FLAMM, W.G., WALKER, P.M.B. & MCCALLUM, M. (1969) Some properties of the single strands isolated from the DNA of the nuclear satellite of the mouse *(Mus musculus). J. Mol. Biol. 40,* 423-443

KAISER-MCCAW, B., HECHT, F., HARNDEN, D.G. & TEPLITZ, R.L. (1975) Somatic rearrangement of chromosome 14 in human lymphocytes. *Proc. Natl. Acad. Sci. U.S.A. 72,* 2071-2075

LIN, M.S. & DAVIDSON, R.L. (1974) Centric fusion, satellite DNA, and DNA polarity in mouse chromosomes. *Science (Wash. D.C.) 185,* 1179-1181

OXFORD, J.M., HARNDEN, D.G., PARRINGTON, J.M. & DELHANTY, J.D.A. (1975) Specific chromosome aberrations in ataxia telangiectasia. *J. Med. Genet. 12,* 251-262

POTTEN, C.S., HUME, W.J., REID, P. & CAIRNS, J. (1978) The segregation of DNA in epithelial stem cells. *Cell 15,* 899-906

ZECH, L., HAGLUND, U., NILSSON, K. & KLEIN, G. (1976) Characteristic chromosomal abnormalities in biopsies and lymphoid-cell lines from patients with Burkitt and non-Burkitt lymphomas. *Int. J. Cancer 17,* 47-56

General discussion I

Siniscalco: I was interested to hear about the potential of the oocyte model for starting reactivation of the X chromosome. How much material does one need? One could use for this purpose a clonal fibroblastic strain (or a clonal tumour line) from a double heterozygote in repulsion at two X-linked loci (e.g. glucose-6-phosphate dehydrogenase and hypoxanthine-guanine phosphoribosyltransferase) so that reactivation of the inactive X could be spotted at the translation level through the characterization of the end-product of the relevant genes.

Brenner: Do frogs, in which females are heterogametic, show any of this regulation?

Gurdon: I don't know whether in the male, which is the homogametic sex in frogs, there is inactivation of one of the sex chromosomes.

I think the experiment Professor Siniscalco has in mind involves injecting nuclei containing genes on a repressed sex chromosome. The limitation is that only about 200 nuclei can be transplanted into a single oocyte (where nuclei do not replicate), and it doesn't help to inject thousands of oocytes because one just raises the background level of any enzyme activity. If too many genes are injected into one oocyte, one can't be sure that in all cases all of them will be activated. The only experiment that bears on that is one where nuclei were transplanted in and proteins were found to be synthesized which represented the activation of genes in those implanted nuclei (De Robertis & Gurdon 1977). What we don't know is whether those were repetitive genes. It is quite possible that it will be much easier to see the production of protein where there are multiple copies of a gene because more message would be made when the gene is turned on.

Brenner: Can the X chromosome of man be purified by cell sorting?

Siniscalco: Perhaps this will eventually be achieved through further refinement of the current techniques of chromosome separation by sedimentation velocity on sucrose gradients, as has recently been attempted with zonal centrifugation (Padgett *et al.* 1977) or sorting (Gray *et al.* 1975).

Henry Harris: The X chromosome of the vole can be purified because it is so big. With this huge heterochromatic region you can actually spin it out fairly simply.

Bodmer: But why would you worry about that? There are lots of other ways.

Brenner: You can put more in.

Bodmer: But then you should use either the vole or some of the marsupials.

Brenner: No, they don't have the genetic markers that would be needed.

Henry Harris: Yes, they do. Where one has looked, the genes on the X chromosome seem much the same in these less well studied species as in the mouse.

Brenner: But do you not need mutant genes?

Henry Harris: For the X chromosome the situation in the vole is, in principle, much the same as in man (Cook 1975).

Bodmer: Hybrids have now been made between marsupials and the mouse. They have large chromosomes which could be purified (Graves & Hope 1977).

Henry Harris: The trouble with hybrids involving the vole chromosome is that the heterochromatin is eliminated at a great rate.

Gurdon: What precisely do you think might come out of that experiment if it were positive, Professor Siniscalco? We have already done experiments where the genes which were not expressed in a certain cell type can be caused to be re-expressed by that particular type of experiment. As I see it, the reactivation of a repressed gene on a repressed X chromosome would be more or less like that. If the experiment you propose were successful, it would only reproduce this other experiment, though it might have the advantage of doing it for a known gene.

Siniscalco: I am interested in knowing whether the X inactivation is a different kind of phenomenon altogether from the switching off of genes in normal somatic differentiation.

Bodmer: The X chromosome is different because, although there have been a couple of reports of possible activation, a lot of people have looked at this and not found activation. So X-inactivation is one of the most stable inactivation states that we know.

Gurdon: It is not obvious that this is different from the other examples that we are discussing.

Bodmer: It is different in terms of somatic cell hybrids. In other cases it has been possible to switch on almost any differentiated gene that you want.

Gurdon: I can see no reason to think that the inactivation of genes on an X chromosome is different in principle from the inactivation of cell-type specific genes in specialized cells.

Brenner: It depends which principles you hold.

Henry Harris: I absolutely accept that in cell fusion experiments the X chromosome hasn't been reactivated except for these one or two problematical cases. But I don't think the ideal experiment has ever been done. The whole point about X-inactivation is to maintain a certain kind of gene-dosage for genes on the X. If you simply fuse cells together without putting on any kind of selection pressure to get rid of active X chromosomes, there may be no reason why you should reactivate the inactive X.

Bodmer: I don't think that is the answer. I think the hybrids change the balance of the number of X chromosomes, though even if you changed that whole balance it would probably make no difference. There is something fundamentally different between X inactivation and other sorts of genetic control.

Milstein: We are talking about extreme cases of gene activation and inactivation. The only other example that I know which involves allelic exclusion involves DNA changes. I am referring to immunoglobulins.

Crick: You said that what characterizes these phenomena is that DNA moves around, Cesar. Here we have a hypothesis that in the active X one alters the DNA and moves it around. That would be different from the cases that John Gurdon has been studying. So at least we can have a reasonable hypothesis that says the active X may be different.

Fincham: Also, in X-autosome segmental rearrangements, the heterochromatization of the X can be spread along the chromosome into a connected autosomal segment. This looks like something special to the X chromosome. We have no reason to suppose that heterochromatization in the same sense is involved in the inactivation of all the scattered genes which may be inactive in a particularly differentiated cell type.

Bodmer: It adds to the argument for the uniqueness of DNA.

Henry Harris: One interesting experiment would be to see what the smallest piece of an X chromosome might be that can remain inactive. This could be done by breaking up the X. The experiment is not all that easy, but it is technically possible.

Brenner: That has been done in flies. For dosage compensation, each gene carries the knowledge that it is on an X chromosome (Lucchesi 1973).

Henry Harris: Down to what sort of level?

Brenner: Down to quite small translocations.

Crick: If a nucleus from a cell in the immune system were put into an oocyte, what would the corresponding phenotype and the resulting immunity be? Is that technically possible?

Brenner: That has been done half way, has it not, by colonization of frog eggs with lymphocyte nuclei?

Gurdon: That was a nuclear transplant experiment into eggs. The progeny did not get far enough for anyone to see what antibodies they had formed (Wabl *et al.* 1975).

We talk about the difficulty of reactivating genes in an X chromosome which is being repressed, but this is probably no more difficult than reactivating genes that have been developmentally repressed. Malawista & Weiss (1974), for example, cloned cells and turned on genes of the wrong cell type. But that is a very rare event.

Ruddle: Not necessarily. It is still an open question whether the X chromosome is activated at all during the normal life cycle. There may not be a mechanism for reactivation.

Gurdon: Has the reactivation of an X-repressed gene been seen as frequently as the activation of developmentally repressed genes in cell fusion experiments?

Bodmer: It has been done quite often. There is a very strong selective system for HGPRT, and inactivation really is remarkably stable.

Brenner: The question is this: if you just look at the germ line continuity and throw the organism away, is there a line of cells in which X chromosomes are never inactivated?

Gardner: It is not clear from studies to date on early human oogonia. Gartler *et al.* (1975) say that they can find only one X chromosome active in early germ cells and that both become active at the onset of meiosis. However, Migeon & Jelalian (1977) claim that both X chromosomes are active in germ cells at the earliest stage of development at which they can be obtained.

Polani: What about human oocytes?

Gardner: In the mature oocytes both Xs are active (Gartler *et al.* 1972).

Brenner: But is the germ line segregated before this X inactivation in normal development?

Gardner: That is not known.

Brenner: I apologize. I had always assumed that in the germ line the X chromosome had never been inactivated, but this is not established. If this is true, then X chromosome inactivation could be a totally irreversible step in the soma. But if it can go on and off in the germ line, I would find myself moving over to Gurdon's picture of it.

Peto: Another example of where the inactivation is getting very strong pressures to get turned off is within carcinomas. Almost every kind of change happens to the cells in a carcinoma. There is an almost indescribable karyotype, yet you get either G6PD type A or G6PD type B which means that with all this change the X that was turned off never gets turned on again.

Brenner: I merely raised this because there are many cases, especially amongst lower animals, of X ejection in the soma. In these cases it is thrown away and that is irreversible. The germ line remains invariant.

Siniscalco: But maybe it never happens to the germ line.

Brenner: That's right. There may never be any biological necessity for a reactivation process.

Henry Harris: As one looks at more and more hybrid cells, it becomes more common to find a differentiated product being turned on. Most of the early hybrids were made with a fibroblastic cell as one of the parents, and the hybrids were selected for their ability to stick to glass or plastic. If you select for hybrid cells that don't stick to glass, that grow in suspension, you often find that the differentiated traits are maintained or turned on. Steffen Junker in our laboratory has made many crosses between hepatoma cells and non-fibroblastic cells. Turning on a liver cell function in the genes of the non-hepatic parent cell is much more common than was thought originally.

Gurdon: But there is an important distinction between the re-expression of a gene previously active in that cell type and the activation of a gene which had undergone developmental expression.

Henry Harris: I am talking about the expression of a differentiated trait by the gene set that did not express it at the time of cell fusion.

Gurdon: It takes many cell generations.

Henry Harris: I can't give you any numbers for how long it takes. It is difficult to study these hybrids until they have gone through a certain number of generations. One of the things we should really like to do is to develop good differentiated markers that can be studied in the heterokaryon in order to see what happens in the first or second cell division.

Brenner: I assume that in teratocarcinoma cells there is no X inactivation. Is that correct?

Jacob: It depends on the strain. In some strains both X chromosomes are active. In other strains, which probably correspond to a more advanced state of differentiation and are derived from the first one, only one X is active.

Brenner: Can you use those later strains to populate animals, and can you get germ lines?

Jacob: According to the published work there is only one animal with the teratoma genome in the germ line.

Siniscalco: But the presence of two active Xs in cultured tumour cell lines doesn't mean anything. It may just be a doubling of the same active chromosome through endoreduplication or other means.

Brenner: You would have to have genetic markers.

Henry Harris: The doubling of an active X chromosome raises problems. Maybe in cells in culture it doesn't matter, but obviously it must matter *in vivo*, or there wouldn't be a mechanism for keeping only one X active. If one ever found two active Xs roaring away *in vivo*, it would be an important observation.

Siniscalco: This has never been found.

Gardner: Mary Lyon and I have been injecting single cells from mouse embryos exhibiting X-chromosome-dependent coat colour mosaicism into blastocysts of a third coat colour genotype in an attempt to define the time of X inactivation (Gardner & Lyon 1971). When taken from embryos up to the stage of implantation at 4½ days *post coitum*, single donor cells can contribute to both germ line and soma in individual chimeras. However, so far we have had no case of a donor germ line contribution in chimeras in which the female donor cell appeared to have undergone X inactivation before injection (R. L. Gardner and M. F. Lyon, unpublished data).

Brenner: So that is consistent with the idea of a throw-away inactivation?

Gardner: So far, but we don't get a very high proportion of germ line chimeras.

Gurdon: I am impressed by what people said but I still think the more important thing is that all of these inactivations are extremely stable. Hadorn (1968) drew attention to transdetermination but one should be impressed by the extreme rarity of that compared to the extraordinary stability of these inactivations. From the examples that Henry Harris was quoting of the reactivation of these genes, my suspicion is that more cell generations are needed than are normally gone through in the life of an individual.

Henry Harris: I doubt that, but I am also impressed by the stability. It is very difficult to avoid the notion that there must be some evolutionary or functional homology in the various mechanisms that animals use to control the dosage of genes on the X chromosome. Some creatures throw out the whole of the paternal chromosome set, some throw out the paternal X chromosome, and some inactivate one of the X chromosomes but retain it. These are not randomly assorted observations. We do not have anything comparable for differentiated traits coded on autosomes.

Crick: What is the actual molecular mechanism that makes the inactive X condense?

Henry Harris: Nobody knows.

Crick: I ask in case anybody has any bright ideas. If you deleted a bit of DNA, can you think of a neat way of keeping the chromosome compact?

Henry Harris: That is rather like the experiment I suggested earlier. I should like to know how small the little bits of the X chromosome could be and still maintain the inactive state.

Crick: What would happen if you made it circular?

Gurdon: Isn't the simplest explanation that a reasonably large amount of untranscribed DNA looks compacted?

Crick: Is it a unitary mechanism? How is it all done? Are there going to be controlling sites all the way along the chromosome or is it just a trivial matter?

Gurdon: The most trivial explanation is that since there are no promoters in a large region and the DNA never has any transcripts on it, it simply becomes highly compacted and takes up the stains that people use for seeing it with. The first thing is to eliminate that explanation.

Crick: That is Henry's argument—that there must be at least one copy of all the genes that are sex-linked, and therefore one evolves a mechanism which is a bit intricate.

Polani: To take up Professor Crick's point: in the mouse X chromosome isn't there an apparent point mutation that leads to non-random inactivation of one of the two Xs (Ohno 1973; Ohno *et al.* 1974)?

Henry Harris: There are several examples of non-random inactivation of the X chromosome. If there is some kind of large-scale structural abnormality, for example a translocation, involving one of the X chromosomes it is generally the paternal X that is inactivated.

Bodmer: Is it the other way round?

Polani: In some way the human decides which X chromosome to inactivate and the inactivating locus is thought to reside in the long arm of the X close to the centromere, which is why it is alleged that persons with isochromosomes for the short arm of the X do not survive whereas those with isochromosomes of the long arm survive (Therman *et al.* 1974). The putative locus that controls the whole switching-off mechanism of the X is presumably (on this evidence) somewhere in the long arm of the X in the human. In the mouse the locus is, I think, linked to *Tabby*, somewhere near the middle of the X.

Bodmer: If that were true, then in somatic cell hybrids which have generated translocations which remove the short arm of the X one would expect the other part of the X to reactivate. I don't know whether there is any evidence for or against that.

Polani: I can't think of any. In the mouse there certainly are various X-autosome translocations where the translocated X gets activated.

Brenner: They are complementary.

Polani: This is true also of humans, at least if you accept heterochromatization and late labelling as evidence of inactivaton. It is not always the normal X but sometimes the translocated one that is late labelling.

Brenner: John Gurdon suggested that whenever one sees heterochromatin it just means that nothing has happened.

Gurdon: This doesn't explain why one X is activated and the other isn't. Heterochromatin as seen under a microscope may be nothing more than the lack of chromosome-associated transcripts, with the resulting compaction of chromatin.

Crick: But if one copy of every locus has to be inactivated, why do all the inactive ones have to be on the same chromosome?

Klein: How complete is X inactivation? In patients with multiple X chromosomes, all except one are inactivated, yet there are increasing numbers of developmental anomalies, particularly of the skeleton and the nervous system, with increasing numbers of inactive X. Does this mean that they are not completely inactive?

Siniscalco: Not necessarily: the majority of the triple-X females are entirely normal and fertile (Hamerton *et al.* 1972) and, of course, most XXY individuals cannot be called really abnormal, apart from their sterility. The point is that during early embryogenesis all X chromosomes are functionally active. This is perhaps what makes XXYs different from XYs. Why some triple Xs are normal and others are not may depend upon the rest of the genome. Since Klinefelter patients with more than two Xs are usually grossly impaired, at least in their mental capacity, one might conclude that it is especially deleterious to the organism to have more than two Xs active in the early embryo when the Y chromosome is also present.

Henry Harris: We all say that one of the X chromosomes is inactivated but the signal isn't quite that. What the signal says is: 'Leave only one X active, and I don't care which!'.

Crick: I can think of any number of ways of doing that. What I don't understand is how one large block is activated. They are different questions. The difficult part is how to inactivate a whole chromosome.

Fincham: As to why all the inactive genes are on one chromosome, that is surely because the whole thing is a chromosome-wide phenomenon, so to speak. I tend to think of this as a property of the packing of the nucleosomes or something of this sort.

Crick: I agree but I haven't been able to think of a neat method of cloning it.

Bodmer: Something must travel down the chromosome.

Fincham: Something not only travels down the chromosome but perhaps can be transmitted from one chromatid to the daughter chromatid through cycles of replication. About two years ago I heard a very exciting but rather speculative talk by Weintraub in which he put forward a model for this, based on there being different conformations of the nucleosomes and conformational constraints operating between symmetrically-related half-nucleosomes (Weintraub *et al.* 1976).

Crick: The problem is that something has to be done about that chromosome if it is to replicate, and it does replicate at a different time.

Bodmer: One interesting example is the HeLa cell. HeLa is a female cell. It is effectively more or less triploid but it has only two Xs. D98/AH2, which is a 6-thioguanine-resistant derivative of HeLa, only expresses G6PD-A but neither of its chromosomes is late replicating. When we make hybrids with a normal diploid (male) cell that is G6PD-B, we get equal levels of G6PD-A and -B, as if only one of those two Xs of D98/AH2 was active. But we don't actually know whether each X is half active or whether one is inactive. If it is the latter, which is more likely, it contradicts Henry's statement that nobody has ever looked at ratios between the autosomes and the sex chromosomes (Bengtsson *et al.* 1975).

Henry Harris: That isn't what I said.

Bodmer: It also contradicts the relationship between late replication and inactivity. So the relationship between the two is not absolute.

Siniscalco: How can you be sure that each hybrid cell contains the same number of active Xs? One of the two could be lost at random.

Bodmer: No, these are not unstable hybrids but human–human hybrids between D98/AH2 and fibroblasts or lymphocytes.

Henry Harris: The dry mass of heterochromatic regions is much lower than the dry mass of euchromatic regions. Heterochromatic regions have much less protein. The additional protein on the euchromatic regions may be the transcription machinery, but its presence on the genetically active regions might be secondary to other activating events.

Gurdon: Musich *et al.* (1977) claim to have separated heterochromatin from euchromatin and they have described some protein differences.

Henry Harris: That has been done by standard biochemical fractionation techniques, but I am talking of the dry mass measurements made by interferometry on individual cells. The differences between euchromatin and heterochromatin are very big.

Siniscalco: Professor Walker spoke about the possibility of repetitive DNA being involved in regulation. If it is, the X chromosome, which is known to be the repository of genes that have been maintained in a condition of linkage at least throughout mammalian evolution, may also carry specific genetic sequences involved in regulatory phenomena. Work from my laboratory (Balazs *et al.* 1978) suggests, but does not prove, that the X chromosome may have more quasi-repetitive DNA sequences than do other chromosomes.

The referee asked us to remove from the paper our comment that this might be important. How do you react to this possibility?

Walker: I don't think I said that repetitive sequences have much to do with regulation. In fact my view was exactly the opposite. Regulatory sequences may be hidden in them but the repetitive sequences have all the characteristics of things that are not important. If the X has specific repetitive sequences this may reflect how much interchange there is between the chromosomes. The human X certainly has specific sequences which are a sub-population of the ordinary satellite DNAs. I also know that there is cross-reaction between the repetitive DNAs of the X and the autosomes.

Brenner: I think there is a very good function for the long intermediate repetitive DNA. In *Drosophila* enough of this DNA exists to account for about one per band, on the average. Lefevre (1973) reports experiments in which the salivary gland chromosomes of larva from irradiated flies were directly examined for rearrangements. The first remarkable result is that he finds considerable numbers of repeats, so that the distribution of what are called breakpoints is non-random. Secondly, a large fraction of the rearrangements are non-mutant; that is, the interruptions occur in DNA with no important function. Now, I do not think that X-rays could produce breaks non-randomly in DNA. It therefore follows that if the *breaks* cannot be clustered, it must be the joins that are clustered. I suggest that the joining takes place by recombination processes between homologous sequences of intermediate repetitive elements. Since there are different subsets of these sequences, each repeated about 40 times in the genome, the postulated process could explain all the classes of aberrations found. The argument then is that these DNA sequences act to preserve linkage of genes to centromeres.

Henry Harris: I can think of other explanations. The number of chromosome breaks induced by radiation increases as a power function of the radiation dose. You are therefore talking about complex events. If you envisage regions of condensed and dispersed chromatin at different sites along the chromosome, you might get mutational hot spots because of the higher order structure of the DNA. I can give you some other models, if you don't like that one.

Cavalli-Sforza: Is it possible for parts of these spacers to be connected in some way with nucleosomes?

Crick: In general that is not possible. I am not sure for satellites.

Walker: For satellites it is possible. There are clusterings of numbers which are of the right sort of order, but you have to include multiples and submultiples, which makes it less convincing. For example, three mouse satellite repeats coincide with four nucleosome repeats. It is still an open question. A periodicity may be imposed by the nucleosome structure increasing the likelihood of replicating a certain length of satellite. This is the hypothesis of Maio *et al.* (1977).

Gurdon: There is a specific case of the spacing of nucleosomes not being in any way related to the naturally repetitious structure of DNA (Gottesfeld & Melton 1978).

Walker: There are certainly exceptions. It is an interesting idea because it is structural constraint which could explain the sequence lengths found.

Crick: The question was whether there was any connection between the repeat imposed by the nucleosome and the repeat in the satellite. The answer as I understand it is that there is evidence to the contrary. There may be the odd case where it might be true but it is not a general rule.

Walker: It is slightly more than an odd case. The trouble is that with the errors involved in the sequence lengths almost anything would fit into it.

Crick: If you cut with the appropriate nuclease, I don't think you cut in the same place in the nucleosomes every time, so I doubt very much that these things have anything to do with each other. It is also unlikely to tell us anything special about the intervening sequences.

References

BALAZS, I., SZABO, P. & SINISCALCO, M. (1978) Hybridization properties of human X-chromosomal RNA transcripts from murine-human hybrids. *Somatic Cell Genet. 4,* 617-631

BENGTSSON, B. O., NABHOLZ, M., KENNET, R., BODMER, W. F., POWEY, S. & SWALLOW, D. (1975) Human intraspecific cell hybrids: a genetic and karyotypic analysis of crosses between lymphocytes and D98/AH-2. *Somatic Cell Genet. 1,* 41-64

COOK, P. R. (1975) Linkage of the loci for glucose-6-phosphate dehydrogenase and for inosinic acid pyrophosphorylase to the X chromosome of the field-vole *Microtus agrestis. J. Cell Sci. 17,* 95-112

DE ROBERTIS, E. M. & GURDON, J. B. (1977) Gene activation in somatic nuclei after injection into amphibian oocytes. *Proc. Natl. Acad. Sci. U.S.A. 74,* 2470-2474

GARDNER, R. L. & LYON, M. F. (1971) X-chromosome inactivation studied by injection of a single cell into the mouse blastocyst. *Nature (Lond.) 231,* 385-386

GARTLER, S. M., LISKAY, R. M., CAMPBELL, B. K., SPARKES, R. & GANT, N. (1972) Evidence for two functional X-chromosomes in human oocytes. *Cell Differ. 1,* 215-218

GARTLER, S. M., ANDINA, R. & GANT, N. (1975) Ontogeny of X-chromosome inactivation in the female germ line. *Exp. Cell Res. 91,* 454-457

GOTTESFELD, J. M. & MELTON, D. A. (1978) The length of nucleosome-associated DNA in transcribed and nontranscribed regions of chromatin is the same. *Nature (Lond.) 273,* 317-319

GRAVES, J. A. M. & HOPE, R. M. (1977) Fusion and hybridization of marsupial and euthesian cells. II. Fusion of marsupial cells. *Aust. J. Biol. Sci. 30,* 461-469

GRAY, J. W., CARRANO, A. V., MOORE, D. H., STEINMETZ, L. L., MINKLER, J., MAYALL, B. H., MENDELSOHN, M. L. & VAN DILLA, M. A. (1975) High-speed quantitative karyotyping by flow microfluorometry. *Clin. Chem. 21,* 1258-1262

HADORN, E. (1968) Transdetermination in cells. *Sci. Am. 219(5),* 110-120

HAMERTON, J. L., RAY, M., ABBOTT, J., WILLIAMSON, C. & DUCASSE, G. C. (1972) Chromosome studies in neonatal population. *Can. Med. Assoc. J. 106,* 776-779

LEFEVRE, G. JR (1973) The one band-one gene hypothesis: evidence from a cytogenic analysis of mutant and nonmutant rearrangement breakpoints in *Drosophila melanogaster. Cold Spring Harbor Symp. Quant. Biol. 38,* 591-599

LUCCHESI, J. C. (1973) Dosage compensation in *Drosophila. Annu. Rev. Genet. 7,* 225-237

MAIO, J. J., BROWN, F. L. & MUSICH, P. R. (1977) Subunit structure of chromatin and the organization of eukaryotic highly repetitive DNA: recurrent periodicities and models for the evolutionary origins of repetitive DNA. *J. Mol. Biol. 117,* 637-655

MALAWISTA, S. E. & WEISS, M. C. (1974) Expression of differentiated functions in hepatoma cell hybrids: high frequency of induction of mouse albumin production in rat hepatoma-mouse lymphoblast hybrids. *Proc. Natl. Acad. Sci. U.S.A. 71,* 927-931

MIGEON, B. R. & JELALIAN, K. (1977) Evidence for two active X-chromosomes in germ cells of female before meiotic entry. *Nature (Lond.) 269,* 242-243

MUSICH, P. R., BROWN, F. L. & MAIO, J. J. (1977) Subunit structure of chromatin and the organization of eukaryotic highly repetitive DNA: nucleosomal proteins associated with a highly repetitive mammalian DNA. *Proc. Natl. Acad. Sci. U.S.A. 74,* 3297-3301

OHNO, S. (1973) Conservation of ancient linkage groups in evolution and some insight into the genetic regulatory mechanism of X-inactivation. *Cold Spring Harbor Symp. Quant. Biol. 38,* 155-164

OHNO, S., GELLER, L. N. & KAN, J. (1974) The analysis of Lyon's hypothesis through preferential X-activation. *Cell 1,* 175-184

PADGETT, T. G., STUBBLEDIELD, E. & VARMUS, H. E. (1977) Chicken macrochromosomes contain an endogenous provirus and microchromosomes contain sequences related to the transforming gene of ASV. *Cell 10,* 649-657

THERMAN, E., SARTO, G. E. & PATAU, K. (1974) Center for Barr body condensation on the proximal part of the human Xq: a hypothesis. *Chromosoma (Berl.) 44,* 361-366

WABL, M. R., BRUN, R. B. & DU PASQUIER, L. (1975) Lymphocytes of the toad *Xenopus laevis* have the gene set for promoting tadpole development. *Science (Wash. D.C.) 190,* 1310-1312

WEINTRAUB, H., WORCEL, A. & ALBERTS, B. (1976) A model for chromatin based upon two symmetrically paired half-nucleosomes. *Cell 9,* 409

Human haemoglobin genetics

D. J. WEATHERALL, J. B. CLEGG, W. G. WOOD and G. PASVOL

Nuffield Department of Clinical Medicine, The Radcliffe Infirmary, Oxford

Abstract The genes which direct the structure of human fetal and adult haemoglobin consist of a linked pair of α-chain loci on chromosome 16 and the ${}^{G}\gamma$-${}^{A}\gamma$-δ-β loci complex on chromosome 11. The δ- and β-chain genes contain inserts similar to those of the mouse and rabbit globin genes. The structure of the various messenger RNAs transcribed from these loci is now worked out although the function of the non-coding regions is not known. The abnormal haemoglobin disorders and thalassaemias result from a variety of lesions at these loci which include single base substitutions, deletions of one or more bases or entire loci, insertions, frame-shifts, fusion genes caused by abnormal crossing over, chain termination mutations and ill-defined defects which lead to a reduced rate of transcription or abnormal structure of messenger RNA. Some progress has been made towards an understanding of the cellular mechanisms whereby the haemoglobin polymorphisms have been maintained. Very little is known about the regulatory mechanisms involved in the switch from fetal to adult haemoglobin production, although it is likely that certain specific areas of the γ-δ-β gene complex are involved in its control.

Human haemoglobin (Hb) is an important model for the study of both normal and abnormal gene action at the cellular and molecular levels. In this review we will summarize briefly what is known about the genetic control of human haemoglobin, the molecular basis for the abnormal haemoglobin disorders, and the cellular mechanisms which are responsible for maintaining the abnormal haemoglobin polymorphisms. We will also consider the major unsolved problems of the field, the developmental biology of haemoglobin, and the management of the common genetic disorders of haemoglobin synthesis.

THE ORGANIZATION OF THE GENETIC CONTROL OF HUMAN HAEMOGLOBIN

The basic structure of all the normal human haemoglobins is similar and consists of two different pairs of globin chains; each chain is associated with one haem molecule. Different haemoglobins are identified by letters of the alphabet or by the place in which they were first discovered, and the constituent globin chains are designated by Greek letters. Individual amino acid residues are numbered from the N-terminal ends and further identified as occupying helical or non-helical parts of the chains.

Human haemoglobin is heterogeneous at all stages of development (Weatherall & Clegg 1972). In the embryo there are at least three types, called Hbs Gower 1, Gower 2 and Portland. These have the molecular formulae $\zeta_2\epsilon_2$, $\alpha_2\epsilon_2$ and $\zeta_2\gamma_2$ respectively. During fetal life the major haemoglobin component is Hb F. This has the general structure $\alpha_2\gamma_2$. However it consists of two molecular forms which differ in the structure of their γ chains; one has glycine at position 136 ($\alpha_2{}^G\gamma_2$) and the other has alanine at this position ($\alpha_2{}^A\gamma_2$) (Schroeder *et al.* 1968). Examination of fetal haemoglobin variants has shown that the structure of the $^G\gamma$ and $^A\gamma$ chains is determined by distinct structural loci. In adult life there is a major haemoglobin (Hb A, $\alpha_2\beta_2$) and a minor component, HbA_2 ($\alpha_2\delta_2$). During human development there is consecutive activation of the ζ and ϵ loci followed by the γ loci and finally, in adult life, the β and δ loci. Alpha-chain synthesis is activated in early fetal life and α chains are produced under the direction of the same loci throughout the rest of fetal development and adult life.

The results of recent cell fusion experiments suggest that the α-chain genes are on chromosome 16 and the β- and γ-chain genes on chromosome 11 (Deisseroth *et al.* 1978). Studies of the $\delta\beta$ and $\gamma\beta$ cross-over variants, Hbs Lepore and Kenya, suggest that the non-α chain genes are linked in a cluster in the order:

$$^G\gamma\text{-}^A\gamma\text{-}\delta\text{-}\beta$$

although the order

$$^A\gamma\text{-}\delta\text{-}\beta\text{-}^G\gamma$$

cannot be excluded.

The general arrangement of the haemoglobin genes and their expression at different stages of human development, are summarized in Fig. 1.

The number of globin genes (Table 1)

It is likely that in most human populations there are two genes per haploid genome controlling α-chain production (Lehmann & Carrell 1968); evidence for this has been summarized in detail recently (Weatherall & Clegg 1976).

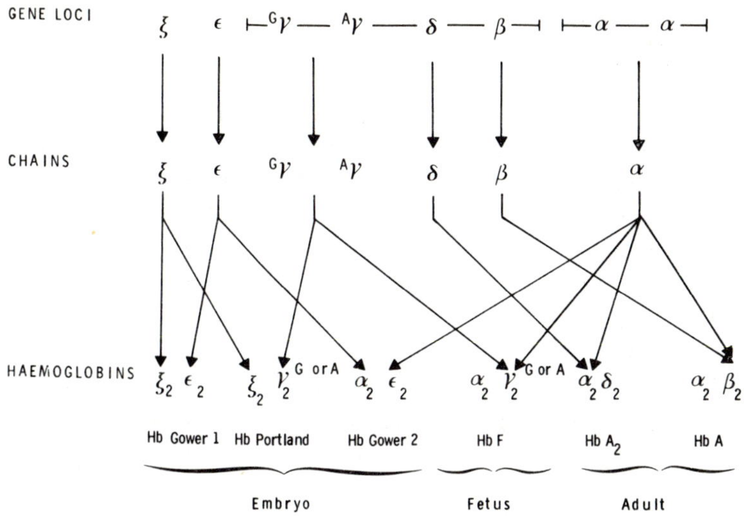

FIG. 1. The genetic control of human haemoglobin at different stages of development.

TABLE 1

Number of human globin genes per haploid genome

Chains	*Number of genes*	
	Genetic analysis	*Hybridization and/or gene mapping*
α	2	2
β	1	1
δ	1	1
$^{G}\gamma$	1 or 2	2–3
$^{A}\gamma$	1 or 2	
$^{T}\gamma^{a}$	?	
ϵ	?	?
ζ	?	?

[a]It is still uncertain where the $^{T}\gamma$ chain is the product of a distinct $^{T}\gamma$ locus

The only finding against there being a total of four α-chain loci is the observation in Melanesian populations that individuals homozygous for the α-chain variant, Hb J Tongariki, have only α^{J} chains (Abramson *et al.* 1970). Recent studies in our laboratory suggest that the Hb J Tongariki homozygotes also have α thalassaemia and that this is the reason they do not produce any Hb A (Old *et al.* 1978*a*). Thus it seems likely that most normal adults have four α-chain loci.

Nothing is known about the number of ϵ or ζ loci. Both genetic evidence and molecular hybridization analysis indicate that there is only one β and one δ chain gene per haploid genome. The number of γ-chain loci remains uncertain. Clearly there must be at least two, coding for the $^{G}\gamma$ and $^{A}\gamma$ chains respectively. Based on analysis of the levels of γ-chain variants in heterozygotes, a model suggesting that there are four γ loci per haploid genome has been derived (Huisman *et al.* 1972*a*). A more recent modification of this model suggests a figure of three genes (Huisman *et al.* 1977). Molecular hybridization analysis has indicated a maximum of two or three γ-chain genes (Old *et al.* 1976; Ramirez *et al.* 1976*b*). The recent discovery of a second ambiguity in γ-chain structure, threonine or isoleucine at position 75, has further complicated the picture (Ricco *et al.* 1976). It is not known whether the threonine-containing γ chain ($^{T}\gamma$) is the product of a distinct $^{T}\gamma$ locus or whether it results from a polymorphism at the $^{G}\gamma$ or $^{A}\gamma$ loci; current genetic evidence is equivocal although if it results from a polymorphism it probably involves the $^{A}\gamma$ locus. Recent restriction mapping analysis indicates that there are only two γ-chain genes, $^{G}\gamma$ and $^{A}\gamma$, and that there is not a separate $^{T}\gamma$ locus (P. Little & R. Williamson, personal communication).

Structure of human messenger RNA (mRNA)

The structure of human α, β and γ messenger RNA is summarized in Fig. 2. The structure of the 5′ and 3′ non-coding regions are known for the α- and β-chain messenger RNAs, as is that for the 5′ non-coding region of the γ-chain mRNA. Each of the human globin messenger RNAs has a poly A tail. The functional significance of the non-coding regions is uncertain. From analysis of conservation of base sequences in these areas it seems possible that certain regions of the 5′ end have critical functions, e.g. ribosome binding, and it has been suggested that these regions may also be involved in stabilizing the molecules in some way. For example the structure is consistent with a hairpin loop configuration of globin messenger RNA (see Proudfoot & Brownlee 1976; Forget 1977).

The fine structure of the human haemoglobin genome

The recently developed techniques for constructing maps of mammalian genes using restriction endonucleases (Jeffreys & Flavell 1977*a, b*) have just started to be applied to the study of human DNA (Orkin *et al.* 1978; Mears *et al.* 1978*a*). This approach has been facilitated by the availability of α, β and γ complementary DNA (cDNA) probes obtained from plasmids although some

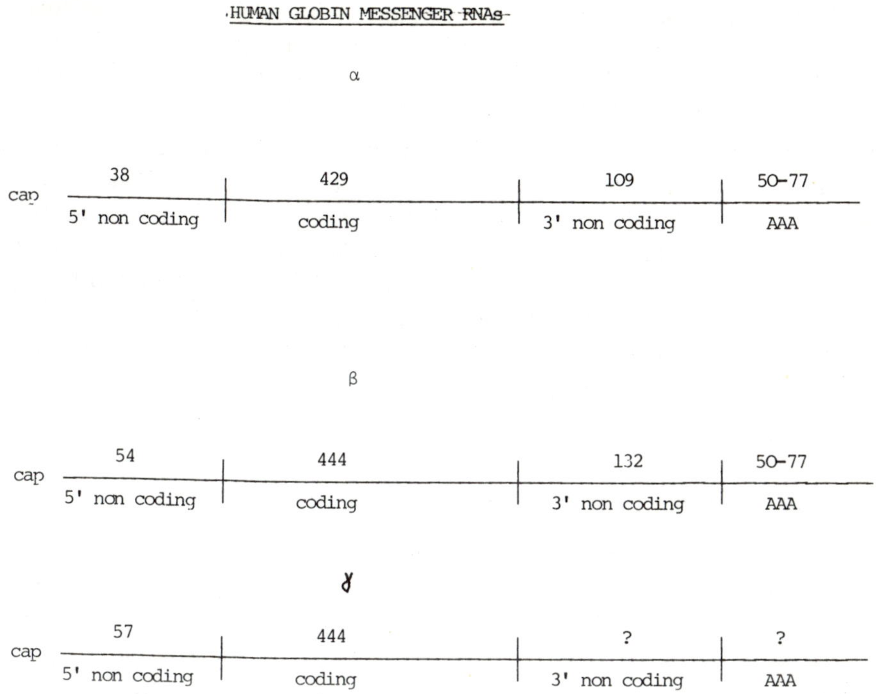

FIG. 2. Diagrammatic representation of human messenger RNA.

preliminary studies of human gene mapping have made use of ^{32}P-labelled cDNA prepared from human reticulocytes (Mears *et al.* 1978*a*; Orkin *et al.* 1978), or rabbit α and β cDNA under non-stringent hybridization conditions (N. J. Proudfoot, personal communication, 1978). The identification of the human haemoglobin genes in individual restriction fragments is facilitated by the existence of certain structural haemoglobin variants in which a particular base substitution provides a new site for restriction endonuclease cleavage. This approach should allow the unequivocal identification of the γ and β loci for example. Furthermore the examination of the regions of the genome containing γ-δ-β gene cluster should be made easier by the availability of DNA from individuals with $\delta\beta$ or $\gamma\beta$ fusion genes (see later section).

Recently information has been obtained regarding the fine structure of the δ and β globin genes (Mears *et al.* 1978*b*; Flavell *et al.* 1978; Lawn *et al.* 1978). Their orientation with respect to the direction of transcription is $5'$-δ-β-$3'$ and the δ and β loci are about 5.4 kilobases apart. The α chain genes are about 2.7 kb apart (S.H. Orkin, personal communication, 1978).

Many mammalian genes have one or more non-coding inserts along their length. This extra material is transcribed and excised during the processing of high molecular weight nuclear RNA to definitive messenger RNA (Tilghman *et al.* 1977, 1978). Inserts containing about 100 and 700 bases have been demonstrated in both mouse and rabbit globin genes (Jeffreys & Flavell 1977*a, b*; Tilghman *et al.* 1978; Leder 1978). In the mouse and rabbit β globin gene the larger insert is the coding sequence between amino acid residues 104 and 105. The smaller insert in mouse β gene lies between residues 29–30 or 30–31. The human δ and β genes contain inserts of 950 and 900 bases, respectively, located between codons for amino acids 104 and 105 (Lawn *et al.* 1978). It is still not clear whether there are inserts in the α genes; if so they must be smaller than in the non-α genes. The functional significance of these inserts is not known.

GENETIC DISORDERS OF HAEMOGLOBIN PRODUCTION

There are four major groups of genetic disorders of haemoglobin production. These are (1) the structural haemoglobin variants, (2) the thalassaemias, which are characterized by unbalanced globin chain synthesis, (3) those structural haemoglobin variants which, because they are inefficiently synthesized, result in globin chain imbalance and the clinical picture of thalassaemia, and (4) hereditary persistence of fetal haemoglobin (HPFH).

STRUCTURAL HAEMOGLOBIN VARIANTS

Well over 200 structural haemoglobin variants have been detected, most of them with abnormal α or β chains. Apart from Hbs S, C and E, which occur at polymorphic frequencies in some populations, these variants are rare. Most of them do not cause disease and hence their detection is dependent on electrophoretic screening techniques. It has been estimated that one in 800 individuals carries an electrophoretic variant and perhaps one in 300 has variants of any sort (Harris 1974). Owing to the degeneracy of the genetic code only four-fifths of the possible mutations in the structural genes for, say, the α or β chains will result in amino acid changes. Of those changes that can result in amino acid substitutions only a third would lead to an alteration in the charge by which the variant could be detected electrophoretically. The remainder would be neutral, detectable only if they result in a functional or structural abnormality in the haemoglobin molecule.

The molecular basis for the production of structural haemoglobin variants is reviewed in detail by Weatherall & Clegg (1976) and current lists of haemoglobin variants will be found in the reviews of Bunn *et al.* (1977) and Winslow & Anderson (1978). The majority of the variants result from single amino acid substitutions which are explicable on the basis of single base substitutions in the coding DNA of the structural genes. Several examples are known of two amino acid substitutions in the same globin chain; individual substitutions in these variants are also found in other abnormal haemoglobins and it has been suggested that such double substitutions result from cross-overs between the respective rare allelic genes. At least three variants, Hbs Sidney, Bristol and Milwaukee, are due to substitutions of alanine, aspartic acid and glutamic acid for valine at position β67. No single valine codeword can accommodate these substitutions, which suggests that either the β-chain gene is polymorphic at this codon or that one of these variants has arisen by a double base mutation.

A number of β chain variants have been described in which sequences from one up to five amino acids have been deleted internally from the normal β-chain sequence. Although no α-chain variants with deletions are known, one with a three-residue insertion has been described. It is presumed that both these types of abnormal globin have arisen by non-homologous crossing-over at meiosis. The sequence of amino acids inserted should be a repeat of an adjacent sequence and in fact this is the case in haemoglobin Grady where the sequence of residues 116 to 118 (Glu-Phe-Thr) is duplicated.

Several variants have been described which result from abnormal crossing-over with the production of fusion genes (Fig. 3). The first of these was haemoglobin Lepore which is made up of normal α chains combined with non-α chains which consist of the N-terminal residues of the δ chain and the C-terminal residues of the β chain. At least three different varieties of Hb Lepore occur, depending on the position of crossing-over between the δ and β genes. Similarly Hb Kenya has resulted from abnormal crossing-over between the γ and β genes: the non-α chains of Hb Kenya consist of the N-terminal residues of the γ chain and the C-terminal residues of the β chain. To date there has been no example of abnormal crossing-over between the embryonic globin chain loci and the γ- or β-chain genes.

Several haemoglobin variants with elongated chains have been described. Those which result from chain termination mutants will be considered later in this paper. There are several variants which are due to alterations of the reading frame. Three frame-shift mutations are known to date, Hbs Tak, Cranston and Wayne, which all have extended C-terminal sequences of the globin chain. The fact that the sequence of these variants extends beyond the normal C terminus indicates that the additional amino acids beyond the C-terminal residue are derived from reading into normally untranslated parts of

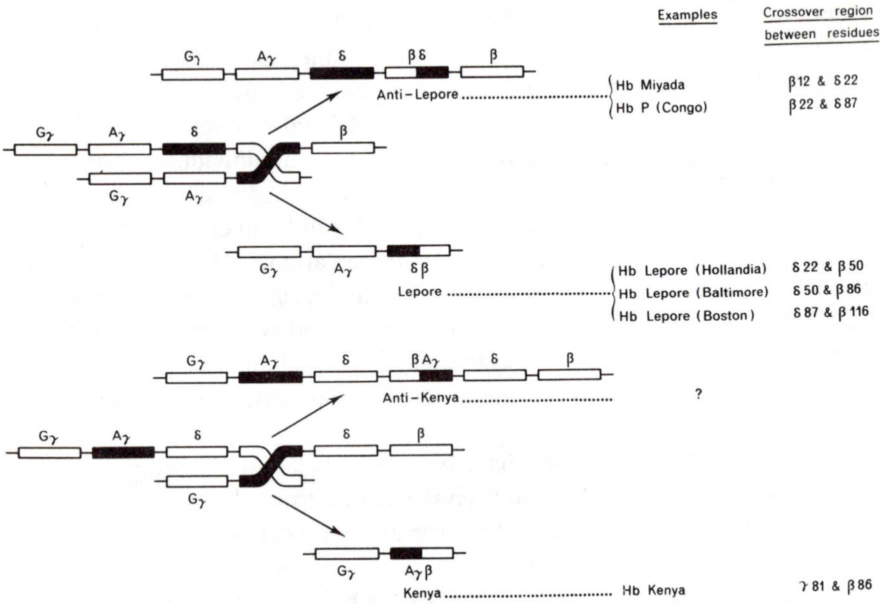

FIG. 3. The production of haemoglobins Lepore and Kenya by unequal crossing-over.

the messenger RNA at the 3′ end until an in-phase termination codon is reached (Fig. 4). This makes it possible to derive hypothetical mRNA sequences which fit the amino acid sequences of the normal α and β chains and the elongated variants. In fact these sequences are entirely consistent with those derived from base analysis of human α and β globin messenger RNA (Proudfoot & Brownlee 1976).

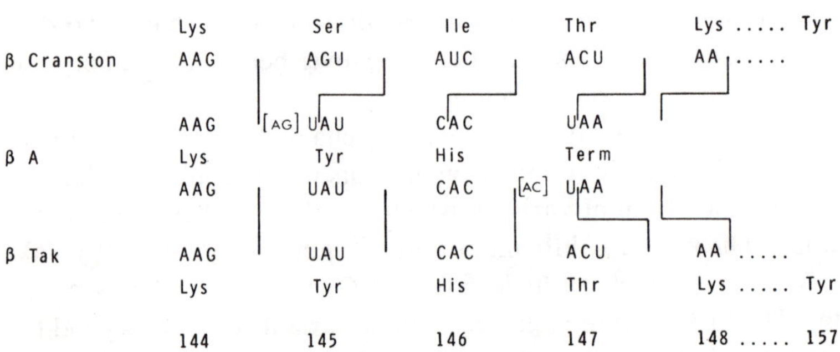

FIG. 4. Human frame-shift mutations.

The molecular basis for the structural haemoglobin variants and for the pathophysiology of those which cause disease are summarized in Tables 2 and 3.

THE THALASSAEMIAS

The thalassaemias are genetic disorders of haemoglobin synthesis characterized by a reduced rate of production of either the α or β chains. They fall into two general groups, the α and β thalassaemias, depending on which pair

TABLE 2

Summary of molecular mechanisms for structural haemoglobin variants

Mechanism	*Examples*
Single base change	Over 200 variants
Deletion of one or more residues—shortened chain (abnormal crossing-over)	Hb Leiden ($\beta^{6 \text{ or } 7}$ Glu→O)
	Gun Hill ($\beta^{92\text{-}96}$ Leu, His, Cys, Asp→O)
Insertion—elongated chain (abnormal crossing-over)	Hb Grady ($\alpha^{116\text{-}118}$ Glu-Phe-Thr duplicated)
Codon polymorphism or double base substitution	Hb Sydney β^{67} Val→Ala
	Hb Bristol β^{67} Val→Asp
	Hb Milwaukee β^{67} Val→Glu
Frame-shift chain elongation	Hb Tak; Hb Cranston
Chain termination mutation	Hb Constant Spring (see Table 4)
Premature termination—short chain	Hb McKees Roçks—terminated 144 Lys
Fusion gene	$\delta\beta$ Hb Lepore } (see Fig. 3)
	$\gamma\beta$ Hb Kenya } (see Fig. 3)

References: Bunn *et al.* 1977; Winslow & Anderson 1978.

TABLE 3

Pathophysiology of structural haemoglobin variants

Clinical disorder	*Mechanism*
Haemolytic anaemia	Hb S—tactoid formation and molecular stacking
	Hb C—reduced solubility
	Unstable Hbs—altered configuration of haem pockets, chains, subunits, etc.
Cyanosis—methaemoglobinaemia	Hbs M—haem fixed in Fe^{3+} state
Polycythaemia	Many variants—increased oxygen affinity due to substitutions at $\alpha_1\beta_2$ contact or 2,3-DPG binding sites
Hypochromic anaemia—thalassaemia	Reduced rate of synthesis of α or β chains (see text)

of globin chains are synthesized inefficiently (Weatherall & Clegg 1972).

The reduced rate of β-chain production in β thalassaemia leads to an excess of α chains which are unstable, precipitate in the developing red cell precursor, and cause its damage and premature destruction in the bone marrow or the blood. As in normal individuals a proportion of the red cell precursors continue to produce γ chains, and these cells are relatively protected against the excess α chains because some of them combine with γ chains to produce Hb F. However, only a relatively small number of red cell precursors produce γ chains in adult life, hence this mechanism is usually unable to compensate for the basic defect in β-chain production. Thus the homozygous state for most forms of β thalassaemia is characterized by a transfusion-dependent life, and death within the second decade from the results of iron loading secondary to repeated blood transfusion (Weatherall & Clegg 1972).

The pathophysiology of α thalassaemia is rather different. Defective α-chain synthesis results in inefficient haemoglobin production in both fetus and adult. In intrauterine life a deficiency of α chains leads to an excess of γ chains, and in adult life to an excess of β chains. Excess γ chains produce a γ_4 molecule or Hb Bart's, and excess β chains produce β_4, or Hb H. These abnormal tetramers, while being soluble in the red cell precursors and hence preventing their destruction in the bone marrow, tend to precipitate in the peripheral blood and cause a haemolytic anaemia. Hence the degree of ineffective erythropoiesis in α thalassaemia is not so severe as that in β thalassaemia. On the other hand Hbs Bart's and H are unstable and, furthermore, are useless as oxygen carriers.

β thalassaemia

The β thalassaemias are remarkably heterogeneous (Table 4). They fall into two major groups: β^+ thalassaemia in which there is a reduced rate of β chain synthesis and β° thalassaemia in which there is no β chain synthesis.

β^+ thalassaemia. Analysis of families with β^+ thalassaemia, particularly where it interacts with β chain haemoglobin variants, suggests that there are probably several different genetic varieties with characteristic rates of β-chain production. They range from a very mild defect in β-chain synthesis, which may well be completely silent in the carrier state, to a severe defect in which less than 10% of β-chain production occurs. The levels of Hb A (i.e. β-chain production) in individuals doubly heterozygous for these different forms of β^+ thalassaemia and structural haemoglobin variants are summarized in Table 4.

TABLE 4

The β thalassaemias

Type	Molecular defect
β^+ thalassaemia[a]	
With 5% Hb A	Reduced β-chain mRNA (? processing defect)
With 30% Hb A	?
With 50% Hb A	?
β° thalassaemia[b]	
Type 1	Globin genes intact. No detectable β globin mRNA
Type 2	Globin genes intact. Abnormal β globin mRNA missing part of 3′ end
Type 3	Globin genes present. Abnormal β globin mRNA—full length with normal non-coding sequences

[a]Classified according to amount of Hb A found in compound heterozygotes for β thalassaemia and β-chain haemoglobin variants.
[b]Based on data in Old *et al.* (1978*b*). Does not include Ferrara variety of β° thalassaemia in which β-chain synthesis is reported to be inducible (Conconi *et al.* 1972).

To date there is very little information about the molecular defect in β^+ thalassaemia. There is a reduced output of β chain messenger RNA, the level of which is usually compatible with the overall degree of β chain synthesis in the red cells (Housman *et al.* 1973; Kacian *et al.* 1973). There is preliminary evidence that some cases of β^+ thalassaemia may result from a defect in processing at the nuclear RNA level but this awaits confirmation (Nienhuis *et al.* 1977).

β° thalassaemia. This condition is characterized by a complete absence of β-chain synthesis. It is also heterogeneous at the clinical and molecular levels.

In all cases of β° thalassaemia examined so far, cDNA/DNA hybridization analysis has indicated that the β globin genes are intact (Tolstoshev *et al.* 1976). However cDNA/mRNA analysis has shown a remarkable heterogeneity of these disorders (Kan *et al.* 1975*c*; Temple *et al.* 1977). Recently three main classes have been defined (Old *et al.* 1978*b*) (Table 4). In group 1 no detectable β-chain mRNA is present. In group 2 hybridizable β-chain mRNA is present which appears to be structurally abnormal. With cDNA probes specific for the 5′ and 3′ ends of β chain messenger RNA it has been shown that at least some of these messenger RNAs are lacking sequences at their 3′ ends. Finally, a third group have full-length β-chain messenger RNA which is not translated. It seems likely that this type of β° thalassaemia results from a structural alteration in the messenger RNA at or near the chain initiation codon.

δβ thalassaemia

In this rarer form of thalassaemia there is a defective synthesis of both the δ and β chains of adult haemoglobin. The homozygous state is characterized by complete absence of δ and β chain production and the haemoglobin consists entirely of Hb F. Heterozygotes carry between 5 and 20% Hb F which is unevenly distributed among the red cells. The degree of chain imbalance in δβ thalassaemia is less than in β thalassaemia, in both homozygotes and heterozygotes, and this suggests that γ-chain production is better able to compensate for the deficiency of β-chain production than in β thalassaemia.

There is evidence of heterogeneity of the δβ thalassaemias. This has been defined in several ways. First there are differences in the structure of the associated Hb F. In most cases both $^{G}\gamma$ and $^{A}\gamma$ chains are produced (Huisman *et al.* 1974). On the other hand some δβ thalassaemias synthesize only $^{G}\gamma$ Hb F (Wood *et al.* 1977*a*). The molecular pathology has been worked out for one family with $^{G}\gamma^{A}\gamma$ δβ thalassaemia (Ottolenghi *et al.* 1976; Ramirez *et al.* 1976*a*). With cDNA/DNA hybridization a deletion of the β and probably of the δ chain loci was defined. This analysis has not yet been carried out in $^{G}\gamma$ δβ thalassaemia although it is tempting to speculate that this may also result from a deletion involving the $^{A}\gamma$ locus as well as the δ and β loci. We will consider the significance of the siting of this deletion and its relationship to persistent γ-chain synthesis later.

The other well defined form of δβ thalassaemia is that associated with the δβ fusion gene variants, the Hb Lepores. As mentioned earlier these have resulted from unequal crossing-over at the δ–β gene complex with the production of δβ fusion genes. The δβ fusion chains are synthesized inefficiently and therefore the clinical picture of δβ thalassaemia is produced. The basis for the defective synthesis of Hb Lepore appears to be instability of δβ messenger RNA (reviewed by Wood *et al.* 1978).

α thalassaemia

The α thalassaemias are also heterogeneous (Weatherall & Clegg 1972) (Table 5). In some populations at least three α thalassaemia genes are found. These are α thalassaemia 1, α thalassaemia 2 and Hb Constant Spring. The latter is an elongated α-chain variant which is inefficiently synthesized and hence produces the clinical picture of α thalassaemia (Clegg *et al.* 1971). In the homozygous state for α thalassaemia 1 there is complete absence of α-chain synthesis and α-chain messenger RNA production. It results in stillbirth of a hydropic infant which produces Hb Bart's and no Hbs F or A. The

TABLE 5

The α thalassaemias

Type	*Molecular defect*
α thalassaemia 1	Deletion of both haploid α chain genes
α thalassaemia 2	Deletion of 1 of pair of haploid α chain genes
Hb Constant Spring	Chain termination mutation UAA→CAA
Hb Icaria	Chain termination mutation UAA→AAA
Hb Seal Rock	Chain termination mutation UAA→GAA
Hb Koya Dora	Chain termination mutation UAA→UCA

The genetics of α thalassaemia is not yet clear in many populations. At least one non-deletion form has been found (Kan *et al.* 1977).

homozygous states for α thalassaemia 2 and Hb Constant Spring are characterized by a mild deficiency of α-chain production but no clinical disability. On the other hand the compound heterozygous states for α thalassaemia 1 and α thalassaemia 2 or Hb Constant Spring are associated with a moderately severe α-chain deficit, the production of variable amounts of Hb H, and the clinical picture of thalassaemia intermedia.

The α thalassaemias have been analysed by cDNA/DNA hybridization (Ottolenghi *et al.* 1974; Taylor *et al.* 1974; Kan *et al.* 1975*a*). Alpha thalassaemia 1 results from a deletion of both pairs of haploid α chain genes; hence the homozygous state is characterized by an absence of all four α chain genes which accounts for the complete absence of α-chain messenger RNA and α-chain synthesis. Alpha thalassaemia 2 results from a deletion of one of the haploid pair of α genes. There is recent evidence, however, that some α thalassaemias are associated with a reduced amount of α chain mRNA but the α-chain genes are intact (Kan *et al.* 1977). The molecular mechanism for this form of the disorder is not yet determined.

The molecular basis for the production of the elongated α chain of Hb Constant Spring is thought to be a single base change in the chain-terminating codon, U to C, such that instead of the α chain being terminated glutamine is inserted and then additional messenger RNA is translated until another in-phase termination codon is reached (Clegg *et al.* 1971; Weatherall & Clegg 1975). The structure of the additional 31 residues at the C-terminal end of the α chain of Hb Constant Spring is entirely compatible with the sequence of the messenger RNA at its 3′ non-coding end (Proudfoot & Brownlee 1976). Indeed Hb Constant Spring is only one of a family of chain termination mutants which have different substitutions at position 142 but in which the rest of the elongated α chain is the same (Weatherall & Clegg 1975) (Fig. 5).

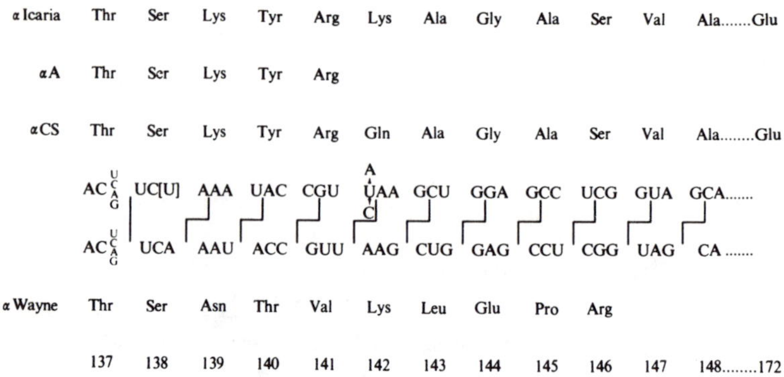

FIG. 5. The C-terminal residues of the chain termination mutants, Hbs Constant Spring (CS) and Icaria, and the frame-shift mutant, Hb Wayne.

It has been suggested that the reason for the very low rate of production of the α Constant Spring chain is the instability of its messenger RNA. Evidence in favour of this notion has been obtained recently because it has been shown that heterozygotes for Hb Constant Spring have a reduced amount of α-chain messenger RNA in their red cells as compared with normal individuals (D.M. Hunt, unpublished work).

The genetics of α thalassaemia as outlined above holds for Orientals and some of the Mediterranean races. However the situation is more complicated in Africans and in the Middle Eastern populations and the molecular basis for α thalassaemia has not yet been defined in these groups.

HEREDITARY PERSISTENCE OF FETAL HAEMOGLOBIN

During normal human development γ-chain synthesis starts to decline from about the 34th week of intrauterine life. During the first few months after birth Hb F is replaced by Hbs A and A_2. In normal adults about 0.5–0.8 Hb F is detectable; this is confined to a small population of red cells called F cells (Boyer *et al.* 1975; Wood *et al.* 1975). The genetic regulatory mechanisms involved in the switch from γ- to β-chain synthesis are unknown.

There is an interesting group of genetic disorders in which Hb F production persists into adult life in the absence of major haematological abnormalities. The name hereditary persistence of fetal haemoglobin (HPFH) is given to these conditions. They are classified broadly into two main groups: pancellular HPFH, in which the Hb F is fairly evenly distributed throughout the red cells, and heterocellular HPFH, in which the Hb F is heterogeneously distributed and seems to be confined to only a proportion of the cells.

Pancellular HPFH

This condition is classified according either to the race in which it was first described or to the structure of the associated Hb F, i.e. whether it contains $^{G}\gamma$ chains, $^{A}\gamma$ chains, or both (Weatherall & Clegg 1972; Huisman *et al.* 1974).

The commonest form is the Negro type. Heterozygotes have about 25% Hb F and homozygotes have no Hb A or A_2 and have 100% Hb F. Hence the condition is characterized by a complete absence of β and δ chain synthesis *cis* to its genetic determinant. Recent studies have shown that the $^{G}\gamma^{A}\gamma$ form of Negro HPFH is due to a gene deletion involving the β chain and probably the δ chain loci (Kan *et al.* 1975*b*; Forget *et al.* 1976; Tolstoshev *et al.* 1976). This allows persistent synthesis of $^{G}\gamma$ and $^{A}\gamma$ chains into adult life such that there is almost complete compensation for the lack of β and δ chains. Compensation is not complete in the homozygotes, however, and they have a mild degree of globin-chain imbalance and hence a clinical picture similar to heterozygous β or $\delta\beta$ thalassaemia. $^{G}\gamma^{A}\gamma$ HPFH in the Negro has been subdivided into several groups according to the level of Hb F in heterozygotes (Huisman *et al.* 1974).

Three Negro families have been described with an HPFH-like condition in which 15% Hb F is associated with both Hbs A and S or C, indicating that β chains are produced *cis* to the HPFH determinant. It appears that these individuals have both persistent γ *and* β synthesis directed by the *same* chromosome (Huisman *et al.* 1975; Friedman & Schwartz 1976; Higgs *et al.* 1979). No hybridization or gene analysis data are available for this condition yet but it is tempting to speculate that it results from a deletion of the $^{A}\gamma$- and δ-chain genes.

Another form of HPFH found in Africans is associated with Hb Kenya (Huisman *et al.* 1972*b*). Heterozygotes carry 7–23% Hb Kenya and 4–9% Hb F which contains only $^{G}\gamma$ chains evenly distributed throughout the red cells. Hb Kenya has non-α chains which consist of the N-terminal end of the γ chains and the C-terminal end of the β chain. The hybrid $\gamma\beta$-chain has probably arisen by unequal crossing-over between the $^{A}\gamma$- and β-chain genes with exclusion of the portion of the $^{A}\gamma$ locus which directs the C-terminal end of the $^{A}\gamma$ chain, the whole of the δ-chain locus, and that part of the β locus which controls the N-terminal end of the β chain (Fig. 3). Thus the chromosome which directs the synthesis of the non-α chain of Hb Kenya contains only $^{G}\gamma$ and $\gamma\beta$ fusion loci.

Clearly each of these conditions involves the deletion of chromosomal material from the $^{G}\gamma$-$^{A}\gamma$-δ-β gene complex. This suggests that certain areas of this complex are involved in the neonatal suppression of the γ genes and presumably in activation of the δ and β genes (Figs. 6, 7).

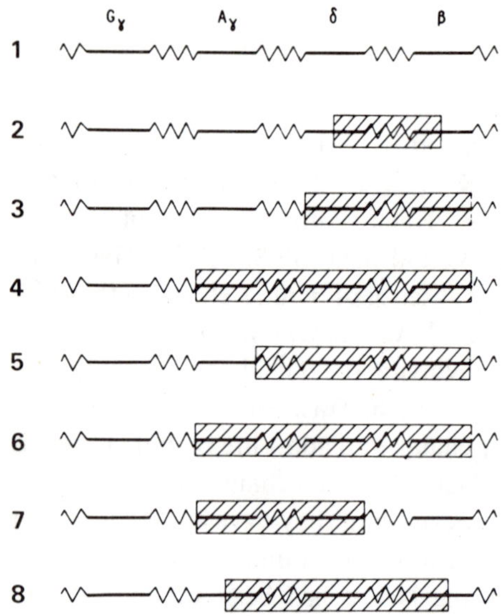

FIG. 6. Proved or assumed deletions of the ${}^{G}\gamma$-${}^{A}\gamma$-δ-β gene complex, causing the disorders summarized in Table 7. Top to bottom: (1) normal genes; (2) Hb Lepore; (3) ${}^{G}\gamma{}^{A}\gamma$ $\delta\beta$ thalassaemia; (4) ${}^{G}\gamma$ $\delta\beta$ thalassaemia; (5) ${}^{G}\gamma{}^{A}\gamma$ Negro HPFH; (6) ${}^{G}\gamma$ Negro HPFH; (7) ${}^{G}\gamma$ β^{+} HPFH; (8) Hb Kenya.

One other variety of pancellular HPFH is documented. This occurs in Greeks and differs from the Negro type in that heterozygotes have a lower level of Hb F which consists mainly of ${}^{A}\gamma$ chains. There is circumstantial evidence that both β and δ chain synthesis is active *cis* to the Greek HPFH determinant. Hence this condition could result from a deletion involving the ${}^{G}\gamma$ gene or from a regulatory mutation.

Heterocellular HPFH

Normal individuals have about 1–5% F cells which can be demonstrated easily by a fluorescent antibody technique using antibody directed against Hb F (Boyer *et al.* 1975; Wood *et al.* 1975). There are several conditions in which there appears to be a genetically determined increase in the number of F cells. The best known of these is Swiss HPFH in which the Hb F level is raised slightly in the range of 1–2% with about 20% F cells. If this condition is inherited together with Hb S or β thalassaemia it results in a higher level of Hb F than is found in sickle cell disease or β thalassaemia alone (Wood *et al.* 1977*b*).

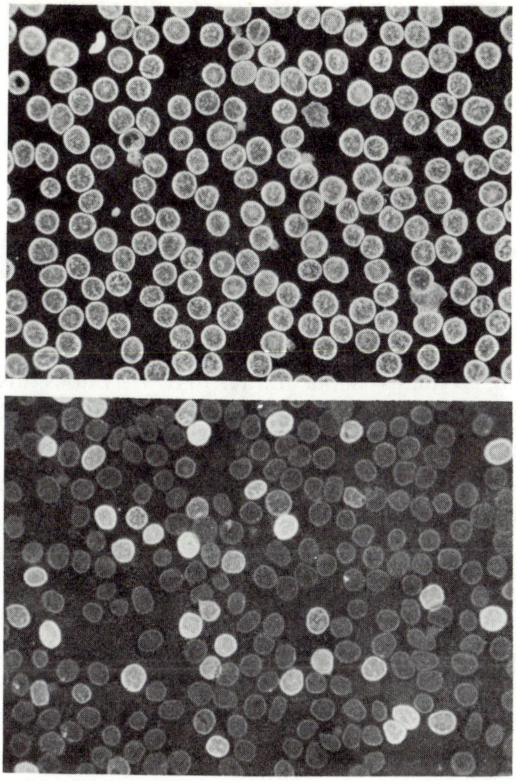

a

b

FIG. 7. (*a*) Blood film from an individual heterozygous for Negro HPFH treated with a fluorescent-labelled anti-Hb F antibody. The Hb F is uniformly distributed among all the cells. (*b*) A similar preparation to (*a*) prepared from the cells of a $^{G}\gamma^{A}\gamma$ $\delta\beta$ thalassaemia heterozygote. The Hb F is unevenly distributed among the cells despite the similarity of the deletion to that which causes HPFH. The reasons for the variability of expression of the γ-chain genes in $\delta\beta$ thalassaemia is not clear.

A British family has been reported with higher levels of Hb F heterogeneously distributed among the red cells in which the inheritance pattern was consistent with the existence of homozygotes and heterozygotes. The heterozygotes had Hb F levels of 4–13% and about 50% F cells while homozygotes had about 20% Hb F and 100% F cells (Weatherall *et al.* 1975*a*). The Hb F was quite heterogeneously distributed among the cells. A similar family to this has been found in an American Negro population (Boyer *et al.* 1977) in which the Hb F levels in heterozygotes ranged from 2.6 to 6%; the Hb F was distributed in about 50% of the cells.

Interestingly the level of Hb F per F cell in these families is similar to that of normal F cells. Thus while it is possible that these conditions result from a regulatory mutation involving the expression of the γ-chain genes it is also feasible that they result from inherited variability of the mechanisms which 'set' the level at which the F-cell population is fixed after the switch from fetal to adult haemoglobin synthesis.

GENETIC SIGNIFICANCE OF FETAL HAEMOGLOBIN PRODUCTION IN SICKLE CELL DISEASE, THALASSAEMIA AND HPFH

Significantly raised levels of Hb F are found in children and adults with sickle cell disease, β thalassaemia, $\delta\beta$ thalassaemia and HPFH. Why does this occur and does this phenomenon tell us anything about the regulation of fetal and adult haemoglobin production?

The proposed mechanisms for increasing Hb F in these disorders are summarized in Table 6 and have been analysed in detail recently (Weatherall *et al.* 1979). We have already mentioned that some red cell precursors in normal adults synthesize Hb F. In health there are about 1–7% F cells in the peripheral blood. In sickle cell anaemia and β thalassaemia these cells will be subject to selection, in the blood in the former condition and in both marrow and blood in the latter. It seems very likely that selection of F cells is the main reason for the apparent increase in Hb F production in these disorders and that the primary genetic lesions which cause them do not have any direct effect on the expression of the γ-chain genes (Weatherall *et al.* 1976). A possible exception is the unusually high levels of Hb F found in Saudi Arabs with sickle cell anaemia. Here there appears to be a genuine increase in γ-chain production in association with the SS genotype (Pembrey *et al.* 1978).

TABLE 6
Mechanisms of Hb F production in adult life

(1) Genetically determined increase in F cells
Heterocellular HPFH
(2) Selection of F cell population
Sickle cell disease
β thalassaemia
(3) Deletions of γ-δ-β gene complex
Pancellular HPFH
$\delta\beta$ thalassaemia
(4) Absolute increase in γ chain output—unknown basis
Saudi sickle cell disease
(5) Reactivation or persistence of fetal erythropoiesis
Juvenile chronic myeloid leukaemia

The situation in $\delta\beta$ thalassaemia and pancellular HPFH is quite different from that in sickle cell anaemia and β thalassaemia. Here we have good evidence for a real increase in γ-chain output. Thus it follows that the deletions of the γ-δ-β gene cluster that cause these conditions somehow lead to a variable failure of repression of the γ-chain loci in adult life. These conditions, which seem to represent a spectrum of disorders of varying severity depending on the output of γ chains, are summarized in Table 7.

In heterocellular HPFH the position is again different. In this case there is no reason for selection of F cells and it is unlikely that there is a major deletion of the δ and β chain genes since Hbs A and F are present in homozygotes. This condition appears to be a genuine regulatory mutation involving loci which are involved in controlling the expression of the γ and β loci in adult life.

It is clear, therefore, that there are several different genetic mechanisms involved in Hb F production in adults. In no case does a condition resembling

TABLE 7

Genetic disorders involving the γ-δ-β gene complex

Designation	*Clinical and haematological characteristics*	*Molecular pathology*
$^G\gamma^A\gamma$ HPFH	Mild thalassaemia with 100% Hb F in homozygotes. Normal haematology with 20–30% Hb F in heterozygotes	Deletions of δ and β genes
$^G\gamma$ HPFH	Normal haematology. Heterozygotes have 5–20% Hb F	?
$^G\gamma$ β^+ HPFH	Normal haematology. Compound heterozygotes with Hbs S or C make Hbs A, F, S or C. Heterozygotes have about 20% Hb F	?
$^A\gamma$ HPFH (Greek HPFH)	Normal haematology. Heterozygotes have 12–15% Hb F	?
Hb Kenya	Normal haematology. Heterozygotes have 4–9% Hb F and 7–24% Hb Kenya	Deletion of part of $^A\gamma$, all δ and part of β genes. $\gamma\beta$ fusion gene
$^G\gamma^A\gamma\delta\beta$ thalassaemia	Homozygotes have variable anaemia and 100% Hb F. Heterozygotes have 5–20% F and mild thalassaemia	Deletion of δ and β genes
$^G\gamma$ $\delta\beta$ thalassaemia	Homozygotes have variable anaemia and 100% Hb F. Heterozygotes have about 20% Hb F	?
Hb Lepore thalassaemia	Homozygotes have severe anaemia with Hbs F and Lepore only. Heterozygotes have Hb Lepore and slight elevation of Hb F	Deletion of part of δ and β genes. $\delta\beta$ fusion gene

true fetal erythropoiesis occur in association with these inherited disorders of haemoglobin synthesis. Thus none of these conditions can be said to provide any real insight into the regulation of Hb F synthesis in normal development although the deletion models do tell us that certain key areas in the γ-δ-β complex are involved in the mechanics of this process. We return to this problem later.

MAINTENANCE OF THE HAEMOGLOBIN POLYMORPHISMS

The structural haemoglobin variants, Hbs S, C and E, occur at high frequencies in certain populations (see Allison 1965; Weatherall & Clegg 1972; Bunn *et al.* 1977). Relative protection against *Plasmodium falciparum* malaria afforded to heterozygous carriers of these disorders is now the accepted mechanism for the high frequency of the genes in areas where malaria is (or was) common (Allison 1965). However, the cellular mechanisms whereby these haemoglobin variants protect the red cells against malarial parasites are still not fully understood. Similarly, population data suggest that the high gene frequencies for β thalassaemia may also have arisen from relative protection of carriers against *P. falciparum* (Siniscalco *et al.* 1966). Again, it is not clear how the thalassaemic erythrocyte protects against the parasite.

The recent development of *in vitro* techniques for short-term culture of *P. falciparum* has enabled the interaction of the parasite with various types of red cells to be examined. Studies along these lines are providing some insight into the way in which genetically abnormal red cells may offer protection against the malarial parasite.

Invasion and growth of *P. falciparum* in red cells containing Hb S has been examined in short-term cultures (Pasvol *et al.* 1978). Under aerobic conditions invasion of the parasites did not differ significantly from normal. Furthermore growth was retarded to only a minor extent as compared with that in normal cells. However, under conditions of reduced oxygen tension there was a significant decrease in the rates of invasion and growth in Hb S-containing cells. These effects seemed to be related to the level of Hb S in the cells but were not dependent on their sickling. These observations suggest a possible mechanism for the protection of sickle cell heterozygotes against *P. falciparum in vitro*. After development of schizonts inside the liver merozoites are liberated which invade red cells. The resultant ring forms of the parasite will mature in well-oxygenated peripheral blood where they normally spend the early stages of the erythrocyte cycle. After this the infected cells move to the deep tissues where the parasites mature to schizonts. During this phase of development the cells are exposed to oxygen tensions very similar to those which cause retardation of parasite development *in vitro*. Hence it seems likely that in red cells containing Hb S this critical stage of parasite maturation may be retarded. Friedman (1978) has described retardation of growth of *P. falciparum* in cells containing Hb S maintained under low oxygen conditions in an entirely different culture system.

The reasons why the red cells of β thalassaemia heterozygotes might protect against *P. falciparum* are less easy to determine. An interesting feature of β thalassaemia heterozygotes is that the rate of decline of Hb F in the first year of life is retarded as compared with normal infants (Beaven *et al.* 1961). It has been shown that cells containing Hb F, while being invaded by *P. falciparum* at the same rate as normal cells, support the growth of the parasite less efficiently than do normal cells (Pasvol *et al.* 1977). These observations have led to the hypothesis that the relatively high levels of Hb F which are found in β-thalassaemia carriers during the first year of life might protect them against *P. falciparum* infection at a critical stage of development when passive immunity is waning but before active immunity has been achieved (Pasvol *et al.* 1977).

It has been shown that *P. falciparum* develops equally well in cells containing Hbs C and E as those containing Hb A. Thus the reason for the high frequency of these variants in some populations is not clear. Early experiments suggested that the rate of decline of Hb F was retarded in carriers of these β-chain variants (Beaven *et al.* 1961). If this is confirmed it raises the interesting possibility that the retardation of growth of *P. falciparum* in Hb F-containing cells may be a protective mechanism common to many of the β-chain haemoglobinopathies.

FUTURE DEVELOPMENTS AND PROBLEMS IN THE HUMAN HAEMOGLOBIN FIELD

The sickling disorders and thalassaemias produce major public health problems in many parts of the world. In the developing countries β thalassaemia causes a tremendous drain on health resources because of the transfusion requirements of homozygotes. There is no doubt that as the anaemias of malnutrition and infection come under control in parts of India and Pakistan, the thalassaemia problem will become one of the major causes of infant morbidity and mortality in these countries as it is already in the Mediterranean region and south-east Asia. With the re-emergence of malaria as a major disease in many parts of the tropics the factors which maintain the high gene frequencies for these conditions will be maintained; hence these common and serious anaemias are likely to be with us for a long time to come. It is clear that, with the availability of the new techniques of gene mapping, the molecular defects in those forms of thalassaemia in which the basis of genetic change has not yet been defined will be worked out within the next few years. Will this sophisticated information be translated into useful forms of clinical management? Unfortunately this seems unlikely! What can be done, or might be done in the future, for affected patients?

The management of sickle cell anaemia and the thalassaemias

The management of sickle anaemia and β thalassaemia has not improved greatly since the diseases were discovered over 50 years ago. The natural history of sickle cell anaemia is better understood and the serious problems produced by infection are now recognized and can be dealt with. However there is no definitive treatment and while many anti-sickling agents have come and gone, none have withstood the rigours of a properly conducted clinical trial. Most homozygous β thalassaemics require regular blood transfusion and if none is available they die in infancy. If they are transfused they die of the effects of iron overload during the second decade. Recent advances in the use of chelating agents (Propper *et al.* 1977; Weatherall *et al.* 1977) have offered some hope for these children but the methods of administration are cumbersome and the pharmaceutical industry has shown little interest or energy in developing new and more easily administered forms of these drugs (Weatherall *et al.* 1977).

Bone marrow transplantation is becoming more feasible with improving technology and experience but the problems of graft-versus-host disease are still serious and at the moment this technique does not seem to be ready to be applied to the problems of sickle cell anaemia and thalassaemia.

For these reasons over the last few years much attention has focused on the prevention of these diseases by genetic counselling and prenatal diagnosis. With the development of methods for studying the relative rates of globin-chain synthesis *in vitro*, and for acquiring small fetal blood samples, it is now possible to accurately diagnose homozygous β thalassaemia and sickle cell anaemia *in utero* (Kan 1977; Alter & Nathan 1978). However, the obstetric techniques are hazardous and the laboratory technology is difficult except in the most experienced hands. Hence these techniques are not yet available for wide-scale use in those countries where they are most needed.

Because it has been observed that patients with sickle cell anaemia or homozygous β thalassaemia who also inherit genes for persistent fetal haemoglobin synthesis are much better off clinically than those who do not inherit the latter, it is clear that the ideal approach to the management of these disorders would be to prevent the neonatal switch from Hb F to Hb A production or to reactivate Hb F production after the neonatal period.

The developmental biology of the human haemoglobins

The factors which control the switch from fetal to adult haemoglobin production are completely unknown and this remains the one area of the human haemoglobin field where relatively little progress has been made to date.

Current information about the developmental biology of the human haemoglobins has been reviewed recently (Wood *et al.* 1977*c*). The pattern of the production of globin chains during human development is summarized in Fig. 8. Although β-chain synthesis is active at a low level throughout early fetal life it is not until about 34–36 weeks that its rate of production increases and this coincides with a decline in production of γ chains. The latter process continues steadily over the first few months after birth until at the end of the first year the small amount of Hb F which continues to be produced into adult life is confined to a few erythrocytes called F cells. Careful analysis of the rates of decline of Hb F in man, and experimental work in the developing sheep, leave little doubt that the switch from γ- to β-chain production is related to the gestational age and stage of development of the fetus. It is not related to the site of erythropoiesis however. It is not absolutely clear whether the switching process occurs uniformly throughout a single cell population or whether it occurs in distinct populations which are programmed to make fetal

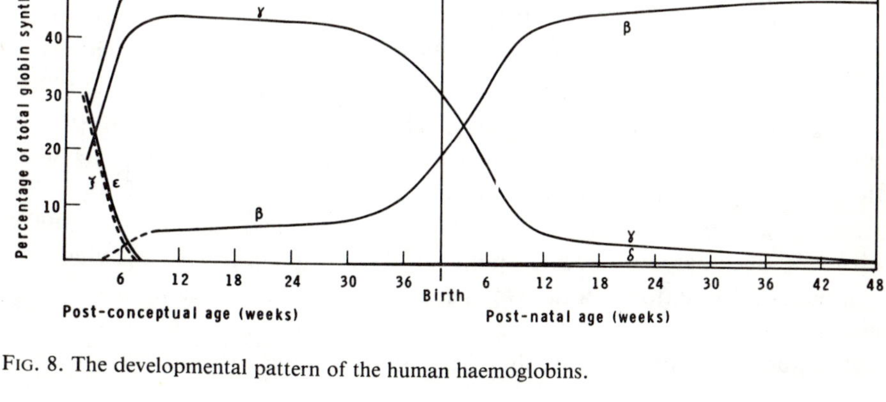

FIG. 8. The developmental pattern of the human haemoglobins.

and adult proteins as appropriate. Certainly with the rather crude methods currently available there is no evidence for the existence of two distinct populations containing Hbs A and F at birth.

The production of Hb F is only one property of the fetal erythrocyte; it has many other features which distinguish it from adult red cells including a low level of Hb A_2, reduced amounts of the carbonic anhydrase isozymes, distinctive differences in the red cell antigens, and a variety of differences in the levels of glycolytic and other enzymes.

From the data obtained from the study of patients with $\delta\beta$ thalassaemia and pancellular HPFH it is clear that deletion of critical areas of the γ-δ-β gene cluster result in persistent γ-chain synthesis in adult life. While this interesting observation tells us that key areas of this gene cluster may be involved in the neonatal suppression of γ-chain production it should be remembered that the cells of individuals with these disorders do not have other fetal properties and therefore the condition is quite different from that seen in the newborn infant. This goes for most of the conditions summarized in Table 7 in which Hb F synthesis occurs in adults. Indeed the only situation in which anything resembling fetal erythropoiesis is seen after birth is in certain rare forms of childhood leukaemia (Weatherall *et al.* 1975*b*). Here there is a genuine reversion to fetal protein pattern in the red cell and a better understanding of the molecular basis for this extraordinary change might provide us with some real understanding about the neonatal switch from fetal to adult erythropoiesis.

Several approaches are being made to the investigation of developmental biology of human haemoglobin. These include analysis of genetic disorders associated with persistence of Hb F production, the study of haemoglobin switching in the sheep and goat (Wood *et al.* 1976), both of which resemble man in the pattern of the change from fetal to adult haemoglobin production, and investigation of haemoglobin production in *in vitro* colony systems (Papayannopoulou *et al.* 1977, 1978). So far none of these systems have given us any real insight into the type of signal which is involved in the switching off of the γ-chain loci and the activation of the β-chain loci at a specific phase during human development. Although these experimental approaches promise to tell us something about the chromosomal mechanisms of activation and suppression of the globin genes during development, and about the stage of erythroid maturation at which genetic programming to make different haemoglobins occurs, none of them seem likely to provide any definitive information about why genes are switched on and off during maturation.

The questions posed in this final section are among the most exciting and challenging in human genetics. Unfortunately at the time of writing it is not clear precisely how the right questions can be asked to answer them.

References

Abramson, R. K., Rucknagel, D. L., Schreffler, D. C. & Saave, J. J. (1970) Homozygous Hb J Tongariki. Evidence for only one alpha chain structural locus in Melanesians. *Science (Wash. D.C.) 169,* 194-196

Allison, A. C. (1965) Population genetics of abnormal haemoglobins and glucose-6-phosphate dehydrogenase deficiency, in *Abnormal Haemoglobins in Africa* (Jonxis, J. H. P., ed.), pp. 365-391, Blackwell Scientific Publications, Oxford

Alter, B. P. & Nathan, D. G. (1978) Antenatal diagnosis of haematological disorders — 1978. *Clin. Haematol. 7,* 195-216

Beaven, G. H., Ellis, M. J. & White, J. C. (1961) Studies in human foetal haemoglobin. III. The hereditary haemoglobinopathies and thalassaemia. *Br. J. Haematol. 7,* 169-186

Boyer, S. H., Belding, T. K., Margolet, L. & Noyes, A. N. (1975) Fetal hemoglobin restriction to a few erythrocytes (F cells) in normal human adults. *Science (Wash. D.C.) 188,* 361-363

Boyer, S. H., Margolet, L., Boyer, M. L., Huisman, T. H. J., Schroeder, W. A., Wood, W. G., Weatherall, D. J., Clegg, J. B. & Cartner, R. (1977) Inheritance of F cell frequency in heterocellular hereditary persistence of fetal hemoglobin: an example of allelic exclusion. *Am. J. Hum. Genet. 29,* 256-271

Bunn, H. F., Forget, B. G. & Ranney, H. M. (1977) *Human Hemoglobins,* W. B. Saunders, Philadelphia

Clegg, J. B., Weatherall, D. J. & Milner, P. F. (1971) Haemoglobin Constant Spring—a chain termination mutant? *Nature (Lond.) 234,* 337-340

Conconi, F., Rowley, P. T., Del Senno, L. & Pontremoli, S. (1972) Induction of β-globin synthesis in the β-thalassaemia of Ferrara. *Nat. New Biol. 238,* 83-87

Deisseroth, A., Nienhuis, A., Lawrence, J., Giles, R., Turner, P. & Ruddle, F. H. (1978) Chromosomal localization of human β globin gene on human chromosome 11 in somatic cell hybrids. *Proc. Natl. Acad. Sci. U.S.A. 75,* 1456-1460

Flavell, R. A., Kooter, J. M., de Boer, E., Little, P. F. R. & Williamson, R. (1978) Analysis of the human δ- and β-globin gene loci in normal and Hb Lepore DNA: direct determination of gene linkage and intergene distance. *Cell 15,* 25-41

Forget, B. G. (1977) Nucleotide sequence of human β globin messenger RNA. *Hemoglobin 1,* 879-881

Forget, B. G., Hillman, D. G., Lazarus, H., Barell, E. F., Benz, E. J. Jr., Caskey, C. T., Huisman, T. H. J., Schroeder, W. A. & Housman, D. (1976) Absence of messenger RNA and gene DNA for β-globin chains in hereditary persistence of fetal hemoglobin. *Cell 7,* 323-329

Friedman, M. J. (1978) Erythrocytic mechanism of sickle cell resistance to malaria. *Proc. Natl. Acad. Sci. U.S.A. 75,* 1994-1997

Friedman, S. & Schwartz, H. (1976) Hereditary persistence of foetal haemoglobin with β chain synthesis in cis position ($^{G}\gamma$-β^{+}-HPFH) in a Negro family. *Nature (Lond.) 259,* 138-140

Harris, H. (1974) Genetic heterogeneity in inherited disease. *J. Clin. Pathol. (Lond.) 27, Suppl. 8,* 32-37

Higgs, D. R., Clegg, J. B., Wood, W. G. & Weatherall, D. J. (1979) $^{G}\gamma\beta^{+}$type of hereditary persistence of fetal haemoglobin in association with Hb C. *J. Med. Genet.* in press

Housman, D., Forget, B. G., Skoultchi, A. & Bentz, E. J. (1973) Quantitative deficiency of chain specific messenger ribonucleic acids in the thalassemia syndromes. *Proc. Natl. Acad. Sci. U.S.A. 70,* 1809-1813

Huisman, T. H. J., Schroeder, W. A., Bannister, W. H. & Grech, J. L. (1972*a*) Evidence for four non-allelic structural genes for the γ-chain of human fetal hemoglobin. *Biochem. Genet. 7,* 131-139

HUISMAN, T. H. J., WRIGHTSTONE, R. N., WILSON, J. B., SCHROEDER, W. A. & KENDALL, A. G. (1972*b*) Hemoglobin Kenya, the product of fusion γ and β polyeptide chains. *Arch. Biochem. Biophys. 153,* 850-853

HUISMAN, T. H. J., SCHROEDER, W. A., EFREMOV, G. D., DUMA, H., MLADENOVSKY, B., HYMAN, C. B., RACHMILEWITZ, E. A., BOUVER, N., MILLER, A., BRODIE, A., SHELTON, J. R., SHELTON, J. B. & APELL, G. (1974) The present status of the heterogeneity of fetal hemoglobin in β-thalassemia: an attempt to unify some observations in thalassemia and related conditions. *Ann. N.Y. Acad. Sci. 232,* 107-124

HUISMAN, T. H. J., MILLER, A. & SCHROEDER, W. A. (1975) A $^G\gamma$ type of the hereditary persistence of fetal hemoglobin with β-chain production in cis. *Am. J. Hum. Genet. 27,* 765-777

HUISMAN, T. H. J., HARRIS, H., GRAVELY, M., SCHROEDER, W. A., SHELTON, J. R., SHELTON, J. B. & EVANS, L. (1977) The chemical heterogeneity of the fetal hemoglobin in normal newborn infants and in adults. *Mol. Cell. Biochem. 17,* 45-55

JEFFREYS, A. J., & FLAVELL, R. A. (1977*a*) A physical map of the DNA regions flanking the rabbit β-globin gene. *Cell 12,* 429-439

JEFFREYS, A. J. & FLAVELL, R. A. (1977*b*) The rabbit β-globin gene contains a large insert in the coding sequence. *Cell 12,* 1097-1108

KACIAN, D. L., GAMBINO, R., DOW, L. W., GROSSBARD, E., NATTA, C., RAMIREZ, F., SPIEGELMAN, S., MARKS, P. A. & BANK, A. (1973) Decreased globin messenger RNA in thalassemia detected by molecular hybridization. *Proc. Natl. Acad. Sci. U.S.A. 70,* 1886-1890

KAN, Y. W. (1977) Prenatal diagnosis of hemoglobin disorders. *Prog. Hematol. 10,* 91-104

KAN, Y. W., DOZY, A. M., VARMUS, H. E., TAYLOR, J. M., HOLLAND, J. P., LIE-INJO, L. E., GANESON, J. & TODD, D. (1975*a*) Deletion of α-globin genes in haemoglobin-H disease demonstrates multiple α-globin structural loci. *Nature (Lond.) 255,* 255-256

KAN, Y. W., HOLLAND, J. P., DOZY, A. M., CHARACHE, S. & KAZAZIAN, H. H. (1975*b*) Deletion of the β-globin structure gene in hereditary persistence of foetal haemoglobin. *Nature (Lond.) 258,* 162-163

KAN, Y. W., HOLLAND, J. P., DOZY, A. M. & VARMUS, H. E. (1975*c*) Demonstration of nonfunctional β-globin mRNA in homozygous β-thalassemia. *Proc. Natl. Acad. Sci. U.S.A. 72,* 5140-5146

KAN, Y. W., DOZY, A. M., TRECARTIN, R. & TODD, D. (1977) Identification of a nondeletion defect in α-thalassemia. *N. Engl. J. Med. 297,* 1081-1084

LAWN, R.M., FRITSCH, E.F., PARKER, R.C., BLAKE, G. & MANIATIS, T. (1978) The isolation and characterization of linked δ- and β-globin genes from a cloned library of human DNA. *Cell 15,* 1157-1174

LEDER, P. (1978) Discontinuous genes. *N. Engl. J. Med. 298,* 1079-1081

LEHMANN, H. & CARRELL, R. W. (1968) Differences between α- and β-chain mutants of human haemoglobin and between α- and β-chain thalassemia. *Br. Med. J. 4,* 748-750

MEARS, J. G., RAMIREZ, F., LEIBOWITZ, D., NAKAMURA, F., BLOOM, A., KONOTEY-AHULU, F. & BANK, A. (1978*a*) Changes in restricted human cellular DNA fragments containing globin gene sequences in thalassemias and related disorders. *Proc. Natl. Acad. Sci. U.S.A. 75,* 1222-1226

MEARS, J. G., RAMIREZ, F., LEIBOWITZ, D. & BANK, A. (1978*b*) Organization of human δ and β globin genes in cellular DNA and the presence of intragenic inserts. *Cell 15,* 15-23

NIENHUIS, A. W., TURNER, P. & BENZ, E. J. Jr. (1977) Relative stability of α- and β-globin messenger RNAs in homozygous β^+ thalassemia. *Proc. Natl. Acad. Sci. U.S.A. 74,* 3960-3964

OLD, J., CLEGG, J. B., WEATHERALL, D. J., OTTOLENGHI, S., COMI, P., GIGLONI, B., MITCHELL, J., TOLSTOSHEV, P. & WILLIAMSON, R. (1976) A direct estimate of the number of human γ-globin genes. *Cell 8,* 13-18

OLD, J. M., CLEGG, J. B., WEATHERALL, D. J. & BOOTH, P. B. (1978*a*) Haemoglobin J Tongariki is associated with α thalassaemia. *Nature (Lond.) 273,* 319-320

OLD, J. M., PROUDFOOT, N. J., WOOD, W. G., LONGLEY, J. I., CLEGG, J. B. & WEATHERALL, D. J. (1978*b*) Characterization of β globin messenger RNA in the β° thalassaemias. *Cell 14,* 289-298

ORKIN, S. H., ALTER, B. P., ALTAY, C. & NATHAN, D. G. (1978) Visualization of human globin genes by endonuclease mapping: a new approach to the prenatal diagnosis of deletion syndromes. *N. Engl. J. Med. 299,* 166-172

OTTOLENGHI, S., LANYON, W.G., PAUL, J., WILLIAMSON, R., WEATHERALL, D.J., CLEGG, J. B., PRITCHARD, J., POOTRAKUL, S. & WONG HOCK BOON (1974) The severe form of a α thalassaemia is caused by a haemoglobin gene deletion. *Nature (Lond.) 251,* 389-392

OTTOLENGHI, S., COMI, P., GIGLIONI, B., TOLSTOSHEV, P., LANYON, W. G., MITCHELL, G. J., WILLIAMSON, R., RUSSO, G., MUSUMECI, S., SCHILIRO, G., TSISTRAKIS, G. A., CHARACHE, S., WOOD, W. G., CLEGG, J. B. & WEATHERALL, D. J. (1976) $\delta\beta$ thalassemia is due to gene deletion. *Cell 9,* 71-80

PAPAYANNOPOULOU, TH., BRICE, M. & STAMATOYANNOPOULOS, G. (1977) Hemoglobin F synthesis *in vitro:* evidence for control at the level of primitive erythroid stem cells. *Proc. Natl. Acad. Sci. U.S.A. 74,* 2923-2927

PAPAYANNOPOULOU, TH., NAKAMOTO, B., BUCKLEY, J., KURACHI, S., NUTE, P. E. & STAMATOYANNOPOULOS, G. (1978) Erythroid progenitors circulating in the blood of adult individuals produce fetal hemoglobin in culture. *Science (Wash. D.C.) 199,* 1349-1350

PASVOL, G., WEATHERALL, D. J. & WILSON, R. J. M. (1977) The effects of fetal haemoglobin susceptibility of human red cells to *P. falciparum. Nature (Lond.) 270,* 171-173

PASVOL, G., WEATHERALL, D. J. & WILSON, R. J. M. (1978) Cellular mechanism for the protective effect of haemoglobin S against *P. falciparum* malaria. *Nature (Lond.) 274,* 701-703

PEMBREY, M. E., WOOD, W. G., WEATHERALL, D. J. & PERRINE, R. D. (1978) Fetal haemoglobin production and the sickle gene in the oases of eastern Saudi Arabia. *Br. J. Haematol. 40,* 415-429

PROPPER, R. D., COOPER, B., RUFEO, R. R., NIENHUIS, A. W., ANDERSON, W. F., BUNN, H. F., ROSENTHAL, A. & NATHAN, D. G. (1977) Continuous subcutaneous administration of desferrioxamine in patients with iron-overload. *N. Engl. J. Med. 297,* 418-421

PROUDFOOT, N. J. & BROWNLEE, G. G. (1976) Nucleotide sequences of globin messenger RNA. *Br. Med. Bull. 32,* 251-256

RAMIREZ, F., O'DONNELL, J. V., MARKS, P. A., BANK, A., MUSUMECI, S., SCHILIRO, G., PIZZARELLI, G., RUSSO, G., LUPPIS, B. & GAMBINO, R. (1976*a*) Abnormal or absent β mRNA in $\beta°$ Ferrara and gene deletion in $\delta\beta$ thalassaemia. *Nature (Lond.) 263,* 471-475

RAMIREZ, F., O'DONNELL, J. V., NATTA, C. & BANK, A. (1976*b*) Quantitation of human gamma globin mRNA with purified gamma globin complementary DNA. *J. Clin. Invest. 58,* 1475-1481

RICCO, G., MAZZA, U., TURI, R. M., PICH, P. G., CAMASCHELLA, C., SAGLIO, G. & BERNINI, L. F. (1976) Significance of a new type of human fetal hemoglobin carrying a replacement isoleucine→threonine at position 75 (E19) of the γ chain. *Hum. Genet. 32,* 305-313

SCHROEDER, W. A., HUISMAN, T. H. J., SHELTON, R., SHELTON, J. B., KLEIHAUER, E. F., DOZY, A. M. & ROBBERSON, B. (1968) Evidence for multiple structural genes for the γ-chain of human fetal hemoglobin. *Proc. Natl. Acad. Sci. U.S.A. 60,* 537-544

SINISCALCO, M., BERNINI, L., FILIPPI, G., LATTE, B., MEERA KHAN, PIOMELLI, S. & RATTAZZI, M. (1966) Population genetics of haemoglobin variants, thalassaemia and glucose-6-phosphate dehydrogenase deficiency, with particular reference to the malaria hypothesis. *Bull. W.H.O. 34,* 379-393

TAYLOR, J. M., DOZY, A., KAN, Y. W., VARMUS, H. E., LIE-INJO, L. E., GANESON, J. & TODD, D. (1974) Genetic lesion in homozygous α thalassaemia (hydrops fetalis). *Nature (Lond.) 251,* 392-393

TEMPLE, G. F., CHANG, J. & KAN, Y. W. (1977) Authentic β globin mRNA sequences in homozygous $\beta°$ thalassemia. *Proc. Natl. Acad. Sci. U.S.A. 75,* 3047-3051

TILGHMAN, S. M., TENNIER, D. C., POLSKY, F., EDGELL, M. H., SEIDMAN, J. G., LEDER, A., ENQUIST, L. W., NORMAN, B. & LEDER, P. (1977) Cloning specific segments of the mammalian genome: bacteriophage λ containing mouse globin and surrounding gene sequences. *Proc. Natl. Acad. Sci. U.S.A. 74,* 4406-4410

TILGHMAN, S. M., TIEMEIER, D. C., SEIDMAN, J. G., PETERLIN, B.M., SULLIVAN, M., MAIZEL, S. V. & LEDER, P. (1978) Intervening sequence of DNA identified in the structural portion of a mouse β-globin gene. *Proc. Natl. Acad. Sci. U.S.A. 75,* 725-729

TOLSTOSHEV, P., MITCHELL, J., LANYON, G., WILLIAMSON, R., OTTOLENGHI, S., COMI, P., GIGLIONI, B., MASERA, G., MODELL, B., WEATHERALL, D. J. & CLEGG, J. B. (1976) Presence of gene for β globin in homozygous β_0 thalassaemia. *Nature (Lond.) 260,* 95-98

WEATHERALL, D. J. & CLEGG, J. B. (1972) *The Thalassaemia Syndromes,* 2nd edn., Blackwell Scientific Publications, Oxford

WEATHERALL, D. J. & CLEGG, J. B. (1975) The α chain termination mutants and their relationship to the α thalassaemias. *Philos. Trans. R. Soc. Lond. B Biol. Sci. 271,* 411-455

WEATHERALL, D. J. & CLEGG, J. B. (1976) Molecular genetics of human hemoglobin. *Annu. Rev. Genet. 10,* 157-178

WEATHERALL, D. J., CARTNER, R., CLEGG, J. B., WOOD, W. G., MACRAE, I. A. & MACKENZIE, A. (1975*a*) A form of hereditary persistence of fetal haemoglobin characterized by uneven cellular distribution of haemoglobin F and the production of haemoglobins A and A_2 in homozygotes. *Br. J. Haematol. 29,* 205-220

WEATHERALL, D. J., CLEGG, J. B., WOOD, W. G., CALLENDER, S. T., SHERIDAN, B. L. & PRITCHARD, J. (1975*b*) Foetal erythropoiesis in human leukaemia. *Nature (Lond.) 257,* 710-712

WEATHERALL, D. J., CLEGG, J. B. & WOOD, W. G. (1976) A model for the persistence of reactivation of fetal haemoglobin production. *Lancet 2,* 660-663

WEATHERALL, D. J., PIPPARD, M. J. & CALLENDER, S. T. (1977) Iron loading and thalassaemia - experimental successes and practical realities. *N. Engl. J. Med. 297,* 445-446

WEATHERALL, D. J., WOOD, W. G. & CLEGG, J. B. (1979) Genetics of fetal hemoglobin production in adult life, in *Cellular and Molecular Regulation of Hemoglobin Switching,* Symposia Specialists, Miami, in press

WINSLOW, R. M. & ANDERSON, W. F. (1978) The hemoglobinopathies, in *The Metabolic Basis of Inherited Disease* (Stanbury, J. B. *et al.*, eds.), pp. 1465-1507, McGraw Hill, New York

WOOD, W. G., STAMATOYANNOPOULOS, G., LIM, G. & NUTE, P. E. (1975) Normal values and levels in individuals with hereditary and acquired elevations of Hb F. *Blood 46,* 671-682

WOOD, W. G., PEARCE, K., CLEGG, J. B., WEATHERALL, D. J., ROBINSON, J. S., THORBURN, G. D. & DAWES, G. S. (1976) Switch from foetal to adult haemoglobin synthesis in normal and hypophysectomised sheep. *Nature (Lond.) 264,* 799-801

WOOD, W. G., CLEGG, J. B., WEATHERALL, D. J., GYDE, O. H. B., OBEID, D. A., TARLOW, M. J., BROWN, M. J. & HEWITT, S. (1977*a*) $^{G}\gamma$ $\delta\beta$ thalassaemia and $^{G}\gamma$ HPFH (Hb Kenya type). Comparison of two new cases. *J. Med. Genet 14,* 237-244

WOOD, W. G., WEATHERALL, D. J., CLEGG, J. B., HAMBLIN, T. J., EDWARDS, J. H. & BARLOW, A. M. (1977*b*) Heterocellular hereditary persistence of fetal haemoglobin (heterocellular HPFH) and its interaction with β thalassaemia. *Br. J. Haematol. 36,* 461-473

WOOD, W. G., CLEGG, J. B. & WEATHERALL, D. J. (1977*c*) Developmental biology of human hemoglobins, in *Progress in Hematology,* vol. 10 (Brown, E. B., ed.), pp. 43-90, Grune & Stratton, New York

WOOD, W. G., OLD, J. M., ROBERTS, A. V. S., CLEGG, J. B., WEATHERALL, D. J. & QUATTRIN, N. (1978) Human globin gene expression: control of β, δ and $\delta\beta$ chain production. *Cell 15,* 437-446

Discussion

Brenner: In the Lepore fusions, in which I assume that there are insertions in the same place, are the cross-overs outside the insertions?

Weatherall: There are three different haemoglobin Lepores which have resulted from unequal crossing-over at different areas between the δ and β chain genes. The cross-over regions are as follows: Hb Lepore Hollandia - δ 22 and β 50, Hb Lepore Baltimore - δ50 and β86, and Hb Lepore Boston - δ87 and β116. Because of the close similarity between the primary structures of the δ and β chains, this is as close as it is possible to be regarding the precise site of crossing-over. The large β insert in the mouse is at residues 104–105. There is a similar large insert in the rabbit β-globin genes somewhere between the coding sequence for amino acid residues 102–120. Thus it appears that the site of unequal crossing-over in Hb Lepore Boston is close to the insert site in the β- and presumably the δ-chain gene. It is also of great interest that Hb Lepore Boston is the only common form of Hb Lepore which has turned up in many different racial groups. The distribution of the Hb Lepores is summarized by Weatherall & Clegg (1972). This observation raises the interesting possibility that the presence of this insert may make it more likely for unequal crossing-over to occur in this region of the $\delta\beta$ genome.

Gardner: Are there any studies on lethally irradiated adult mouse systems? If you repopulate by, say, yolk sac cells, what is the pattern of the haemoglobin synthesis? Are there any problems in distinguishing between fetal and adult haemoglobin in the mouse?

Weatherall: One of the problems of using the mouse as a model for studying haemoglobin switching is that the 'fetal' haemoglobins are confined to embryonic erythropoiesis and as soon as erythropoiesis occurs in the liver adult haemoglobins are synthesized. In other words there is no intermediate haemoglobin like human haemoglobin F during differentiation and development in the mouse. This appears to be the case in the chick and probably the hamster although there is some recent evidence that the hamster produces a short-lived fetal haemoglobin. A further difficulty is that it is becoming clear that the primitive yolk sac cells of these animals do not produce only embryonic haemoglobin but that using sensitive immunological techniques it is possible to demonstrate small amounts of adult haemoglobin in this cell line. Furthermore some of the definitive line cells which are produced slightly later in maturation appear to contain small quantities of embryonic haemoglobin. Hence the haemoglobins produced in these early embryonic yolk sac cells and in the later liver cells are not confined to single molecular species. It would, however, be interesting to see what the products of yolk sac cells injected into a lethally irradiated adult mouse would be. Incidentally we are attempting to ask a similar question by bone marrow transplantation experiments in which we are infusing fetal erythropoietic tissue into newborn sheep after they have switched from fetal to adult haemoglobin production.

Gardner: So you presume that some of the yolk sac cells that were making embryonic haemoglobin switch immediately to the adult pattern.

Polani: If the mouse's marrow is repopulated with yolk sac cells the mouse dies, because the yolk sac cells are aggressive.

Weatherall: What do you mean by aggressive?

Polani: The yolk-sac transplanted mice runt very quickly.

Nossal: The experiment can be done now. Johnson & Metcalf (1977) at the Hall Institute have worked out techniques for cloning fetal liver stem cells in agar. These single cells make both red cells and white cells in a single clone. The trick is to use spleen cells stimulated with pokeweed mitogen to generate a supernatant which contains the right activator(s). It should be possible to develop this technology for human stem cells. Their over-all experience suggests that what works for the mouse can be adapted for man.

You definitely need to know, when you talk about switching in haemoglobin synthesis, whether you are talking about switching within a single clone, or about a shifting population genetics, with each clone being fixed. Cloning technologies should be able to work that out for you quite nicely.

Gurdon: You mentioned that in β thalassaemia the sequence of the transcript of the active gene is normal. What happens when you go upstream from that area? That is, assuming that there is some defect, some base change, somewhere in the DNA, is it possible to get some indication of how near that must be to that gene?

Weatherall: Not really. One is always frightened about talking about linkage data in human populations! One can only say that the β thalassaemia mutation always appears to be at or fairly close to the β structural locus. There have been no recorded cross-overs between any of the β thalassaemia genes and those for the β haemoglobin structural variants. However, this statement is based on something like 100 to 150 possibilities for crossing-over. Another approach to this problem is to look at the relationship between β and δ structural variants and the β thalassaemia genes. The whole problem of linkage disequilibrium between the β and δ structural variants and β thalassaemia has been discussed recently (Weatherall *et al.* 1976; Stamatoyannopoulos *et al.* 1977). One possible interpretation of the data derived from studies of this type is that the β thalassaemia locus is further from the β structural locus than is the δ structural locus. However, this information is based largely on apparent linkage disequilibrium between various loci and there are other possible explanations for this other than the distances between the various genes which have been analysed.

Bodmer: Can you set a limit to that recombination fraction by the number of individuals in the denominator?

Weatherall: This is the problem. We are talking of about 100 or 150 possibilities of crossing-over so it could be some distance away.

Gurdon: What does some distance mean in terms of base pairs? Is it an enormous number?

Weatherall: Yes.

Bodmer: One per cent is about 3 million.

Neel: The red cell has a life expectancy of 120 days so there will be a significant lag period between the switch from fetal to adult haemoglobin synthesis and the manifestation of the switch. Has anybody yet corrected the curve showing the decline in haemoglobin F production for red cell survival? The switch might be more abrupt than the curve you showed indicated.

Weatherall: This is an extremely complicated question. Not only does one have to consider the differential survival of fetal erythrocytes as compared with those produced after birth but one has to consider complications arising from the fact that erythropoiesis almost shuts down immediately after birth and doesn't take off again for several weeks. Hence one is examining the decline of haemoglobin F at a period when there is a marked change in the relative rates of erythropoiesis and probably in the survival of any erythrocytes which are being produced. Peggy Shepard and I tried to look at the problem of whether a new population of red cells was appearing by doing sequential acid elution analyses of the red cells of newborn infants, starting at birth and continuing up to the third month of life (Shepard *et al.* 1962). At birth most of the red cells stained for Hb F but a few adult 'ghost' cells were present. The population is heterogeneous, but at about two months after birth the population is clearly bimodal, with a new population of cells which contain very little haemoglobin F as judged by this rather insensitive technique. More recently Bard *et al.* (1970) analysed the synthesis of fetal haemoglobin from about 26 weeks of gestation until the 20th week of postnatal life. Surprisingly there is a fairly smooth curve of disappearance of γ-chain synthesis although some γ chains are being produced right up to about 30 weeks postnatally. I interpret this result as indicating that even the cell populations that are being synthesized after erythropoiesis takes off again are producing small but definite amounts of haemoglobin F. All these observations suggest that both Hb F and Hb A are being produced in the same cell before, during and after birth, and there is no evidence for the existence of clear-cut populations producing either fetal or adult haemoglobin. The findings do not, however, exclude the possibility that there are different populations producing mainly haemoglobin F or haemoglobin A and the whole question remains undecided. Certainly recent work on the newborn sheep suggests that there may be distinct populations of red cells, as judged by their size and potassium content (Valet *et al.* 1978).

Siniscalco: Is the Koya Dora elongated α chain present in everybody?

Weatherall: Haemoglobin Koya Dora was found in 20 out of 200 individuals from the Koya Dora tribe from Andhra Pradesh (De Jong *et al.* 1975). Haematological and genetic studies suggested that carriers of Hb Koya Dora have a mild α-thalassaemia-like disorder and that the genes for this variant can interact with α thalassaemia to produce Hb H disease in exactly the same way that Hb Constant Spring can interact with α thalassaemia to produce this condition.

Polani: Has anybody looked carefully at the chromosomes in juvenile myeloid leukaemia to see whether there is anything peculiar on banding and so on?

Weatherall: Yes; no abnormality has been found.

Polani: What happens in pregnancy when the fetal haemoglobin is switched on 'a little bit'?

Weatherall: We have recently re-examined the problem of the production of Hb F in normal pregnancy (Popat *et al.* 1977). We found, using a fluorescent anti-Hb F antibody technique, that the number of F cells was raised in all normal pregnancies at about mid-term. The number of F cells started to increase at about 16 weeks, reached a maximum between 20 and 24 weeks and started to decrease in the last trimester. The reason for this phenomenon is still far from clear. There has been much interest in whether it is related to the level of chorionic gonadotropin production which rises a little earlier in pregnancy than the F cells and then falls before the F cell level starts to fall. However, we have been unable to find any increase in Hb F and F-cell production in women with choriocarcinoma with very high levels of chorionic gonadotropin. Furthermore, the administration of large amounts of chorionic gonadotropin to monkeys failed to cause any increase in fetal haemoglobin production or in the number of F cells.

There is another possible explanation for the increase in F cells and Hb F production in mid-pregnancy. There is increasing evidence that under any state of sudden increase in erythropoiesis there may be a transient increase in F cell production. This is well documented in baboons with drug-induced haemolysis (De Simone *et al.* 1978), after recovery from bone marrow transplantation (Alter *et al.* 1976) and after bone marrow aplasia induced by anti-leukaemic agents (Sheridan *et al.* 1976). Pregnant women start to increase their red cell mass from about the 15th week of pregnancy onwards and it may well be that the increase in F cells which is observed at this time is simply a reflection of the 'on/off' of F cell production after erythropoietic stress of this type. If this is the case then the maternal increase in haemoglobin F synthesis tells us nothing about the regulation of fetal haemoglobin production.

Bodmer: Are the fetal cells that fluoresce with your antiserum also making normal adult haemoglobin?

Weatherall: Yes. In the F cells which are found in normal adults there is about 5 pg of Hb F out of a total cellular haemoglobin of about 30 pg. Interestingly, in conditions where there is a genetically determined increase in the number of F cells, the amount of haemoglobin F per F cell remains fairly constant at about 4 pg per cell. Thus in these conditions there appears to be a genetically determined increase in the number of F cells rather than an increase in the amount of Hb F per F cell.

Bodmer: How many complex variants, such as deletions and thalassaemia-type variants, are there compared with the single substitutions? All infants are subject to the selective sieve of survival but one gets the impression that the frequency of bizarre variants is going up and up. One wants to get some idea of the initial rate of production for these sorts of variants.

Weatherall: This is a very difficult problem. One's immediate guess is that since well over 200 haemoglobin variants with single amino acid substitutions have been described, these outnumber what you call the 'complex variants'. The worry is, however, that although the numbers of complex variants seem to be rapidly increasing one has a real problem of ascertainment bias since the majority of such variants as the chain elongation mutants, frame-shift mutations, deletion and insertion mutants and abnormal cross-over variants all tend to cause some clinical abnormality and therefore they are much more likely to be identified in a population. Furthermore it is becoming clear that at least some of these variants, such as the chain termination mutants, have been subject to the same selective advantages as the thalassaemias.

Bodmer: If you stick to the rare ones that are clearly not polymorphic, what do you think the ratio is?

Weatherall: I really don't believe that it is possible to say. Assuming that one is more likely to pick up the complex variants, and assuming that some of them are subject to selection then one would guess that the variants with single amino acid substitutions are very much more common.

Neel: There is a serious ascertainment bias here. The more extreme variants come to clinical attention more often than the minor ones.

Harry Harris: There is an indirect way of getting at this. Quite a lot of the rare haemoglobin variants that give unstable haemoglobin, so producing a clinical condition, have now been looked at. One can now say what proportion of those is due to single amino acid substitutions which alter the configuration of haemoglobin and make it unstable, and what proportion is due to frank amino acid deletions. At the last count about 20% were due to deletions. This seems to imply that these things in general are a commoner type of mutation than one had thought.

Weatherall: But again there is a real ascertainment bias.

Harry Harris: Yes, but the point is that the number of amino acids deleted in the unstable haemoglobins which have been identified is never much over five or six. If larger numbers of amino acids were deleted the variants might be so unstable that they would not come to be identified. This would tend to increase the proportion of deletions versus single amino acid substitutions in this class of mutant proteins.

Brenner: Miller in Geneva has been analysing the frequency and the nature of spontaneous mutations in a bacterial gene. Many of the localized deletions, and also the symmetric duplications seem to depend on the sequence homology. It is not certain whether these arise by anomalies of classical recombination.

Henry Harris: In the cases where you delete two genes and the other two over-produce, is the over-production measured on a cell population basis or within individual cells?

Weatherall: In general the best model for answering this question is the Negro form of hereditary persistence of fetal haemoglobin. In this there is a deletion of the δ and β chain loci with persistent activity of the ${}^{G}\gamma$ and ${}^{A}\gamma$ loci. This occurs in all the red cells and the haemoglobin F is relatively uniformly distributed among them. Compensation for the deficient δ and β chains is almost although not absolutely complete and there is an over-all mild degree of globin chain imbalance in homozygotes.

Henry Harris: There is a human G6PD-defect which is due to a single amino acid substitution and which results in the defective protein being produced at four times the normal concentration.

Lachmann: Is it clear that it is over-production and not hypocatabolism?

Henry Harris: It looks like over-production. The paper is by Yoshida (1970).

Weatherall: At the moment very little is known about the size of these deletions and how far they extend. It is clear, however, that if the δ and β loci are deleted, an area of the γ-δ-β gene complex which is normally involved in the 'shutting down' of the γ chain synthesis in adult life is also removed and this is presumably why affected individuals go on producing large amounts of γ chain throughout adult life. It is really an extremely well compensated form of $\delta\beta$ thalassaemia in which the output of γ chains almost entirely compensates for the absence of δ and β chains. The output of γ chains in absolute terms is almost equal to the normal output of β chains. There is slight chain imbalance in homozygotes for hereditary persistence of fetal haemoglobin. Indeed one can look at the Negro form of hereditary persistence of fetal haemoglobin at one end of a spectrum of thalassaemias; at the other end there are the β and $\delta\beta$ thalassaemias in which there is a severe deficiency of δ and β chains in which γ chain synthesis is inadequate to compensate for the deficiency of δ and β chain production. These people have normal haemoglobin levels.

Henry Harris: Do they sometimes over-produce?

Weatherall: The only example that I know of, of a human haemoglobin disorder in which there appears to be an excessive production of chains, is a very interesting Irish family which we studied in detail a few years ago (Weatherall *et al.* 1973). In this family there were three individuals who had a moderate degree of anaemia with ineffective erythropoiesis. Their red cells were relatively well filled but both in the bone marrow and in the peripheral blood after splenectomy there were large haemoglobin inclusions. We found that these individuals had excessive α chain over β chain production in their bone marrow and reticulocytes of a similar magnitude to that of β thalassaemia heterozygotes. Unlike the latter, however, their red cells were well filled. We suggested therefore that this disorder was due to a primary overproduction of α chains due to an increase in the number of α chain loci per haploid genome, perhaps due to abnormal crossing-over. However, in the last year we have obtained further information about this fascinating family. Dr John Old, in our laboratory, has measured the ratio of α/β messenger RNA in the reticulocytes of one of the family members and shown that this ratio is normal. In other words, we have excessive α chain production in the presence of a normal ratio of α/β messenger RNA. This strongly suggests that the reason for the excess of α-chain synthesis in this family is an increased rate of α-chain translation. We are currently attempting to study the translation of this messenger RNA in a cell-free system and plan to go on to try to sequence the 5′ end, assuming that the defect may well lie at the initiation codon, so leading to an increased rate of initiation of α chain. I was particularly interested in Dr Weissmann's remarks about the effects of artificially producing base substitutions in this region.

Pontecorvo: It is very interesting to note that the α gene doesn't have inserts while the other two do. The α gene is the most primitive one and though a single comparison is not much, it may give ideas for the future on the evolution of splicing, inserts and so on.

Could part of the excess of one or the other chains in your various cases be due to the final processing of the haemoglobin? In other words, are there changed polypeptide chains which do not combine in the proper way?

Weatherall: This kind of model exists with the unstable haemoglobin variants. Many of them are so unstable that they appear to fall apart in the red cell. The unstable α or β chains precipitate, leaving an excess of either α

or β partner subunits. It is quite possible that some of the variants are so unstable that they never form a viable tetramer and one is left simply with the normal subunit in excess.

Pontecorvo: So you could have an *in vitro* test for this sort of change, and use the site-directed mutagenesis for identifying the critical sequences?

Weatherall: Yes.

Milstein: We have been looking at structural mutants of immunoglobulin chains spontaneously occurring in tissue culture (Adetugbo *et al.* 1977). Of over 5000 clones studied, four mutants were investigated in detail; three involved partial heavy chain deletions, and all three included full domains. The immunoglobulin chains are made of pseudo-subunits which are folded independently of each other. The deletions always involve the whole of one of these subunits. That points strongly to the fact that the stability of the product of the mutant cell must play an important role. In one of the cases the deletion was internal to the heavy chain and digestion studies indicated that the Ig molecule was assembled in a different way from the wild type. The product of digestion with proteolytic enzymes indicated that the exposed regions were different from the wild-type molecule (Secher *et al.* 1977).

Bodmer: Couldn't it also be related to sequence, if there is unequal crossing-over? The ends of the domains could give the right configuration needed for unequal crossing-over.

Milstein: In one case (IF3) the deletion was clearly due to a frame shift leading to premature termination. In another the deletion is consistent with a point mutation resulting in a chain termination sequence. It is only the third case (IF2) where the deletion is internal and which could be due to something special. We don't think it is unequal crossing-over.

Ruddle: Does the γ-chain persistence mutation which you said might ameliorate sickle cell anaemia behave as a dominant or a recessive? Is there any suggestion that it might map into the γ-chain region?

Weatherall: This is a complicated field. There are really three varieties of hereditary persistence of increase in haemoglobin production in adults. I have already described the pancellular form of the condition, which is presumably due to deletion of the δ- and β-chain genes with the persistent γ-chain production in every cell. The second group is hereditary persistence of F cells or heterocellular hereditary persistence of fetal haemoglobin. A slight increase in F cells occurs in about 2% of North Europeans. If this condition is inherited together with the β thalassaemia or sickle cell gene it tends to ameliorate the condition.

In Saudi Arabia the situation is different again. Here we find that individuals who are homozygous for the sickle cell gene produce large amounts of Hb F but we now have good evidence that this form of persistent fetal haemoglobin production is different from pancellular or heterocellular hereditary persistence of fetal haemoglobin, at least as far as we can tell. In this population it looks as if there is either a genetic determinant which 'goes along' with the sickle cell gene and behaves like a recessive, or that for some reason the Saudi population increase their output of haemoglobin F when the rate of erythropoiesis is increased. However the latter explanation seems unlikely, and it does appear as though there may be a genetic factor related to the β^S gene which in some way allows the production of large amounts of haemoglobin F in sickle cell homozygotes.

In answer to your second question about the mapping of these various forms of hereditary persistence of fetal haemoglobin, it is clear that the pancellular form maps as an allele of the β structural locus. There is less information about the genetic location of the heterocellular hereditary persistence of fetal haemoglobin gene; what evidence there is suggests that it may be linked to the γ-δ-β gene complex, although at some distance, as some cross-overs have been described. However, there is a real difficulty in this type of genetic analysis because it is not absolutely certain that one can identify all carriers of pancellular hereditary persistence of fetal haemoglobin (Wood *et al.* 1976).

Ruddle: But this affects the differentiation of the haemopoietic cells—it operates on a cellular level.

Weatherall: Yes. This is very interesting. It appears that the action of the heterocellular hereditary persistence of fetal haemoglobin gene is to raise the number of F cells. As I mentioned earlier (p. 164), the amount of haemoglobin F per F cell remains normal. Furthermore there seems to be a genuine reciprocal relationship between the output of δ- and β-, and γ-chain genes in the cells of individuals with this condition. Those cells that produce most γ chains produce least β and δ chains, i.e. there is almost a 'seesawing' between the activities of γ and δ and β loci. This is seen generally in conditions where haemoglobin F is raised in adult life.

Ruddle: You said that perhaps a genetic engineering trick or intervention could be used to improve the condition of individuals with thalassaemia. Have you any specific ideas or speculations on how that might be done?

Weatherall: I really have no idea how one might do this. It is clear that one would only need to increase the F cell population by a relatively small amount to improve the clinical status of patients with β thalassaemia and sickle cell anaemia. This is clear from studies of patients with these disorders who have also inherited the gene for heterocellular hereditary persistence of fetal haemoglobin (Wood *et al.* 1977).

Brenner: Weren't you implying cellular engineering rather than genetic?

Weatherall: That's right.

Ruddle: But gene transfer is a possibility.

Weatherall: Yes, I suppose it is a possibility. The problem is that at the moment we know nothing about the cellular origins or control mechanisms involving the size of the F cell population. Recent studies using the methylcellulose erythroid colony technique have suggested that F cells come from the same stem cell pool as adult haemoglobin-containing cells (Papayannopoulou *et al.* 1977). Furthermore, studies of the distribution of F cells in clonal disorders such as polycythaemia vera are in keeping with this idea. However, apart from the fact that the numbers of F cells are genetically determined and that they are increased in acute erythropoietic stress, we know nothing about the factors which maintain their level in the peripheral blood.

Fincham: You described two kinds of β-thalassaemia cases where the $\beta\delta$ genes were deleted. In one kind there was full compensation and in the other much less than that. The difference between these, as I recall, was in the extent of the deletion. How well defined were the limits of the deletions? How was this affecting the output of the γ genes?

Weatherall: Data on the size of the deletions of the γ-δ-β gene cluster are limited. Several groups have shown by cDNA/DNA hybridization techniques that both the Negro form of hereditary persistence of fetal haemoglobin and $\delta\beta$ thalassaemia results from a gene deletion and that the size of the deletion in the former condition is larger. This suggests that the greater output of γ chains in hereditary persistence of fetal haemoglobin is due to loss of genetic material somewhere between the δ and $^{A}\gamma$ loci. At the moment we have no real idea about the size of these deletions but once it is possible to clone the appropriate pieces of DNA from these patients it should be possible to sequence these areas of the genome and to get an accurate assessment of the size of the deletions.

Brenner: I venture to predict that if it has not already happened, differential excision models will be produced to explain those results.

Siniscalco: Wasn't this confirmed by restriction analysis studies?

Weatherall: There are only very preliminary restriction map data on $\delta\beta$ thalassaemia (Mears *et al.* 1978; N.J. Proudfoot & J. Old, unpublished data). These studies simply confirm that there is probably a deletion of the β and δ genes but they don't tell us anything about the size of the deletion.

Neel: Is failure of transcription absolutely ruled out as an explanation for the loss of the δ and β chains?

Weatherall: No. The two families that have been looked at so far have both shown deletions but it is quite possible that there will be forms of $\delta\beta$ thalassaemia due to failure of transcription of the δ- and β-globin genes. All the β° thalassaemias that have been looked at so far have intact β-globin genes and are presumably due to a defect in transcription or processing of β messenger RNA. Certainly the genes are intact (Old *et al.* 1978).

References

ADETUGBO, K., MILSTEIN, C. & SECHER, D. S. (1977) Molecular analysis of spontaneous somatic mutants. *Nature (Lond.) 265,* 299-304

ALTER, B. P., RAPPEPORT, J. M., HUISMAN, T. H. J., SCHROEDER, W. A. & NATHAN, D. G. (1976) Fetal erythropoiesis following bone marrow transplantation. *Blood 48,* 843-854

BARD, H., MAKOWSKI, E. L., MESCHIE, G. & BATTAGLIA, F. C. (1970) The relative rates of synthesis of hemoglobins A and F in immature red cells of newborn infants. *Pediatrics 45,* 766-772

DE JONG, W. W., KHAN, P. M. & BERNINI, L. F. (1975) Hemoglobin Koya Dora; high frequency of a chain termination mutant. *Am. J. Hum. Genet. 27,* 81-90

DESIMONE, J., BIEL, S. I. & HELLER, P. (1978) Stimulation of fetal hemoglobin synthesis in baboons by hemolysis and hypoxia. *Proc. Natl. Acad. Sci. U.S.A. 75,* 2937-2940

JOHNSON, G. R. & METCALF, D. (1977) Pure and mixed erythroid colony formation in vitro stimulated by spleen conditioned medium with no detectable erythropoietin. *Proc. Natl. Acad. Sci. U.S.A. 74,* 3879

MEARS, J. G., RAMIREZ, F., LEIBOWITZ, D., NAKAMURA, F., BLOOM, A., KONOTEY-AHULU, F. & BANK, A. (1978) Changes in restricted human cellular DNA fragments containing globin gene sequences in thalassemias and related disorders. *Proc. Natl. Acad. Sci. U.S.A. 75,* 1222-1226

OLD, J. M., PROUDFOOT, N. J., WOOD, W. G., LONGLEY, J. I., CLEGG, J. B. & WEATHERALL, D. J. (1978) Characterization of β-globin mRNA in the β° thalassemias. *Cell 14,* 289-298

PAPAYANNOPOULOU, TH., BRICE, M. & STAMATOYANNOPOULOS, G. (1977) Hemoglobin F synthesis *in vitro*: evidence for control at the level of primitive erythroid stem cells. *Proc. Natl. Acad. Sci. U.S.A. 74,* 2923-2927

POPAT, N., WOOD, W. G., WEATHERALL, D. J. & TURNBULL, A. C. (1977) The pattern of maternal F-cell production during pregnancy. *Lancet 2,* 377-379

SECHER, D. S., MILSTEIN, C. & ADETUGBO, K. (1977) Somatic mutants and antibody diversity, in *Molecular Aspects of Antibody Genes (Immunol. Rev. 36)* (Möller, G., ed.), pp. 51-72, Munksgaard, Copenhagen

SHEPARD, M. K., WEATHERALL, D. J. & CONLEY, C. L. (1962) Semi-quantitative estimation of the distribution of fetal hemoglobin in red cell populations. *Bull. Hopkins Hosp. 110,* 293-310

SHERIDAN, B. L., WEATHERALL, D. J., CLEGG, J. B., PRITCHARD, J., WOOD, W. G., CALLENDER, S. T., DURRANT, I. J., MCWHIRTER, W. R., ALI, M., PARTRIDGE, J. W. & THOMPSON, E. N. (1976) The patterns of fetal haemoglobin production in leukaemia. *Br. J. Haematol. 32,* 487-506

STAMATOYANNOPOULOS, G., WEITKEMP, L. R., KOTSAKIS, P. & AKRICAKIS, A. (1977) The linkage relationships of the β and δ hemoglobin genes. *Hemoglobin 1,* 561-570

VALET, G., FRANZ, G. & HAUF, P. K. (1978) Different red cell populations in newborn, genetically low potassium sheep: relation to hemopoietic, immunologic and physiologic differentiation. *J. Cell. Physiol. 94,* 215-228

WEATHERALL, D. J. & CLEGG, J. B. (1972) *The Thalassaemia Syndromes,* 2nd edn., Blackwell Scientific Publications, Oxford

WEATHERALL, D. J., CLEGG, J. B., KNOX-MACAULAY, H. H. M., BUNCH, C., HOPKINS, C. R. & TEMPERLEY, I. J. (1973) A genetically determined disorder with features both of thalassaemia and congenital dyserythropoietic anaemia. *Br. J. Haematol. 24,* 679-702

WEATHERALL, D. J., CLEGG, J. B., MILNER, P. F., MARSH, G. W., BOLTON, F. G. & SERJEANT, G. R. (1976) Linkage relationships between β- and δ-structural loci and African forms of β thalassaemia. *J. Med. Genet. 13,* 20-26

WOOD, W. G., WEATHERALL, D. J. & CLEGG, J. B. (1976) Interaction of heterocellular hereditary persistence of foetal haemoglobin with β thalassaemia and sickle cell anaemia. *Nature (Lond.) 264,* 247-249

WOOD, W. G., WEATHERALL, D. J., CLEGG, J. B., HAMBLIN, T. J., EDWARDS, J. H. & BARLOW, A. M. (1977) Heterocellular hereditary persistence of fetal haemoglobin (heterocellular HPFH) and its interaction with β thalassaemia. *Br. J. Haematol. 36,* 461-473

YOSHIDA, A. (1970) Amino acid substitution (histidine to tyrosine) in a glucose-6-phosphate dehydrogenase variant (G6PD Hektoen) associated with over-production. *J. Mol. Biol. 52,* 483-490

Multilocus enzymes in man

HARRY HARRIS

Department of Human Genetics, University of Pennsylvania, Philadelphia

Abstract Two or more separate loci may be concerned in coding for enzyme proteins with the same or very similar catalytic properties. The phenomenon appears to be remarkably common and rough estimates suggest that perhaps 50% or more of structural loci coding for enzyme proteins are concerned with these so-called multilocus enzymes. The separate enzymes in such a multilocus set resemble each other not only in their enzyme properties, but also in subunit structures and in molecular sizes. However there are often remarkable differences in expression of the several loci of a set from tissue to tissue. Such sets of loci probably arose in evolution by gene duplications with subsequent divergence by point mutations. Tissue differentiation in expression suggests that, *pari passu* with the divergence of the structural genes, there was an evolution of genetic regulating systems controlling their tissue expression.

The genome of each individual contains a large number of so-called 'structural' genes, each of which determines the amino acid sequence of a polypeptide chain of a particular enzyme. Mutations at such loci involve some change in the base sequence of the DNA and will often result in a change in the amino acid sequence of the polypeptide, and hence in the structure and the properties of the enzyme. And during the last two decades a tremendous amount of work has been done on the nature of such enzyme variants and on their distribution among individual members of various human populations and also, of course, on natural populations of other animal species.

This work, much of which has involved the technique of electrophoresis, has embraced in various ways more than a hundred different enzymes. But as the work progressed, it became increasingly apparent that although many enzymes appear to be determined by single loci—as one had expected on the basis of the classical one gene–one enzyme hypothesis, as developed by Beadle, Tatum and others in the early 1940s—there are also quite a number of enzymes whose structures are coded by two or more different gene loci. To be more precise, there often appear to be two or more different loci which determine the structures of a set of enzyme proteins which, though structurally distinct, have the same or very similar catalytic activities. For lack of a better term, I will refer to such sets of apparently closely related enzymes determined by two or more loci as multilocus enzymes. In the present paper I propose to consider some of the main characteristics of these multilocus enzymes as they are now emerging. And I want to focus attention on some of the fundamental problems they pose in the future development of enzyme genetics.

The phenomenon as such is not new; its occurrence in the case of particular enzymes such as lactate dehydrogenase has been known for more than 15 years and some of these cases have been widely studied and extensively discussed (Markert 1968). However, it is only relatively recently, with the accumulation of data on a wide range of different enzymes, that some of its general features have begun to emerge. The phenomenon, of course, has its counterpart in non-enzyme proteins. The best known example is haemoglobin where it now appears that at least eight structural loci are involved.

The topic is of some general importance from several different points of view:

(1) It is telling us something about the organization of structural genes in the genome as a whole.

(2) Because of this, it is of considerable interest in relation to the problem of enzyme evolution and molecular evolution in general.

(3) It has a not unimportant bearing on the general issue of enzyme regulation and more specifically on the problems of tissue differentiation.

(4) It is also relevant when one is trying to sort out the relationship of phenotype to genotype in a variety of genetically determined diseases.

INCIDENCE

How often do such multilocus enzymes occur among enzymes as a whole? To get some idea of this, data on some 83 different enzymes have been reviewed. The enzymes were selected simply by the fact that each had been studied electrophoretically in man and there was some information about the number of loci involved. In fact, electrophoresis has proved to be a very powerful tool in elucidating this question. There are probably few or no other examples of enzymes which have not been studied by electrophoresis where it is definitely known how many loci are involved. The data are summarized in Table 1.

In 10 enzymes there is evidence for three loci and in 19, for two. Thus some 35% of these enzymes are coded by two or more loci and about 56% of the loci are involved in coding for such multilocus enzymes. I found this a somewhat surprising result because although I had been well aware of the phenomenon before, I had hardly appreciated how extensive it was.

In fact, Table 1 probably underestimates the frequency of the phenomenon. A rather conservative position had been taken in deciding whether multiple loci were involved in any particular case and only unequivocal examples were classified as multilocus enzymes. All the others were included in the single locus category and it is quite probable that at least some of these on further study will turn out to involve more than one locus.

There is no reason to think that because the series included only enzymes for which electrophoretic methods had been devised there was unwitting selection for those with multilocus causation. However, certain types of enzymes are undoubtedly represented disproportionately in the sample and one may well ask whether the phenomenon is more common in some types of enzymes than in others. One way of approaching this question is to classify the enzymes by catalytic function as set out by the International Enzyme Commission, as shown in Table 2. Multilocus enzymes occur in each of the enzyme classes where three or more enzymes had been looked at. The differences in incidence between the classes are not statistically significant.

TABLE 1

Incidence of multilocus enzymes (based on data given in Harris & Hopkinson 1976)

No. of loci	*No. of enzymes*		*Total loci*	
3	10	29 (35%)	30	68 (56%)
2	19		38	
1	54		54	
	83		122	

TABLE 2

Incidence of multilocus and single locus enzymes among enzymes of different types

	Multiple loci	*Single loci*	*Total*
Oxidoreductases	8	11	19
Transferases	9	11	20
Hydrolases	7	26	33
Lyases	4	3	7
Isomerases	1	2	3
Ligases	—	1	1
	29	54	83

The classification in Table 2 is simply based on type of catalytic activity. However it would be of interest to know whether the phenomenon occurs more commonly among the enzymes of particular metabolic pathways. This question is rather difficult to analyse in any systematic way with the present data, but it is perhaps worth noting that the phenomenon appears to be particularly common among the sequence of enzymes concerned in glycolysis (Table 3). Out of 11 enzymes, as many as nine involve two or more loci and there is evidence for a total of at least 25 loci coding for the enzymes of this pathway.

SUBUNIT STRUCTURE

So far, in identifying these multilocus enzymes we have simply used as our criterion the close similarity in catalytic activity of the enzyme protein products of the different loci. This presumably implies some similarity in the nature of the active sites of the isozymes. It is pertinent, however, to ask how closely sets of isozymes classified in this way resemble each other in their overall protein structures.

One basic feature of an enzyme protein is its subunit structure. Some enzymes are monomers, others dimers, others trimers, and so on. We can ask—how far do the enzyme protein products of loci which are involved in particular multilocus enzymes resemble each other in the number of polypeptide subunits which make up their structures?

Recently a study of the distribution of subunit numbers among the enzyme products of 100 different loci was carried out (Hopkinson *et al.* 1976). There

TABLE 3

Enzymes involved in glycolysis

Glycolysis	*No. of loci*
Hexokinase	3
Glucose phosphate isomerase	1
Phosphofructokinase	2 (or 3)
Aldolase	3
Triose phosphate isomerase	2
Glyceraldehyde-phosphate dehydrogenase	1
Phosphoglycerate kinase	2
Phosphoglyceromutase	2
Enolase	3
Pyruvate kinase	3
Lactate dehydrogenase	3
Total	25 (or 26)

appeared to be 28 monomers, 43 dimers, four trimers, 24 tetramers and one probable octamer. Among these there were 25 multilocus enzymes in which the subunit number had been determined on the enzyme products of at least two of the loci. In seven cases they were each monomers, in 10 cases they were each dimers, and in seven cases they were each tetramers. There was only one exception to the general rule that the subunit structures of the enzyme products are the same for each member of a set of loci coding for a multilocus enzyme. The exception is superoxide dismutase, which occurs in two forms, each coded by a separate locus: one is the so-called soluble or cytoplasmic form which is a dimer, the other is the mitochondrial form which is a tetramer.

We can also ask to what extent the polypeptide subunits coded by members of a set of loci determining a particular enzyme resemble each other in molecular size. Estimates of subunit size were obtained for two (or three) of the enzyme products coded by the separate loci determining 22 multilocus enzymes (Hopkinson *et al.* 1976). In all, there were 52 loci. The mean subunit size was 50 600, the standard deviation 21 800 and the range 16 000 to 116 000, values not significantly different from those obtained for the subunit sizes of 46 enzymes classified as 'single locus'.

Table 4 shows an analysis of variance for the subunit sizes of the products of the loci between and within multilocus enzymes. The mean square between is very much larger than the mean square within and the ratio is highly significant ($F = 43$, $P \ll 0.001$). This indicates that the enzyme products of loci determining a particular multilocus enzyme tend to resemble each other closely in subunit size. The overall correlation coefficient r is about 0.95.

Since the size of a polypeptide subunit depends on the number of amino acids it contains and this in turn is determined by the number of bases in the

TABLE 4

Analysis of variance of subunit sizes of products of loci between and within 22 multilocus enzymes

Source	*d.f.*	*Sum of squares* ($\times 10^6$)	*Mean square* ($\times 10^6$)
Between enzymes	21	23 433	1116
Within enzymes	30	772	26
Total	51	24 205	

$$F = \frac{\text{mean square between}}{\text{mean square within}} = 43$$

$P \ll 0.001$

coding region of the DNA in the corresponding gene, the result indicates that the length of the coding regions must be very similar for the several loci determining a particular multilocus enzyme.

It is interesting to note that in the soluble and mitochondrial forms of superoxide dismutase, which in this series was the one exception to the rule with regard to similarity in subunit number, the subunit sizes are very similar, 16 000 and 18 000, and lie at the extreme lower end of the over-all subunit size distribution.

TISSUE DIFFERENCES IN EXPRESSION

A striking feature of certain of these multilocus enzymes is the remarkable degree of variation which occurs in the expression of the different loci from tissue to tissue.

For example, in the well-known case of lactate dehydrogenase there are three loci determining the polypeptide subunits known as A (or M), B (or H) and C. A and B are both expressed in most tissues of the body. But there are marked differences in their relative expression from one tissue to another. In skeletal muscle and liver, A predominates over B in a ratio of about 10:1, whereas in heart and kidney, B predominates over A in about the same ratio. In contrast, C is found only in post-pubertal testis and sperm.

Marked tissue differences in locus expression are also found in alcohol dehydrogenase (between liver, intestine, lung and kidney); creatine kinase (between brain, skeletal muscle and heart muscle); pyruvate kinase (between skeletal muscle, liver and kidney); enolase (between muscle, liver and brain); aldolase (between skeletal muscle, liver and brain); phosphoglyceromutase (between brain, skeletal muscle and heart muscle); alkaline phosphatase (between placenta, intestine, liver, bone and kidney); amylase (between salivary gland and pancreas). There are also in many of these cases other tissues where none of the loci are expressed to a detectable degree. For further details and references to these and other examples see Harris & Hopkinson (1976) and Edwards & Hopkinson (1977).

Although in general the different isozymes which make up these various multilocus sets are very similar to one another in their catalytic functions, differences in their kinetics, their inhibition characteristics and other properties such as stability have been noted in quite a number of cases. It is difficult, in viewing the striking tissue differentiations which occur, not to conclude that the detailed enzymic properties of the different isozymes of the set have been tailored in the course of evolution so that they are appropriate for the specific metabolic roles they subserve in the particular intracellular environment of the tissues in which they are found. However, in most cases the exact nature of such presumed functional differences has not been clearly defined and this is a major task for the future.

In one important group of multilocus enzymes marked tissue differences are not usually seen. The products of both loci are generally present in virtually all tissues. However, within the cells of a particular tissue, they are sharply compartmentalized. These are the enzymes which occur in so-called soluble and mitochondrial forms, the mitochondrial form being localized in the mitochondrion and the soluble form in the cytosol. Quite a number of these soluble and mitochondrial enzyme pairs have now been studied. They include oxidoreductases such as malate dehydrogenase, transferases such as glutamate-oxalate transaminase and lyases such as aconitase.

In general, the mitochondrial form appears to be coded by a chromosomal gene and not in mitochondrial DNA. So presumably both the soluble and mitochondrial polypeptides are synthesized on ribosomes in the same cell, and somehow the mitochondrial enzyme finds its way into the mitochondrion, whereas the soluble form is kept in the cytosol. How this comes about is not known. Presumably there must be some receptor or recognition sites in the mitochondrial membrane and in the enzyme itself. This is another problem for the future.

Possibly related to this is the curious finding that heteromeric isozymes containing both soluble and mitochondrial type polypeptides are never seen even when, as is often the case, the pairs of soluble and mitochondrial enzymes are multimeric. This absence of heteromeric isozymes containing the polypeptide products of the two loci is in striking contrast to the situation observed with other multilocus enzymes in which the products of the different loci are formed in the same cell but are retained in the cytosol.

CHROMOSOMAL ASSIGNMENTS

A question of some importance is whether the separate loci in a multilocus enzyme system are located together on the same chromosome or are on different chromosomes, and whether if they are on the same chromosome they are closely linked. The great advances in chromosomal assignment which have been made in the last few years, largely as a result of the technique of interspecific somatic cell hybridization, make it possible to begin to answer this question.

Table 5 lists the main results so far obtained for multilocus enzymes where at least two of the loci have been chromosomally assigned (McKusick & Ruddle 1977; Human Gene Mapping *4* 1978). Out of the 11 cases, assignment

TABLE 5

Human chromosomal assignments for loci of some multilocus enzymes (McKusick & Ruddle 1977; Human Gene Mapping *4* 1978).

Multilocus enzymes	*Chromosome assignments*
Phosphoglucomutase	1, 4, 6
Amylase	1, 1
Hexosaminidase	5, 15
Lactate dehydrogenase	11, 12
Enolase	1, 12
Malate dehydrogenase	2, 7
Isocitrate hydrogenase	2, 15
Glutamate-oxalate transaminase	10, 6
Superoxide dismutase	21, 6
Aconitase	9, 22
Adenylate kinase	9, 1, 9

of members of the set to the same chromosome occurs only twice. In one of these cases, amylase on chromosome 1, the two loci are known to be closely linked. In the other case, AK_1 and AK_3 on chromosome 9, they are possibly not closely linked. Also the third AK locus is on a different chromosome, No. 1.

GENE DUPLICATION

How can we account for the apparently widespread occurrence of these multilocus enzymes? The most plausible explanation in evolutionary terms is that in each case the two or three genes involved are descended from a common ancestral gene which at some point in the course of evolution was duplicated. A later duplication of one or other of the first pair would account for the occurrence of three loci, and so on. The close similarity in enzyme function and structure exhibited by the isozymes derived from these sets of loci appears to provide strong evidence in favour of this hypothesis. That the products of the different loci now differ in some degree in their primary structures and catalytic properties can be readily accounted for in terms of subsequent point mutations and consequent divergence of the products of the original duplicate genes.

The alternative hypothesis is that they are the consequence of convergent evolution from two or more quite unrelated ancestral genes. This hardly seems plausible as a general explanation, though of course it possibly accounts for occasional cases.

The phenomenon of gene duplication has long been recognized in genetics. The first case was the Bar locus in *Drosophila* described in the 1920s. And in the 1930s and 1940s the potential significance of gene duplication in evolution was clearly recognized. It was important because it provided a simple explanation for the origin of new genes and new functions by resolving the dilemma that if evolutionary change was a consequence of gene mutation, then emergence in a gene of a new function would be associated with the loss of its old function and this hardly seemed to be a satisfactory way of accounting for the evolutionary changes which appeared to have occurred. However, gene duplication followed by mutation would allow the emergence of a new function from one of the duplicated products, with the retention of the old function by the other.

This concept was developed well before there were clear ideas about the relationships of genes to the structure of proteins, and certainly before their molecular bases had been elucidated. However in the 1960s, as structural and genetic work on a variety of proteins over a range of different species progressed, the results were found to fit neatly into this older concept. The most striking example perhaps is haemoglobin, where it emerged that several different loci coding for the structurally distinct polypeptide chains, alpha, beta, gamma, delta, and so on, are involved. And the primary sequences of the chains showed obvious homologies which clearly implied that the various genes had evolved from a common ancestral form by a series of successive gene duplications. Thus there are strong precedents for thinking that the sets of gene loci involved in determining these multilocus enzymes are in each case derived from a common ancestral gene by a process involving gene duplication.

We should note, of course, that the fact that we recognize these multilocus enzymes at all must on this view imply that the degree of divergence by mutation subsequent to the duplication event has still left the enzyme products of the different loci closely similar in enzyme function. The A, B and C forms of lactate dehydrogenase are still recognizably lactate dehydrogenases.

It is obviously possible that certain of the so-called single locus enzymes may have a similarly common origin, but that the process of mutational divergence has progressed so far that we no longer recognize the final products as related to one another because their enzyme functions have become so different. Ultimately, when amino acid sequences and three-dimensional structures of many different enzymes are worked out, we may see a whole hierarchy of interrelationships which are at present quite obscured.

Accepting that multilocus enzymes originated from gene duplications, we can ask how such duplications actually arose.

It is usually thought that most gene duplications are so-called tandem duplications which arise from random breaks at different, though not very distant, sites in two homologous chromosomes or sister-chromatids, followed by aberrant reunions which result in exchange of material between the two original chromosomes or chromatids. According to the position of the breakage sites, the duplicated genes may be very close together on the chromosome or separated by a number of other duplicated genes.

For the chromosome with the duplicated region eventually to become fixed in the species, it is necessary that it must first be transmitted to the offspring by the individual in whom it first occurred, and then progressively spread through the population and the species. For many generations it would mainly occur in individuals who are heterozygous for the chromosome carrying the duplication and for another chromosome of the original type. Such individuals would thus be trisomic for the duplicated piece of chromosome. And it would therefore seem necessary that the duplicated area should not be so large that it produces grossly deleterious phenotypic effects, since then the chromosome would simply be eliminated by natural selection.

Once such a primary duplication has occurred and the chromosome with the duplicated region has spread through the population, then further secondary tandem duplications can occur as a result of what has been called homologous, but unequal, crossing-over.

If the sets of loci involved in our series of multiple locus enzymes have indeed originated in tandem duplications of this sort, then one might expect the loci in each case to occur together on the same chromosomes and, usually, to be closely linked. In fact, as we have seen, what data there are bearing on this question suggest that this is not usually the case. Out of the 15 pairs of such loci so far assigned in man, only two appear to be on the same chromosome, and only one of the pairs appears to be closely linked.

So if tandem gene duplication was in fact the process by which these multilocus enzymes originated, it must be postulated that in most cases the pair of loci were subsequently separated by other chromosomal breakage events, resulting in the exchange of segments between non-homologous chromosomes—so-called translocations.

However it is also possible that the primary gene duplication itself was not a tandem duplication but involved breaks and aberrant reunions of two different chromosomes and resulted in the placement of the two loci on separate chromosomes. Indeed, a whole variety of different types of chromosomal aberrations resulting from breaks and aberrant reunions between non-homologous chromosomes have been described which could bring about this effect.

Of course the vast majority of such translocations which result in gene duplication are unlikely to survive and become common or fixed throughout the species, simply because of the adverse phenotypic effects they produce or because of disturbances in meiosis. So it is presumably only the very rare example of such a phenomenon which could have survived and these examples would presumably be cases where the duplicated chromosomal region was very small. The extensive studies in cytogenetics in human populations which have been carried out in recent years make it clear that such aberrations occur with a significant frequency in natural populations. And even if only a minute fraction of them were to be viable and to give rise to the gene duplication phenomenon, it is perhaps not implausible that viewed against the extensive time-scale of evolution, this could itself account for the apparently high incidence of duplicated loci which we observe.

However, there is a quite different hypothesis which could account for the extensive degree of gene duplication which we appear to observe. Ohno (1970) has argued that instead of this resulting from the piecemeal accumulation of single gene duplications during evolution, much of the phenomenon may have occurred in one fell swoop by tetraploidization. He suggests that at an early stage of vertebrate evolution, possibly at the time when our ancestors were fish or amphibia, there was a complete doubling of the chromosome complement, so that in effect each gene locus was duplicated and the two members of the pair came to be present on separate chromosomes. Such tetraploidization, he points out, could hardly have occurred later than our fish or amphibian ancestors because here the chromosome sex determination mechanism was still at a quite primitive stage, while later when fully developed it would have probably precluded tetraploidization.

This hypothesis has, of course, the great attraction of simplicity. It immediately accounts for the rather high proportion of multilocus enzymes which we appear to observe and also for the finding that, in the majority of cases, the loci are on different chromosomes rather than closely linked on the same chromosome.

The hypothesis does not preclude the possibility that in the course of evolution subsequent to tetraploidization, single gene duplications of the standard sort could have occurred, and this would account for the multilocus enzymes where three loci appear to be involved. For haemoglobin, where there are even more loci, it is known that the γ, δ and β loci are closely linked on one chromosome and that there are two α loci probably closely linked on another. The suggestion is that each of these gene clusters arose by tandem duplication but that the progenitors of the two clusters may have been produced and separated on different chromosomes by the original tetraploidization. The matter is further complicated by the evidence that the gene coding for myoglobin also has a common ancestor with the various haemoglobin genes. However, its chromosomal location is not yet known.

The tetraploidization hypothesis also needs to account in some way for the various single locus enzymes which appear to occur. This could be done perhaps either by suggesting that in some cases the second form has not yet been identified, and, in other cases, that the second form, though originally present after the tetraploidization event, subsequently lost its functional activity as a result of a so-called null mutation.

I think it is hardly possible with the data at present available to adjudicate between these various hypotheses of duplication. However, in the future the matter may very well become susceptible to more critical analysis.

Finally in this evolutionary context one should discuss, if only very briefly, what is perhaps the most remarkable feature of many of these multilocus enzymes. This is the high degree of tissue differentiation in the expression of the two or more loci involved in many of the cases. In some tissues one of the loci may be expressed, in other tissues another, and in yet other tissues, both may be expressed. And when both loci are expressed in a variety of tissues, as in the case of lactate dehydrogenase, the relative degree of their expression may vary in a quite characteristic manner from tissue to tissue.

These characteristic patterns of tissue expression must imply that during the course of evolution not only has there been some divergence in the structures and properties of the enzymic products, presumably consequent on point mutations and natural selection, but there has also been an evolution of the genetic elements which regulate the expression of the loci in various tissues.

Unfortunately we have as yet no satisfactory general theory of genetic regulation with a basis in firm experimental observations which would be the counterpart of the general theory we do possess—that relating gene structure and mutation to enzyme protein structure and function. Consequently, with regard to these regulatory phenomena we are still at a purely descriptive stage. And the impasse is only likely to be broken when we can see how to devise experimental genetic systems which will enable us to begin a serious analysis at a more basic level. It is possible that certain of these multilocus enzymes may provide the appropriate tools to work with.

CONCLUSIONS

In summary, I have suggested:

(1) That multilocus enzymes, far from being unusual, are a very common phenomenon, and perhaps 50% or more of all structural genes coding for enzymes are involved in their determination.

(2) That the enzyme protein products of the members of such a set of loci, besides being very similar, though usually not identical in their enzymic properties, also resemble each other quite closely in their over-all structures, in particular, in their subunit compositions and in the molecular sizes of the subunits.

(3) That usually, though not invariably, the different loci are located on different chromosomes.

(4) That the gene loci involved in any particular case have generally been derived from a common ancestral form by gene duplication. Though whether they have all been generated separately in different duplication events, or whether much of the duplication occurred in one fell swoop by tetraploidization, is still an open question.

(5) That in many cases there is a marked degree of tissue differentiation in the expression of the different loci in a set. This implies that there has been an evolution of specific genetic regulatory elements *pari passu* with the evolution of the structural genes themselves. The nature of these regulatory elements is however still quite obscure.

ACKNOWLEDGEMENT

This work was supported in part by a grant CA-20296 from the National Cancer Institute, Department of Health, Education and Welfare.

REFERENCES

EDWARDS, Y. H. & HOPKINSON, D. A. (1977) Developmental changes in the electrophoretic patterns of human enzymes and other proteins, in *Isozymes: Current Topics in Biological and Medical Research*, vol. 1 (Rattazzi, M. C. *et al.*, eds.), pp. 19-78, Alan R. Liss, New York

HARRIS, H. & HOPKINSON, D. A. (1976) *Handbook of Enzyme Electrophoresis in Human Genetics*, North-Holland, Amsterdam

HOPKINSON, D. A., EDWARDS, Y. H. & HARRIS, H. (1976) The distributions of subunit numbers and subunit sizes of enzymes: a study of the products of 100 gene loci. *Ann. Hum. Genet. 39*, 383-411

HUMAN GENE MAPPING 4 (1978) 4th International Workshop 1977. Birth Defects, Original Article Series 14, 4. National Foundation, N.Y.

MARKERT, C. L. (1968) The molecular basis for isozymes. *Ann. N.Y. Acad. Sci. 151*, 14-40

MCKUSICK, V. A. & RUDDLE, F. H. (1977) The status of the gene map of the human chromosomes. *Science (Wash. D.C.) 196*, 390-405

OHNO, S. (1970) *Evolution by Gene Duplication*, Springer, Heidelberg

Discussion

Henry Harris: Are the Michaelis constants for a group of enzymes within a multilocus system all much the same?

Harry Harris: They vary. Usually they seem very close; occasionally there are quite significant differences, e.g. the phosphoglucomutases.

Henry Harris: In how many cases could one plausibly think that maybe one hasn't got the right substrate?

Harry Harris: Some perhaps, but probably only a minority.

Henry Harris: Where there are two enzymes, say on different chromosomes, that happen to cope with a particular substrate, but have very different K_m values for it, is it possible that although you give them the same name, they may have quite different natural substrates?

Harry Harris: Yes, this could obviously be a source of difficulty but I don't think it can alter the over-all picture.

Henry Harris: Is there enough sequence information to rule this out for any particular case?

Harry Harris: No.

Fincham: I take it that in cases where the two enzymes formed a hybrid or family of hybrids, that would be good evidence for fairly close homology between them.

Harry Harris: That is another important point.

Milstein: But the activation by glucose-1,6-diphosphate strongly implies that both enzymes are phosphoglucomutases.

Harry Harris: There are situations where one can't see any catalytic difference, yet the enzyme products of the separate loci have localized in different tissues. In other cases, differences in K_m or other properties such as thermostability have been detected, but it is generally quite unclear how these are related to the particular tissues in which the enzymes are expressed.

Henry Harris: You are probably right. Are the inactivation temperatures of the different variants widely different?

Harry Harris: There are usually no very large differences.

Ruddle: We think about the multiplication of genes and their distribution within genomes as being due to tetraploidization and translocation. But perhaps this could also be accomplished by excision and reinsertion of genes or very small parts of the chromosomes, and we should start looking for evidence to support that view. By looking at the nucleotide sequence composition in the flanking regions we should be able to get some insight into that possibility.

Harry Harris: Yes, indeed.

Brenner: Hence my short dissertation earlier (p. 144) on the intermediate repetitive DNA.

Gurdon: Some kinds of genes apparently make very small numbers of transcripts, such that a transcript is being made only once every five or six hours. The genes that show this behaviour might be ones that are not regulated but which always make transcripts at this very low rate. Have you worked out, from the stability and abundance of the enzyme (which may not be directly related to its enzyme activity), what rate of transcription you would expect for these genes in the cell types concerned?

Harry Harris: No; we are nowhere near being able to do that since the methods are still fairly crude. However, the large differences that I am pointing to rather exclude that possibility in most cases.

Gurdon: What percentage of the total protein do these enzymes constitute?

Harry Harris: We don't know, but it may vary a lot from enzyme to enzyme.

Peto: How thoroughly has Ohno's hypothesis been investigated? For example, if lactic dehydrogenase is multilocus in one species, is it multilocus in all species of mammals? If that hypothesis is true, it sounds almost as though the changes with the fish and the amphibians are too recent.

Harry Harris: The hypothesis hasn't been investigated in great depth, but I think Ohno's ideas and speculations usually deserve to be taken seriously. In general, if one of several loci coding for a particular enzyme is expressed in a particular tissue such as muscle or brain, that seems to apply in all vertebrates.

Peto: Is this before mammals split off?

Harry Harris: There are certainly three lactate dehydrogenase loci in fish. A very interesting case here supports Ohno's idea. In certain fish—salmon and trout, I believe—there is strong evidence for tetraploidization from enzyme and other studies. Here, instead of finding one A and one B locus for lactate dehydrogenase, two As and two Bs can be identified. Unfortunately, there is only one C, which creates a bit of a problem.

Peto: So the changes ought to be even further back than the fish?

Harry Harris: Yes.

Bodmer: There are surely a lot of cases where tetraploidization is not true. In the immunoglobulins, for example, there are two unlinked light-chain loci and the heavy-chain loci, and then there is the relation between β_2-microglobulin and immunoglobulins. So there are families of related things which don't fall into twos. One has to accept the much more likely possibility that duplications of substantial parts of chromosomes evolved at a time when chromosome imbalance was much less damaging to the organism, quite apart from translocations and insertions. To focus on tetraploidization as the main argument is in my view almost certainly going to be wrong.

Harry Harris: I am inclined to agree. I am not sold on tetraploidization as the main explanation, but I don't see that it can at present be dismissed out of hand.

Peto: At what sort of time in the distant past do you mean that that wasn't harmful? Are you talking about a single cellular stage of primitive life, or what?

Bodmer: Plants, for example, can tolerate a much wider range of chromosome variation than mammals. Mammals have come a long way along the evolutionary path.

Milstein: To what extent should we take the necessity of putting duplicated genes on different chromosomes as a dramatic event? Isn't it happening all the time? Perhaps the multilocus enzymes are a specific case of a more general situation in which large gene families are found, in some cases very large indeed. Sometimes they are linked, sometimes they are unlinked, depending on the function. The histones for instance are linked but in some cases (like 5S RNA) they are not all in the same chromosome. They occur in bits of families distributed all over the place. Movement from one chromosome to another is perhaps something which the animal can deal with in a simple way rather than in a difficult way.

Harry Harris: You might be right.

Henry Harris: I am prepared to offer you an evolutionary argument that it is a selective advantage, or another argument that it is a selective disadvantage. They would both be perfectly plausible arguments and you could take your pick.

Milstein: But both may come into the evolutionary events. In some cases one thing happens and in other cases something else happens. Both arguments have been used.

Siniscalco: Sister chromatid exchange might perhaps lead to inheritable gene duplication if it takes place in early pregonial stem cells. For a duplication to occur, the exchange between sister chromatids should of course be slightly inaccurate, like the non-homologous crossing-over that is postulated to occur during meiosis. At any rate, SCE might be held responsible for gene duplication at least in somatic cell lineages (e.g. immunoglobulin genes).

Crick: The general point is, how fast in evolution do things get moved around? Obviously there may be different sorts of mechanisms. If we look at later vertebrate evolution, we should eventually be able to get enough information to say how fast they move. There are complications. For example, we don't know, as is implied for the haemoglobin genes, whether they are kept together for some functional reason. We don't know whether selection keeps them together. Presumably in the future that will be illuminated as we get to know more about the controls and whether these related genes are together because they haven't been separated or whether they are together for some particular reason. Once we have some idea of the rate at which things are moving around, which I don't think we have at the moment, and once we know something about the mechanisms, we can begin to address the question.

Tandem duplication in big gene families does not affect the issue of whether there was early tetraploidization. That is a quite separate issue. If the rate of moving around and taking things in and out was rather fast, then even if tetraploidization occurred it would be lost. But if it isn't too fast, one would expect after tetraploidization to find more genes of different sorts together and unseparated than one would expect otherwise. Until we have a lot more taxonomic data we are not going to get very far.

Brenner: Could you comment on the conservation of linkage amongst mammalian species?

Ruddle: Lalley *et al.* (1978) have recently claimed that tightly linked genes in one species are more likely to reside on a single chromosome than in another species, and vice versa. However, the number of genes which are clearly homologous is small and the associations have not yet been subjected to statistical analysis.

Crick: We are discussing the separation of genes which are very close together. Things which are a reasonable distance apart and are in large chunks are not relevant to the argument of whether one can separate tandem duplications.

Bodmer: From chromosome-banding studies in different species one can get gross estimates of the rate at which break-points have been substituted during evolution. One has to allow for the things that one can't see, but the number is relatively small—about 6 to 12 per chromosome arm—compared to the size of the gene cluster. The whole question of what keeps gene clusters together and what puts them apart is a matter of selective advantages and disadvantages.

Milstein: It is sometimes very difficult to know whether a gene contains a duplicated form or not. The immunoglobulin Cλ gene in humans is a good case in point. We know that there must be at least three genes linked together in the genome, but they differ from each other by only one amino acid. This was easy to spot only because of the intensive screening of myeloma and Bence-Jones protein variants.

Crick: Nobody is arguing that tandem duplications don't occur. The point is that it is not easy to separate them. If Walter is going to discuss that in his paper we should perhaps leave it there.

Polani: It is nice to look at a chromosome such as the X chromosome that suffers from isolation and see what happens. On very circumstantial evidence, it looks as if there are two loci for many things in some sort of tandem on this chromosome—for example, colour blindness, muscular dystrophy and clotting.

Brenner: All these isozymes or similar proteins, when thoroughly investigated, always have physiological significance, as Henry was hinting. Lactic dehydrogenases are really two different enzymes that just happen to have the same substrate.

Reference

LALLEY, P. A., FRANCKE, U. & MINNA, J. D. (1978) Homologous genes for enolase, phosphogluconate dehydrogenase, phosphoglucomutase, and adenylate kinase are syntenic on mouse chromosome 4 and human chromosome lp. *Proc. Natl. Acad. Sci. U.S.A. 75,* 2382-2386

Gene clusters and the HLA system

WALTER F. BODMER

Genetics Laboratory, Department of Biochemistry, University of Oxford

Abstract The major human histocompatibility, or HLA system, is now known to include a number of closely linked loci controlling cell surface specificities, components of the complement system and immune response. Three loci, *HLA-A, B* and *C,* control serological determinants found on nearly all cells; a fourth locus, *HLA-D,* codes for determinants controlling the mixed lymphocyte culture reaction that are present only on B lymphocytes and monocytes.

The products of the *HLA-A, B* and *C* loci are very similar to each other, as expected if they have arisen by duplication from a common ancestor. The *HLA-D* products however, show no homology with these and each in turn is different from the complement components (C2, C4 and Bf) coded for in the HLA region.

If this complex genetic region has arisen by a series of duplications, one must explain how four different structural gene products can arise from a single duplicated nucleic acid sequence. Different proteins could be translated from the same nucleotide sequence through the use either of different reading frames or of different, not necessarily adjacent, parts of a sequence in different combinations.

It seems likely that the general organization of the genetic material of higher organisms is into clusters of duplicatd genes like the HLA region and that the numbers of such clusters, rather than the number of individual genes, reflects the genetic complexity of higher organisms.

The HLA system, developed initially out of a search for white blood cell groups that could form the basis for transplantation matching, is now known to include a number of closely linked loci controlling a variety of cell surface specificities, certain components of the complement system and immune responsiveness. This system has its counterpart in the mouse, H-2, and in the many similar systems defined in other species (see e.g. Götze 1977). These major histocompatibility systems, so-called because they are the most relevant for transplantation matching, are complex genetic regions containing many genes with interrelated functions, some, if not most, of which are likely to have arisen from a common ancestor by tandem gene duplication. There are now many examples in higher organisms of such duplicated gene clusters, some like the immunoglobulins perhaps of comparable complexity to the HLA system, and others, such as the haemoglobins, apparently much less complex.

Ever since the discovery by Bridges in 1919 of duplication, there has been discussion of its role in gene evolution. The possible significance of gene clusters in the general organization of the genetic material of higher organisms has gradually been clarified by our increasing understanding of the molecular structures of genes and their products. This started with the simple gene/protein relationship exemplified by the earliest work on the sickle cell haemoglobin and has continued to the most recent revelations on the genetic structure of organisms at the DNA level (see e.g. Gilbert 1978). It now seems likely that gene clusters of greater or lesser complexity are a characteristic and basic feature of the genetic structure of higher organisms, and that a genetic classification of cellular systems in terms of such clusters may be more fundamental than other more conventional functional classifications. The aim of this paper is to discuss these issues, with the HLA system as a model.

The HLA system will first be defined, including a description of its gene products and their interrelationships. This leads naturally to a discussion of the evolutionary origins of such gene clusters and the nature of their structure in the light of the most recent advances in understanding of gene structure at the DNA level. Finally, the relevance of gene clusters to arguments about the number of functional genes in higher organisms, in relation to total DNA content, will be emphasized.

THE HLA SYSTEM

The HLA system was first defined by a series of antigenic specificities identified by serological techniques using white blood cells, and especially lymphocytes, from the peripheral blood. Early studies by Dausset and others made use of white cell agglutinins in the serum of poly-transfused individuals. These, however, turned out to be too complex to be generally useful, and further developments depended on the independent discovery of van Rood & Payne in 1958 that white cell antibodies were produced quite frequently by fetal–maternal stimulation. Since the introduction by Terasaki & McClelland (1964) of a microlymphocytotoxicity assay, based on specific killing by antibodies in the presence of complement and using only microlitres of reagents and a few thousand cells, this has become the standard serological assay.

It soon became apparent that the specificities defined at first could be assigned to two separate series, originally called *LA* and *FOUR*, which correspond to two closely linked loci each with multiple alleles, now called *A* and *B* respectively. The combined system was, by international agreement, given the name HLA. As the complexity of the system has increased and more constituent loci of the HLA system were defined, the prefix HLA has been reserved to describe the whole system including in particular the *HLA-A* and *B* loci. Subsequently a third locus, *HLA-C*, has been defined which is closely analogous to the *HLA-A* and *HLA-B* loci. The products of these three loci are found on nearly all tissues, except for red blood cells. These three loci are remarkably polymorphic, as is illustrated by the European Caucasoid gene frequency data shown in Table 1. The *HLA-A* and *HLA-B* loci, in particular, both have a large number of alleles and each gives rise to an average heterozygosity of about 90%.

The fourth major polymorphic locus in the HLA region, *HLA-D*, originates from the discovery by Bach & Amos (1967) that mixed lymphocyte culture reactions (MLC as measured by the proliferative response of lymphocytes from different individuals when cultured together) are controlled by genetic differences in the HLA region. Subsequent work established a series

TABLE 1

HLA-A, -B and -C gene frequencies in European Caucasoids (in per cent)

HLA-*A1*		15.8	HLA-*B5*		5.9	HLA-*CW1*	4.8
A2		27.0	*B7*		10.4	*CW2*	5.4
A3		12.6	*B8*		9.2	*CW3*	9.4
AW23	} *A9*	2.4	*B12*		16.6	*CW4*	12.6
AW24		8.8	*B13*		3.2	*CW5*	8.4
A25	} *A10*	2.0	*B14*		2.4	*CW6*	12.6
A26		3.9	*B18*		6.2	*Blank*	46.7
A11		5.1	*B27*		4.6		
A28		4.4	*B15*		4.8		
A29		5.8	*BW38*	} *BW16*	2.0		
AW30		3.9	*BW39*		3.5		
AW31		2.3	*B17*		5.7		
AW32		2.9	*BW21*		2.2		
AW33		0.7	*BW22*		3.6		
Blank		2.2	*BW35*		9.9		
			B37		1.1		
			B40		8.1		
			Blank		3.6		

Data from Bodmer *et al.* (1978), based on 228 individuals. The brackets joining pairs of alleles, as for example *AW23* and *AW24*, related to *A9*, indicate alleles closely related by cross-reaction which were originally not separated.

of determinants identified by the MLC reaction that were controlled by alleles at a locus genetically separable from *HLA-A, -B, -C* although closely linked to them.

More recently, with serological techniques, several antigens have been defined that are closely associated with the *HLA-D* locus determinants identified by the MLC reaction and which are most probably the same gene products. These serological determinants are distinguished from those controlled by the *HLA-A, -B* and *-C* loci primarily by the fact that they have a much more restricted tissue distribution, occurring mainly on B lymphocytes and monocytes. These antigens have been called DR, for D-related, and are identified by the same numbers as the *HLA-D* locus determinant to which they correspond. The same sources of sera are used to define the HLA-DR determinants as for HLA-A-B-C typing and the main assay is the same microlymphocytotoxicity assay, but used on appropriate sources of B lymphocytes, often purified from peripheral blood. Reactions to HLA-DR determinants are most easily distinguished from those to HLA-A, -B and -C determinants by the absorption of sera with platelets, which express the latter but not the former. The evidence for the close relationship between the HLA-D and HLA-DR determinants comes both from their very close association in family and population data, and from the fact that HLA-DR antisera block corresponding MLC reactions. The close relationship between the HLA-D and -DR determinants is illustrated in Table 2, which shows 2 × 2 associations in family data between corresponding determinants. The *HLA-D* locus is also very polymorphic, as illustrated by the European Caucasoid gene frequency data shown in Table 3.

TABLE 2

Association between HLA-DRw and -Dw types in 189 haplotypes from healthy Caucasoids

DRw	*Dw*	+ +	+ −	− +	− −	r^a
1	1	22	1	2	164	0.92
2	2	39	3	0	147	0.95
3	3	28	4	0	157	0.92
4	4	9	10	0	170	0.66
5	5	9	2	0	178	0.89
6	6	8	2	1	178	0.83
7	7	18	3	4	164	0.81

Data from Bodmer *et al.* (1978).

[a]Correlation coefficient measure of association between the DRw and Dw types. All values are highly significant.

+ +, + − etc.: positive for DRw and Dw, positive for DRw and negative for Dw, etc.

TABLE 3

HLA-D and -DR gene frequencies in European Caucasoids (%)

HLA-*Dw1*	7.9	HLA-*DRw1*	6.2
Dw2	9.5	*DRw2*	11.2
Dw3	9.5	*DRw3*	8.9
Dw4	5.1	*DRw4*	7.8
Dw5	9.0	*DRw5*	15.1
Dw6	11.5	*DRw6*	8.6
Dw7	5.8	*DRw7*	15.6
Dw8	2.5	*W1A8*	5.6
Blank	39.1	*Blank*	21.1
No. tested	99		334

Data from Bodmer *et al.* (1978).
W1A8 is an unofficial description of a determinant that seems to correspond to Dw8.

The development of the HLA system has been greatly stimulated by a series of international collaborative workshops started by Amos in 1964 and organized subsequently in 1965, 1967, 1970, 1972, 1975 and 1977 by van Rood, Ceppellini, Terasaki, Dausset, Kissmeyer-Nielsen and Bodmer and their colleagues. Proceedings of these workshops provide the best documentation of the development of the system and the most up-to-date account of this will be found in the most recent publication (Bodmer *et al.* 1978; see also Barnstable *et al.* 1979 and Bodmer 1978 for further reviews of the HLA system).

The recombination fraction between the *HLA-A* and *-B* loci has been shown by family studies to be just under 1%. In the mouse the genetic distance between the *H-2D* and *H-2K* loci, which are thought to be the mouse equivalents of *HLA-A* and *B*, is about the same. The mouse H-2 system, originally discovered by Gorer in 1937, has in many respects been an important model for studies of the HLA system in man. There is no space here to describe the H-2 system in any detail, and for this further reference must be made, for example, to the comprehensive review by Klein (1975). Much of our knowledge about the functions of the many genes in major histocompatibility regions has come from work in the mouse. This has, in particular, shown that these regions include genes controlling specific immune responses, and has shown how the equivalents of the *HLA-A, -B, -C* and *-D* loci are involved in specific cellular interactions in the immune system.

Another major category of genes in the HLA and H-2 regions consists of those controlling certain of the complement components. The first clue to this came from studies by Demant *et al.* (1973) that showed a correlation between complement levels in the mouse and variants of the Ss protein, a gene product of the H-2 region with at that time no assigned function and no obvious connection with the immune system. Subsequently it has been shown that this protein is part of, or at least shares some molecular structures with, C4, the fourth component of complement, and data from human material have clearly shown that a structural gene for C4 is located in the HLA region. The two other complement components that are located in the HLA region are Factor B (Bf) of the alternative complement pathway, and its close relative C2, the second component of the classical complement pathway (for review see Lachmann & Hobart 1978 and Lachmann, this symposium).

A simplified comparative schematic map of the H-2 and HLA regions is shown in Fig. 1. The main difference between the maps of mouse and man is that the genes of the mouse I region, which control immune response, the mixed lymphocyte culture reaction and the so-called Ia, or immune associated antigens (which include the equivalents of the HLA-DR antigens), lie between the *H-2K* and *H-2D* loci adjacent to *H-2K*, while their human equivalents, thought to be in the neighbourhood of *HLA-D*, lie outside the interval *HLA-A* to *-B*. So far, the *HLA-C* locus seems to have no clear equivalent in the mouse, but other loci have been defined in the mouse which control, for example, red cell and thymus-specific antigens, and no doubt parallels to these may be expected in man. The evidence for the existence of immune response genes in man, probably in the neighbourhood of the *HLA-D* locus, is mainly indirect and comes, in particular, from HLA and disease association studies. A number of striking associations have now been described between particular

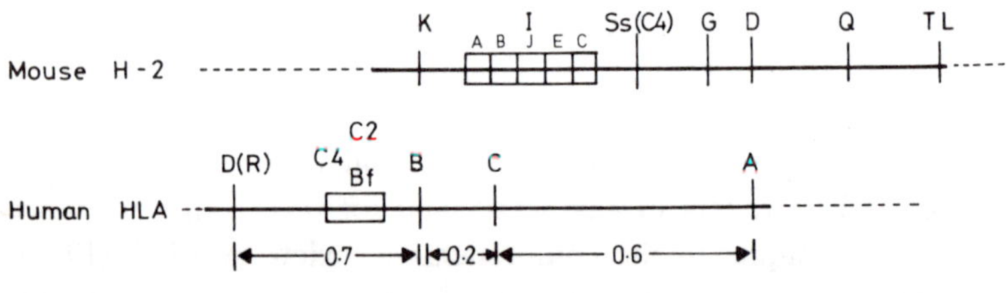

FIG. 1. A schematic comparison of the genetic maps of the mouse H-2 and human HLA regions. The letters above the line indicate loci or regions. Ss stands for serum substance and is now known to be a gene for the fourth component of complement C4. H-2G, Q and TL are respectively loci for red cell blood groups, antigens expressed on thymocytes and some lymph node cells, and a series of antigens expressed also on thymocytes and certain thymic leukaemias. I represents the region containing several immune response loci as well as loci controlling the mixed lymphocyte culture response and the Ia (immune-associated) antigens. C2, C4 and Bf refer to the second and fourth factors of the classical complement pathway and factor B of the alternative pathway. The numbers underneath the line for HLA are the approximate recombination fractions in per cent.

HLA region determinants and a variety of diseases, many of which have a presumptive or suspected immune aetiology, for example coeliac disease, active chronic hepatitis, myasthenia gravis, juvenile onset diabetes, rheumatoid arthritis, multiple sclerosis and ankylosing spondylitis (see Dausset & Svejgaard 1977; Bodmer 1978). The main explanation for many of these associations, as suggested by McDevitt & Bodmer (1974) and others, is that they are due to immune response genes in the HLA region near to the *HLA-D* locus. Other explanations must however be borne in mind, especially for diseases such as psoriasis and haemochromatosis, for which there is no known or suspected immune aetiology. More direct evidence for the existence of specific immune response genes comes from associations between HLA and ragweed hay fever (Marsh *et al.* 1977), and for some associations with differences in antibody levels (see Law *et al.* 1979).

The molecular structures of the four classes of gene products from the HLA region, namely those controlled by the *HLA-A, -B, -C* loci, by the *HLA-D* locus, by C2, Bf, and by C4, are at least to some extent known. Products of the *HLA-A, -B, -C* and *-D* loci (in the latter case actually DR) have been purified using as an assay the inhibition of an appropriate specific serological reaction. Molecules recognized by antisera to *HLA-A, -B* and *-C* locus specificities are composed of a 43 000 molecular weight glycosylated polypeptide, which carries the polymorphic specificities, and which is non-covalently linked to a non-glycosylated polypeptide of molecular weight 12 000, which is β_2-microglobulin (β_2m). This latter protein, which was first identified as a urinary protein excreted in certain cases of kidney malfunction in a context totally unrelated to work on the HLA system, is not known to be polymorphic, but has a primary structure which shows a significant homology with certain regions of the immunoglobulin molecules (see Möller 1974). The HLA region has been shown to lie on the short arm of chromosome 6 (Jongsma *et al.* 1973; Franke & Pellegrino 1977). Genetic studies using human–mouse somatic cell hybrids have, however, shown that the gene for β_2m is on chromosome 15 and so not in the HLA region (Goodfellow *et al.* 1975), although other evidence indicates a clear relationship between the genetic control of the *HLA-A, -B* antigens and β_2m (Jones *et al.* 1976; Arce-Gomez *et al.* 1978). Molecular characterization of the HLA-A, -B and -C products, including some amino acid sequence data, indicates a close similarity between them, strongly suggesting that the corresponding genes are related by gene duplication (for review, see Crumpton *et al.* 1978).

The molecules recognized by antisera to the HLA-DR products also have a two-chain structure. These antigens do not contain β_2m, but comprise two non-covalently linked glycosylated polypeptides of molecular weights 33 000 and 28 000 (see Crumpton *et al.* 1978). Studies with somatic cell hybrids have indicated that it is the 33 000 molecular weight chain which carries the polymorphic determinants and so is probably coded for by the *HLA-D* locus, while the 28 000 molecular weight chain is probably coded for by a gene on a different chromosome. A comparison of the fingerprints of the separated 33 000 and 28 000 molecular weight chains clearly indicates that they differ substantially in primary structure (M. J. Crumpton, personal communication).

Factor B and C2 have very similar structures, again indicating a probable common origin by gene duplication. They are each single polypeptide chains with molecular weights of 90 000 for Factor B and 100 000 for C2. The fourth complement component, C4, has a total molecular weight of 200 000 and is composed of three polypeptide chains of respective molecular weights 30 000, 80 000 and 90 000 (for review see Porter 1979). It is not known which of these might be coded for in the HLA region. Moreover, a recent report by Hall & Colten (1977) suggests that the three chains of C4 may first be synthesized as a single polypeptide, which further complicates the issue, as it then, from the genetic data, has to be encoded for in its entirety by a gene in the HLA region.

These structural studies clearly indicate that the genes of the HLA region code for at least four apparently distinct polypeptide chains, as summarized in Table 4. This obviously poses a problem for the suggestion that all the genes in the HLA region are derived from a common precursor by duplication.

EVOLUTION OF GENE CLUSTERS BY DUPLICATION

The significance of gene duplication for the evolution of new gene functions

TABLE 4

Protein products of, and associated with, the HLA region

HLA region product	*Molecular weight*	*Associated product*	*Molecular weight(s)*
HLA-A, -B, -C	43 000	β_2-microglobulin	12 000
HLA-D(R)	33 000	unnamed	28 000
C2	100 000	none	
Bf	90 000	none	
C4	200 000	3 chains	30 000
			80 000
			90 000

The HLA-A, -B and -C products are clearly structurally homologous, as are C2 and Bf. For C4 it is not known which chain, or indeed whether all chains, might be coded for in the HLA region.

has been discussed ever since Bridges' original definition of the phenomenon (Bridges 1919). Horowitz in 1945 suggested how gene duplication could be the basis of the evolution of those genes involved in sequential steps in biosynthetic pathways, and Lewis in 1951 discussed the whole concept of gene evolution by duplication in relation to the complex allelic series that had by that time been described in *Drosophila*. The first clear evidence at the molecular level for duplicate genes arranged in tandem came from studies of the haemoglobin β, γ and δ chains (see Weatherall *et al.*, this symposium). The role of gene duplication in evolution has more recently been re-emphasized by Ohno (1970) and others.

Interest in the evolution of linked interacting gene systems also developed from another direction, through Fisher's (1930) discussion on the interaction between linkage and selection. He proposed a specific model, involving selective interactions between loci, which could favour closer linkage between interacting genes. This notion was subsequently greatly expanded by population geneticists, such as E. B. Ford and P. M. Sheppard, at the experimental level (see Ford 1971) and by Kimura, Lewontin, Bodmer and others at the theoretical level (see Bodmer & Parsons 1962; Bodmer & Felsenstein 1967; Lewontin 1974). These evolutionary discussions referred to systems such as the plant incompatibility systems and the complex human blood groups as possible examples of gene clusters involving interacting gene sets with related functions, long before there was any possible understanding of these functions at a biochemical level. Out of these discussions grew the concept of the 'super-gene', a term coined by Darlington & Mather in 1949 to refer to 'a group of linked genes mechanically held together on a chromosome and usually held together as a unit'. The word system is defined by the Oxford Dictionary as 'a set of connected things or parts', while the term complex derives from Dunn's (1954) use of complex-locus to refer to a cluster of two or more closely linked, and functionally related, genes particularly in connection with the *T/t* gene complex in the mouse. The concepts of system, complex, and super-gene coincide when, as with the HLA system, one is dealing with sets of closely linked genes controlling functions that are to a considerable extent interrelated and for which, in at least some cases, there is evidence of selective interactions occurring in such a way as to favour closer linkage (see Bodmer & Thomson 1977).

The simple theory of gene evolution by duplication is that once a gene has been duplicated, one of the copies can diverge from the original and so acquire new functions while not jeopardizing those functions fulfilled by the original gene before it duplicated. Once duplication has occurred, further duplications are much more likely to arise by the process of unequal crossing over. Naturally, such a model leads to the expectations that, first, there should be readily detectable structural evolutionary homologies between the products of the duplicated genes, and second that the set of functions evolved will be closely related. Both of these are clearly features of the haemoglobin and immunoglobin systems. Another particular feature of these two gene clusters is that genes within them, for example those that specify the β, γ and δ chains of haemoglobin, tend to be mutually exclusively expressed, in the sense that a cell at any given time is likely to be expressing only one member of the gene set within the cluster. This feature of the control of expression of genes within a cluster may turn out to be fairly general for higher organisms, and it contrasts markedly with the bacterial operon systems, whose rationale seems to be mutually coordinated simultaneous expression of a set of closely linked and functionally interrelated genes. This pattern of control may, perhaps, also be related to the fact that, for example, in haemoglobin the two sets of chains, namely α and β, γ, δ or ϵ which between them form the normal complete molecule, are controlled by unlinked sets of genes. In fact, to my knowledge, in all those cases where the genetics of dimeric proteins that are made up of different subunits is sufficiently well advanced (e.g. lactate dehydrogenase, hexoseaminidase A and B, haemoglobin, immunoglobulins: see *Cytogenetics and Cell Genetics* 1976), the genes for the two different chains are on different chromosomes, or at least very loosely linked. This situation also, of course, applies to the relationship between the HLA-A, B and C products and β_2m and, probably, also to the relationship between the *HLA-D* locus product and the gene controlling its 28 000 molecular weight partner. For haemoglobin, where the α chains are clearly related evolutionarily to the β, γ and δ set, it is an obvious suggestion that the two gene sets are themselves related by duplication. This must have been followed by separation to different chromosomes, either by translocation or re-duplication of a whole or part of a chromosome, or even, according to Ohno (1970), by tetraploidization. Whatever its origins, such an arrangement must have evolved because of some selective advantage, which could well be related to its advantages for mutually related control of gene expression. This, of course, is in addition to the presumed greater variety of functional adaptations that become possible on the transition from a single chain structure to one involving various combinations of two different chains.

The genes coding for ribosomal, transfer and 5S RNAs were the first to be clearly shown to be highly tandemly repeated at the molecular level (see *Cold Spring Harbor Symposia on Quantitative Biology* 1974). In these cases the repetition may mainly serve the needs for very high rates of synthesis of the corresponding gene products, though for the ribosomal RNA the arrangement clearly also serves the needs for coordinate control of production of the two major RNA subunits. The histone genes are another highly repeated gene family where, again, the repetition may be because of high synthesis rates and where, in contrast to the haemoglobins β, γ, δ and ϵ, and to the immunoglobulins, control of synthesis is coordinate rather than mutually exclusive (see Kedes 1976; Lifton *et al.* 1978).

The theory originating from Fisher (1930), by which selective interaction between genes favours closer linkage, can equally well be used to explain why tandemly duplicated genes, which by their nature arise as a closely linked set, are likely, on the whole, to be held together throughout evolution. If there are selective interactions between these genes, for example connected with coordinated control of gene expression, then a partition of the gene cluster into two sections by translocation may disrupt this control, and so be at a selective disadvantage.

The paradox posed by the HLA system for these models of the evolution of gene clusters is that it seems to consist not of one but of at least four different gene clusters, namely *HLA-A, -B, -C; HLA-D;* Bf, C2; and C4. Evidence already exists for homology between the products of the *HLA-A, -B* and *-C* loci, as well as between Factor B and C2. Data in the mouse on the products of the H-2 I region certainly suggest that the *HLA-D* locus will be one member of a cluster of related genes. Though the evidence is against structural evolutionary homologies between these four different sets of gene products, nevertheless they do seem to have somewhat interrelated functions. Thus, the relationship between the HLA-A, -B, -C and -D products is clear, as is that between the complement components. Complement components moreover can act on the cell surface and Barnstable *et al.* (1977) have suggested that the HLA-A, -B, -C and -D products should be looked on as having complement-like functions on the cell surface, and in relation to cellular interactions. It is possible, of course, that a diverse collection of genes has been trapped in the HLA region and that these have subsequently been duplicated to different extents. It is still, however, much more appealing to try and explain the origin of the system in terms of duplication of a single, perhaps admittedly complex, nucleotide sequence. Recent dramatic advances in our understanding of the genetic organization of higher organisms at the DNA level do seem to make this possible, as discussed in the next section.

COMPLEX ORGANIZATION OF GENE CLUSTERS

If the HLA system has evolved by duplication of a single nucleotide sequence, the basic problem is to explain how this sequence could give rise, apparently, to several structurally unrelated polypeptide products. The first clue to the way this might have happened came from the elucidation by Sanger *et al.* (1977) of the complete nucleotide sequence of the bacteriophage ϕX174, which showed that it contained overlapping genes, namely that there were nucleotide sequences which could contribute to the determination of more than one polypeptide product. This arose because the same sequence could be read by the protein-synthesizing machinery in three different reading frames and, also, in two directions. Such an explanation for the production of different gene products from the same nucleotide sequence would lead to the prediction that the corresponding amino acid sequences could be related by an overall frame shift or an inversion at the nucleotide level. Even more recently two further striking discoveries have considerably further expanded the possible mechanisms by which a single polynucleotide sequence could give rise to a variety of polypeptides. The first arises from work with the animal viruses, adenovirus, SV40 and polyoma, which shows that messenger RNAs producing a given polypeptide may be constructed from non-contiguous nucleotide sequences (for a brief review see Sambrook 1977). The assumption is that the whole of a DNA sequence may first be transcribed into messenger RNA, but that, subsequently, segments of this initial RNA transcript are spliced out, so bringing together nucleotide sequences that are not next to each other at the DNA level. The corresponding polypeptide is then produced by the normal translation mechanisms from this processed messenger RNA. This possibility was foreshadowed by explanations for the bringing together of the variable and constant regions of immunoglobulin chains, though here it was generally assumed that the primary processing was likely to be at the DNA level (see Tonegawa *et al.* 1978). The second striking discovery is of the existence of apparently silent intervening nucleotide sequences in the middle of a protein-determining nucleotide sequence (see Jeffreys & Flavell 1977; for a brief review see Gilbert 1978). Here also it seems that the intervening untranslated sequence is originally transcribed into messenger RNA, but then spliced out of the messenger before this is translated into a polypeptide product. Both of these discoveries show that it is possible that a given nucleotide sequence may not be translated as a whole into a polypeptide product, but that different sub-segments of it may contribute, perhaps even in different combinations, to a variety of different polypeptide products. It seems entirely possible that the sections of a given nucleotide sequence which determine one polypeptide derived from it may either overlap, or not, with sections giving rise to another product, and that various sections may also be read in any one of the three different possible reading frames and in either direction. Given these possibilities, one long polynucleotide sequence could easily give rise to a considerable variety of polypeptide products that are, apparently, structurally unrelated. The only requirement on the sequence would, of course, be that it contained all the information needed for producing the various classes of polypeptide products, together, presumably, with appropriate control sequences. So long as the primordial sequence from which the whole of the gene cluster was derived contained all the necessary information, the evolution of a region such as HLA by duplication from one sequence can now, it appears, be readily explained. It is, of course, not necessary to assume that the duplicates all retain the original information. Subsequent evolutionary divergence could easily lead to different classes of duplicates, defined perhaps in part by segments of the original sequence which may have been lost by internal deletion, duplicated by internal duplication or inverted, leading to further complications in the assessment of the homologies between the different classes of genes within the clusters.

In this context it is worth recalling the evolutionary change in the enzyme tryptophan synthetase from *E.coli* to *Neurospora*. In *E.coli* the enzyme is composed of two separate polypeptides, A and B, catalysing slightly different reactions. In *Neurospora,* on the other hand, the enzyme is a single polypeptide consisting of the equivalents of the A and B subunits of *E.coli*, joined by a small third segment (Bonner *et al.* 1965). This evolutionary transition suggests how two different gene segments, coding for separate proteins, may be joined together to form a new protein that combines, semi-independently, the properties of the original pair. As Blake (1978) has succinctly put it: genes-in-pieces may imply proteins-in-pieces. Changes like these may lie behind the evolution of the primordial sequence from which a complex gene cluster, such as HLA, may have evolved by duplication. Since duplicate gene clusters are not, apparently, a feature of prokaryotes, this stage of evolution of eukaryotes may date back to a prokaryote phase before duplication and gene clustering became major features of genetic organization.

Yet another complication to the evolution of a cluster may be the existence within clusters of transposable genetic elements, or what have been called 'insertion sequences' in bacteria (for review see Nevers & Saedler 1977). The existence of such elements was predicted many years ago by McClintock in her extraordinary work on jumping genes in maize (see McClintock 1956). These elements are apparently DNA sequences that can be inserted and excised from many regions of the DNA, without apparently any need for significant sequence homology between the transposable element and the DNA region into which it is being inserted. This phenomenon means that it may be possible that there will exist miscellaneous apparently unrelated sequences in the midst of a gene cluster whose origin in other respects is by gene duplication.

An intriguing example of a dispersed duplicated gene that appears itself to be such a transposable element has recently been described in *Drosophila* (Finnegan *et al.* 1978).

A further complication to the structure of a complex gene cluster arises from the suggestion made by Bodmer (1973) that some of the apparent allelism in complex genetic systems, such as HLA, may really reflect polymorphism for control of the expression of tandemly duplicated genes. The simplest model situation is where there are just two apparently allelically controlled antigens, say 1 and 2. The genes for both the antigens 1 and 2 are always present on the relevant chromosome, but a chromosome carrying the apparent allele 1 has only gene 1 expressed, with gene 2 silent, while apparent allele 2 corresponds to the expression of gene 2 with gene 1 switched off. The genetic control of which gene is expressed must be transmitted through the germ line. This model can be elaborated in at least two main ways. First there may also be true alleles, and, second, more than one member of such a gene set may be switched on at any given time.

The origin of this model was the need to explain the fact that allo-antisera made to detect polymorphic differences within one species are also sometimes apparently able to pick up polymorphic differences within another species. This implies that the polymorphic differences now found within these various species pre-date their evolutionary separation, a possibility not easily accommodated by present concepts in population genetics, or by data on protein evolution. In other words, the normal expectation is that two allelic products would differ by at most a few amino acids. On an evolutionary time scale, the particular different versions of a given product present in a population will change, so that it is unlikely that two species that have been separated evolutionarily for a considerable period of time will contain two or more versions of a gene which differ in the same amino acid positions. Where this is apparently found to be the case, and where in addition apparent alleles differ by considerable numbers of amino acids, as occurs for example in some cases of immunoglobulin allotypes (see Mole *et al.* 1971), then it is much more likely that the variation being seen is due to polymorphism for control of the expression of duplicated genes. Such duplicates may be related to each other in a way analogous, for example, to the relation between the β and δ chain genes of haemoglobin. The duplicates are likely to have diverged in evolution during speciation, retaining common features which account for the apparent relatedness of allelic differences in widely separated species. An important prediction of this model, therefore, is that some apparent alleles could be related to each other in complex ways, and are likely to differ in a relatively large number of amino acids. Certainly, data on the mouse H-2 gene products from Coligan *et al.* (1978) support this notion. It is also much easier to explain the complex patterns of cross-reaction found between the various allelic products of the HLA loci on the assumption that they are at least in part the products of different, but related, tandemly duplicated genes. A similar hypothesis was independently put forward to explain certain anomalies in the immunoglobulin allotypes by Rivat *et al.* (1973), and this has recently been further expanded on by Hood and his collaborators and by others (see Hood *et al.* 1975).

Undoubtedly the sort of complex genetic regions envisaged by this discussion (see Fig. 2) may provide the opportunity for a relatively high rate of spontaneous genetic changes involving complex events, such as inversion and deletion, that are mediated by intrachromosomal unequal crossing over. Indeed, it might well be predicted that such events will be more frequent than simple point mutations occurring at the DNA level. Furthermore, it seems likely that such complex mutational events will lead to switching off of genes, rather than to the production of an altered product. This prediction agrees with the impression that a high proportion of spontaneous mutations in higher eukaryotes are indeed loss mutations. Certainly, these concepts may have a profound influence on the whole notion of what a gene is and also on the interpretation of mutations, especially in relation to human inherited abnormalities.

AMOUNT OF DNA AND NUMBER OF GENETIC FUNCTIONS

There is much argument about the number of functional genes in higher organisms. The existence of the highly repetitive satellite DNA sequences (see

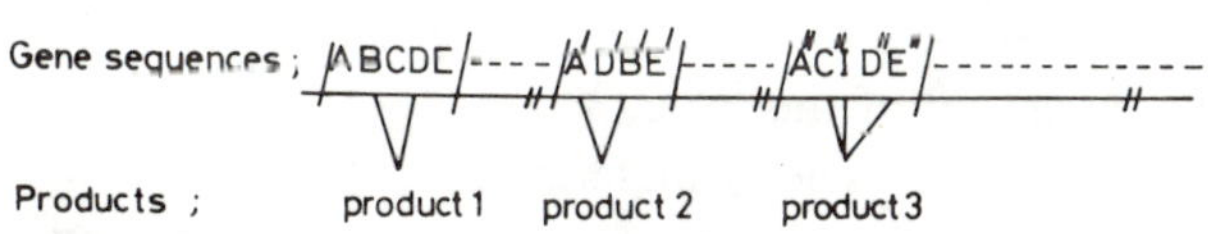

FIG. 2. Schematic example of a complex duplicated gene cluster coding for three different classes of product. A, B etc. represent nucleotide sub-segments. The primes, e.g. A, A′, A′′ indicate segments which have a common origin by duplication, but which have diverged and which may be translated in different reading frames, and in different directions. A B C D E, A′ D′ B′ E′, A′′ C′′ I D′′ E′′ are representatives of three basic subclusters of genes, each of which may be duplicated as a whole many times, and not all of which are necessarily expressed. For product 1, C represents a silent intervening sequence. For product 2, D′ is silent; this sequence has lost C by deletion and then had a subsequent inversion of B and D. For product 3, D′′ is silent, while I represents an unrelated insertion sequence or transposable element. The sequences contributing to a product may be coding for control regions as well as specifying a protein product. The example is highly schematic and is only meant to illustrate some of the possible complexities which may arise in the evolution of a duplicated gene cluster.

Walker, this symposium) clearly shows that much of higher organism DNA may have no obvious function. Nevertheless, up to 40% or more of the DNA may represent more or less unique sequences, leaving room for nearly half the DNA to be functional in the normal sense. Estimates of the proportion of higher organism DNA which is functional range from this upper limit of about 40% to a lower limit of at most a few per cent. The argument is sometimes made that there is no need for mammals to have a thousand times as many genes as bacteria, this being the approximate ratio of their DNA contents. The average haploid mammalian genome contains about 3×10^9 nucleotide pairs of DNA. Even if we assume that only 10% of this is functional, then, with an average gene size of a thousand nucleotide pairs, and allowing for an average size protein and for some controlling sequences, this would give a total of some 300 000 genes. It is clear, however, that our notions of what a gene is at the DNA level have, until quite recently, been much too simple. The new discoveries discussed in the previous section imply that, even for the single gene that is not a member of a duplicated cluster, there may be a considerable number of nucleotide pairs not involved directly in the specification of a polypeptide product. If, however, in addition we assume that most genes occur in duplicated clusters of greater or lesser complexity, then the whole notion of what a gene is, and so of the number of basic genetic functions, is radically altered. The idea of polymorphism for gene control, namely the existence of tandem gene duplicates only one of which may be expressed in a given individual, suggests that even simple enzyme functions that show no apparent evidence of gene duplication may, nevertheless, be members of small gene clusters. Regions such as HLA, which from the fact that it encompasses a recombination fraction of up to 2 or 3% and therefore may be as much as one thousandth of the total human genome and so may contain the equivalent of a few hundred if not up to a thousand genes, are probably at the upper end of the scale of complexity.

If the general organization of the genetic material of higher organisms involves clustering of tandemly duplicated genes as a general phenomenon, then it seems only right that the basic number of genetic functions should be counted in terms of the number of gene clusters, rather than the number of individual genes. This view, though it does not lead to a reduction in the total number of genes below the figure of a few hundred thousand, may well reduce the functional complexity they represent to something of the order of 5000 to 10 000. This is a number often quoted as plausible for the number of functional genes, especially in *Drosophila* (see e.g. Garcia-Bellido & Ripoll 1978). In this case the problem of unravelling the genetic organization of higher organisms may be considerably simplified. The problem becomes not so much that of studying individual genes, as of studying gene clusters of greater or lesser complexity. The recent discoveries about genetic organization at the DNA level certainly provide plenty of scope, as discussed above, for the generation of considerable complexity in the evolution of a gene cluster by duplication and related events. Now, our hope must be that recent developments in DNA technology, including in particular recombinant DNA techniques (see Weissmann, this symposium) together with the now established techniques of somatic cell genetics, will lead to a detailed understanding of the organization of gene clusters, even of such complexity as the HLA system, and through this to a major advance in our general understanding of genetic organization and its control.

References

ARCE-GOMEZ, B., JONES, E. A., BARNSTABLE, C. J., SOLOMON, E. & BODMER, W. F. (1978) The genetic control of HLA-A and B antigens in somatic cell hybrids: requirement for β_2 microglobulin. *Tissue Antigens 11,* 96-112

BACH, F. H. & AMOS, D. B. (1967) Hu-1: major histocompatibility locus in man. *Science (Wash. D.C.) 156,* 1506-1509

BARNSTABLE, C. J., JONES, E. A., BODMER, W. F., BODMER, J. G., ARCE-GOMEZ, B., SNARY, D. & CRUMPTON, M. (1977) Genetics and serology of HLA linked human Ia antigens. *Cold Spring Harbor Symp. Quant. Biol. 41,* 443-455

BARNSTABLE, C. J., JONES, E. A. & BODMER, W. F. (1979) The genetic structure of major histocompatibility regions, in *Defence and Recognition,* 2nd edn. (Lennox, E.S., ed.), MTP Press, Lancaster (in press)

BLAKE, C. C. F. (1978) Do genes-in-pieces imply proteins-in-pieces? *Nature (Lond.) 273,* 267

BODMER, W. F. (1973) A new genetic model for allelism at histocompatibility and other complex loci: polymorphism for control of gene expression. *Transplant. Proc. 4,* 1471-1475

BODMER, W. F. (ed.) (1978) The HLA system. *Br. Med. Bull. 34,* 213-321

BODMER, W. F. & FELSENSTEIN, J. (1967) Linkage and selection: theoretical analysis of the deterministic two locus random mating model. *Genetics 57,* 237-265

BODMER, W. F. & PARSONS, P. A. (1962) Linkage and recombination in evolution. *Adv. Genet. 11,* 1-100

BODMER, W. F. & THOMSON, G. (1977) Population genetics and evolution of the HLA system, in *HLA and Disease* (Dausset, J. & Svejgaard, A., eds.), pp. 280-295, Munksgaard, Copenhagen

BODMER, W. F., BATCHELOR, J. R., BODMER, J. G., FESTENSTEIN, H. & MORRIS, P. J. (eds.) (1978) *Histocompatibility Testing 1977 (Report 7th Int. Histocompatibility Testing Workshop Conf.),* Munksgaard, Copenhagen

BONNER, D. M., DEMOSS, J. A. & MILLS, S. E. (1965) The evolution of an enzyme, in *Evolving Genes and Proteins* (Bryson, V. & Vogel, H. J., eds.), pp. 305-318, Academic Press, New York

BRIDGES, C. B. (1919) Duplication. *Anat. Rec. 15,* 357

COLIGAN, J. E., KINDT, T. J., EWENSTEIN, B. M., UEHARA, H., NISIZAWA, T. & NATHENSON, S. G. (1978) Primary structure of murine MHC alloantigens. II. Amino acid sequence studies of the cyanogen bromide fragments of the H-2K^b glycoprotein. *Proc. Natl. Acad. Sci. U.S.A. 75,* 3390-3394

CRUMPTON, M. J., SNARY, D., WALSH, F. S., BARNSTABLE, C. J., GOODFELLOW, P. N., JONES, E. A. & BODMER, W. F. (1978) Molecular structure of the gene products of the human HLA system: isolation and characterization of HLA-A, -B, -C and Ia antigens. *Proc. R. Soc. Lond. B Biol. Sci. 202,* 159-175

Cold Spring Harbor Symp. Quant. Biol. 38 (1974) *Chromosome Structure and Function*

Cytogenet. Cell Genet. 16 (1976) *Proc. 3rd Int. Workshop Human Gene Mapping*

DARLINGTON, C. D. & MATHER, K. (1949) *The Elements of Genetics,* Allen & Unwin, London

DAUSSET, J. & SVEJGAARD, A. (eds.) (1977) *HLA and Disease,* Munksgaard, Copenhagen

DEMANT, P., CAPKOVA, J., HINZOVA, E. & VORACOVA, B. (1973) The role of the histocompatibility-2 linked Ss-Slp region in the control of mouse complement. *Proc. Natl. Acad. Sci. U.S.A. 70,* 863-864

DUNN, L. C. (1954) The study of complex loci. *Proc. 9th Int. Congr. Genetics,* vol. 1, p. 156. *Caryologia Suppl. 6*

FINNEGAN, D. J., RUBIN, G. M., YOUNG, M. W. & HOGNESS, D. S. (1978) Repeated gene families in *Drosophila melanogaster. Cold Spring Harbor Symp. Quant. Biol. 42,* 1053-1063

FISHER, R. A. (1930) *Genetical Theory of Natural Selection,* Oxford University Press, London

FORD, E. B. (1971) *Ecological Genetics,* 3rd edn., Chapman & Hall, London

FRANKE, U. & PELLEGRINO, M. A. (1977) Assignment of the major histocompatibility complex to a region of the short arm of human chromosome 6. *Proc. Natl. Acad. Sci. U.S.A. 74,* 1147-1151

GARCIA-BELLIDO, A. & RIPOLL, P. (1978) The number of genes in *Drosophila melanogaster. Nature (Lond.) 273,* 399-400

GILBERT, W. (1978) Why genes in pieces? *Nature (Lond.) 271,* 501

GOODFELLOW, P. N., JONES, E. A., VAN HEYNINGEN, V., SOLOMON, E., BOBROW, M., MIGGIANO, V. & BODMER, W. F. (1975) The beta-2 microglobulin gene is on chromosome 15 and not in the HLA region. *Nature (Lond.) 254,* 267-269

GORER, P. A. (1937) Further studies on antigenic differences in mouse erythrocytes. *Br. J. Exp. Pathol. 18,* 31-36

GÖTZE, D. (ed.) (1977) *The Major Histocompatibility System in Man and Animals,* Springer, Berlin

HALL, R. E. & COLTEN, H. R. (1977) Cell-free synthesis of the fourth component of guinea pig complement (C4): identification of a precursor of serum C4 (pro-C4). *Proc. Natl. Acad. Sci. U.S.A. 74,* 1707-1710

HOOD, L., CAMPBELL, J. H. & ELGIN, S. C. R. (1975) The organization, expression and evolution of antibodies and other multigene families. *Annu. Rev. Genet. 9,* 305-353

HOROWITZ, M. H. (1945) On the evolution of biochemical synthesis. *Proc. Natl. Acad. Sci. U.S.A. 31,* 153-157

JEFFREYS, A. J. & FLAVELL, R. A. (1977) The rabbit β-globin gene contains a large insert in the coding sequence. *Cell 12,* 1097-1108

JONES, E. A., GOODFELLOW, P. N., KENNETT, R. H. & BODMER, W. F. (1976) The independent expression of HLA and β_2-microglobulin on human mouse hybrids. *Somatic Cell Genet. 2,* 483-496

JONGSMA, A., VAN SOMEREN, H., WESTERVELD, A., HAGEMEIJER, A. & PEARSON, P. (1973) Localization of genes on human chromosomes by studies on human-chinese hamster somatic cell hybrids. *Humangenetik 20,* 195-202

KEDES, L. H. (1976) Histone and messengers and histone and genes. *Cell 8,* 321-331

KLEIN, L. H. (1975) *Biology of the Mouse Histocompatibility 2 Complex,* Springer, Berlin

LACHMANN, P. J. & HOBART, M. J. (1978) Complement genetics in relation to the HLA. *Br. Med. Bull. 34,* 247-252

LACHMANN, P. J. & HOBART, M. J. (1979) The genetics of the complement system, in this symposium, pp. 231-245

LAW, H. Y., BODMER, W. F., MATHEWS, J. D. & SKEGG, D. C. G. (1979) The immune response to vasectomy and its relation to the HLA system. *Tissue Antigens,* in press

LEWIS, E. B. (1951) Pseudoallelism and gene evolution. *Cold Spring Harbor Symp. Quant. Biol. 16,* 159-174

LEWONTIN, R. C. (1974) *The Genetic Basis of Evolutionary Change,* Columbia University Press, New York

LIFTON, R. P., GOLDBERG, M. L., KARP, R. W. & HOGNESS, D. S. (1978) The organization of the histone genes in *Drosophila melanogaster:* functional and evolutionary implications. *Cold Spring Harbor Symp. Quant. Biol. 42,* 1047-1051

MARSH, D. G., CHASE, G. A., GOODFRIEND, L. & BIAS, W. (1977) 'Mapping' of postulated Ir genes within HLA by studies in allergic populations. *Monogr. Allergy 11,* 106-115

McCLINTOCK, B. (1956) Controlling elements and the gene. *Cold Spring Harbor Symp. Quant. Biol. 21,* 197-216

McDEVITT, H. O. & BODMER, W. F. (1974) HLA, immune response genes and disease. *Lancet 1,* 1269-1275

MOLE, L. E., JACKSON, S. A., PORTER, R. R. & WILKINSON, J. M. (1971) Allotypically related sequences in the Fd fragment of rabbit immunoglobulin heavy chains. *Biochem. J. 124,* 301-318

MÖLLER, G. (ed.) (1974) β_2-Microglobulin and HLA antigens. *Transplant. Rev. 21*

NEVERS, P. & SAEDLER, H. (1977) Transposable genetic elements as agents of gene instability and chromosomal rearrangements. *Nature (Lond.) 268,* 109-115

OHNO, S. (1970) *Evolution by Gene Duplication,* Springer, Berlin

PAYNE, R. & ROLFS, M. R. (1958) Fetomaternal leukocyte incompatibility. *J. Clin. Invest. 37,* 1756

PORTER, R. R. (1979) Complement, in *Defence and Recognition* (Lennox, E. B., ed), MTP Press, Lancaster, in press

RIVAT, L., GILBERT, D. & ROPARTZ, C. (1973) Immunoglobin allotypic specificities in mixed leucocyte structures. *Immunology 24,* 1041-1049

SAMBROOK, J. (1977) Adenovirus amazes at Cold Spring Harbor. *Nature (Lond.) 268,* 101-104

SANGER, F., AIR, G. M., BARRELL, B. G., BROWN, N. L., COULSON, A. R., FIDDES, J. C., HUTCHINSON III, C. A., SLOCOMBE, P. M. & SMITH, M. (1977) Nucleotide sequence of bacteriophage ΦX174 DNA. *Nature (Lond.) 265,* 687-695

TERASAKI, P. I. & McCLELLAND, J. D. (1964) Microdroplet assay of human serum cytotoxins. *Nature (Lond.) 204,* 998-1000

TONEGAWA, S., MAXAM, A. M., TIZARD, R., BERNARD, O. & GILBERT, W. (1978) Sequence of a mouse germ-line gene for a variable region of an immunoglobin light chain. *Proc. Natl. Acad. Sci. U.S.A. 75,* 1485-1489

VAN ROOD, J. J., EERNISSE, J. G. & VAN LEEUWEN, A. (1958) Leukocyte antibodies in sera from pregnant women. *Nature (Lond.) 181,* 1735

WALKER, P. B. M. (1979) Genes and non-coding DNA sequences, in this symposium, pp. 25-39

WEATHERALL, D. *et al.* (1979) Human haemoglobin genetics, in this symposium, pp. 147-174

WEISSMANN, C. (1979) Reversed genetics: a new approach to the elucidation of structure-function relationship, in this symposium, pp. 47-58

Discussion

Milstein: You referred to the control functions of alleles and mentioned the rabbit allotypes as one of them.

Bodmer: I don't think they are controlling genes in the usual sense of this term. I was referring to the idea that there are duplicate genes for the product being studied and that the genetic polymorphism observed for this product is for the control of expression of which of these diverged duplicates is switched on in an individual, in contrast to different tissues of a given individual.

Brenner: The extreme case is that all of us have all the haemoglobin variants.

Bodmer: No, I only said that it could apply. There are sequence data on the allotypes that Mole *et al.* (1971) got, for example, which showed that there were large numbers of amino acid differences between apparent allotypes.

Milstein: There have been two ideas about this. One is the one you have just described. The other is that they originate from what we call different sub-groups and are pseudo-alleles which have been deleted or moved around during evolution. There is very strong positive evidence for this second view. It appears that in hares the rabbit allotypes are not alleles since they all appear in all animals. In that case there is no control and the controlling elements must have arisen after speciation.

Bodmer: I don't claim that not more than one duplicate can be expressed, nor does expression have to precede speciation. The evolution of the duplicates is pre-speciation.

Milstein: But once the three markers have been established, the control for their expression is present in the hare and in the rabbit. The homozygous rabbit expresses only one allotype while the hare seems to express all three.

Bodmer: The hares just switch all on at the same time.

Milstein: The alternative which to me is much more obvious is that some of the alleles have been deleted. Evolution by deletion and expansion is fairly well established by now.

Bodmer: That is just saying that the differences are not due to two alleles but to expanded or contracted genes.

Crick: It is a matter of fact whether they are there or not, but how is it explained by your hypothesis?

Bodmer: In my model mostly they are there, but I wouldn't exclude some polymorphisms due to a mutation having been lost.

Milstein: You gave another example of a marker in the rat in the histocompatibility system which is present in humans. But as I understand it such a marker is produced when antisera to the rat are raised in another species. The sequence in the rat may have an ancestral sequence which is preserved in the human, and nothing strange happens except that in both cases variants may arise which allow the expression of the 'antiancestral' antiserum.

Bodmer: But that marker is polymorphic in the rat.

Milstein: It doesn't matter—one of them may be ancestral.

Bodmer: You can do the population kinetics, however imperfectly. You do not expect from normal population genetic models that alleles will show complex differences.

Milstein: You also have to take into account expansion and deletion.

Bodmer: As soon as you accept expansion or deletion, you are accepting my ideas. You are accepting that you are not simply looking at one sequence with variants within that sequence.

Brenner: Milstein is allowing silent expansion which becomes audible later.

Henry Harris: It is very clear in *E.coli* that although some related groups of genes are clustered, clustering is not essential for coordinate regulation. I have always thought that a good evolutionary reason for clustering of related genes in *E.coli* is provided by the unusual nature of its sex life. *E.coli* transmits a single long thread of DNA from one cell to another and, if related groups of essential genes were clustered, this would minimize the chance of a chromosome break producing a lethal disruption of a concatenated system.

Bodmer: Sex life is a rare event for *E.coli*!

Henry Harris: That doesn't affect the argument. If you are thinking of a form of regulation comparable to the operon in *E.coli*, then it seems to me you gain nothing if the two genes are a million bases away from each other. They might as well be on different chromosomes.

Bodmer: First one has to accept the fact that there is some coordinate regulation within a very large region here. 'Coordinate' is perhaps the wrong word; it is mutually related.

Crick: Did you say *fact*?

Bodmer: Yes, there is some mutually related expression. β_2-Microglobulin controls the expression of HLA-A, B and C together, and they can be quite far apart.

Henry Harris: Clusters of genes are sometimes coordinately regulated, but it is clear that clustering is not essential for coordinate regulation.

Bodmer: Not coordinate—but mutually related. In most clusters in mammals, such as in the haemoglobins, it is not a question of having them all on together but of knowing that one is on and not the others. It is a mutually related control.

Your other point was about distance. The population genetic theory of linked loci, which we shouldn't forget, tells us the magnitude of selection interaction with linkage and is quite adequate to explain, for the distances we are discussing, the need to hold things together.

Crick: There is no actual reason why they should be held together. It is just an abstract idea.

Bodmer: That is not quite true. There are interrelations of the functions here. The mouse, for instance, may have pairs of immune response loci that interact.

Crick: That isn't a strong argument.

Henry Harris: The thing about clustering of genes in *E.coli* is that it sometimes achieves the concatenated transcription of a group of genes into one piece of RNA. If you are not thinking of this kind of gene regulation, I find it difficult to see that it can make much difference, in the regulation of the activity of two related genes, whether they are separated by a length of DNA equivalent to one *E.coli* chromosome or five *E.coli* chromosomes.

Bodmer: I think it is wrong to assume that it doesn't make any difference if they are as far as two million nucleotides apart. It will be necessary to have related control of genes that are that far apart.

Crick: There are two arguments, Walter. I would be surprised if an actual control mechanism extended over so big a region, but Fisher's argument that you gave *would* extend over a bigger region. The difficulty is that in that case you have to show why it is a selective advantage for the components not to be separated. I would accept your arguments about small tandem duplications, and I agree with Cesar that it is not unimportant whether you have a deletion or not, but let's leave that on one side. It is a valid point but it is not the main thing. What I am worried about is that you call it a region, and as far I can see it is just because A, B and C are separated. Suppose A, B and C were next to each other, what would your justification be for saying that that was one region?

Bodmer: You then have to go to the population genetics and you certainly go to notions which are not hard.

Crick: But why is it considered a region?

Nossal: It is linkage disequilibrium that defines the extent of the 'region'.

Crick: That is not what it is. That would explain why the region was made that way.

Bodmer: That is what lies behind the population genetics. One reason is because these are interrelated functions.

Crick: But you are only asserting that—that is what I am arguing.

Bodmer: They are definitely interrelated.

Milstein: What about immunoglobulin $\varkappa$ and λ chains? I think these are the best examples of interrelated functions which are not genetically linked.

Bodmer: I pointed out that when $\varkappa$ and λ go together with a heavy chain, there is a different sort of control, just as the α chain set in haemoglobin is not linked to β, γ, δ.

Crick: May I sharpen my question a little? I understand why some of them are interrelated, but why is D related to A, B and C?

Bodmer: Because functionally it seems to be involved in the same sort of restrictions between cells, in this case between macrophages and T cells.

Crick: But they just happen to be on the surface.

Bodmer: No, HLA-D appears to be serving a similar function in the interrelationship between the cells in the immune system as A, B and C do for the T cell type of recognition. The D locus does it for macrophages in relation to T cells.

Crick: So what you are trying to say is that they are similar functions but on different sorts of cells?

Bodmer: In fact A, B and C are found on the same cells as D. I agree that many of these ideas are still speculative. The reason for relating these to complement is a question of one's view of the nature of the whole restriction phenomenon, which I don't want to get into here.

Crick: If you looked at all the polypeptide chains expressed in those regions, would you expect to find unrelated ones which apparently are functioning quite disparately?

Bodmer: No, I would not.

Crick: So you make a firm prediction that all genes found in that region will be related. Now what do you mean by related?

Bodmer: They have some functional interrelationship, which may be hard to define! In this case I would say that the functional interrelationship is with respect to their effect on cell interactions. Complement plays a role like that and these HLA-A, B, C and D products play a role that is like complement but on the cell surface. That is my view of the role of these products in restriction.

Brenner: Could you expand that? I take it you can account for why all the other complement factors are elsewhere?

Bodmer: No, not entirely. We aren't meant to discuss complement yet! There is after all clear evidence for C4 on the cell surface.

Linkage disequilibrium is very important because it is through linkage disequilibrium that to a large extent one defines what goes together. We know that there is very strong linkage disequilibrium, for example, between certain alleles at the *A* and *B* loci and between the *B* and *D* and *B* and *C* loci. And although in some cases, like the *B* and *C* loci, it is hard to argue that this is due to selection—it could just be due to very close linkage—I think that where it involves *A* and *B, B* and *D,* and probably *D* and the complement factors, its nature in terms of persistence and population distribution argues that it is likely to be maintained by selection. My prediction is that if this is right, in a functionally interrelated group there would be linkage disequilibrium amongst the alleles of the constituent loci. But another gene which could be quite close by, like glyoxylase, whose distance from HLA is not much farther than the length of the HLA region, should not show linkage disequilibrium because it has no functional interrelationship. That is a firm prediction. Some evidence for it exists but we don't have enough information yet. I would expect that linkage disequilibrium of this sort of order would help to define the boundaries of the region, and that it is an essential concept to interrelate the products of that region. Obviously we have a lot to learn about the real explanations for those interrelationships and how they work.

Crick: So the origin of the region wasn't due to tandem duplication in macro as opposed to micro? That is a completely separate subsidiary hypothesis.

Bodmer: There are some problems there because of the way the A, B and C loci are distributed. One can speculate just so far, but in the end we have to find out what the facts are.

Pontecorvo: How far back in evolution does the main structure go?

Bodmer: It is found in all mammals that have been studied sufficiently. It is found in chickens and certainly in reptiles. I don't know how good the data in fish are.

Fincham: What is known about the molecular homology between these?

Bodmer: In mammals and birds they are clearly homologous products.

Ruddle: Is there any significance in the linkage of glyoxylase and H-2?

Bodmer: The question of linkage homology at the chromosomal level is a very difficult one. Ohno (1973) has made two suggestions which are very interesting. One is very right and one is very wrong. The one that is right concerns the conservation of the X chromosome. The one that I think is clearly wrong is the conservation of linkage in other chromosomes. One has to ask what is the significance of the conserved linkage which is beyond the cluster. Ohno said that the rate of substitution of break-points was so slow that there is every reason for expecting chromosome linkages to be maintained. I think the data simply contradict that idea. Even the crude data on chromosome-banding patterns and their changes, allowing for ascertainment problems, permit the production of break-points at a rate which would scramble up the genome. If the figure I once gave (Bodmer 1975) of about 6 to 12 per chromosome arm is right, then, since this is probably just one little band in the chromosome, the fact that the glyoxylase–H-2, glyoxylase–HLA linkage was conserved might not be significant. If we find that the T locus equivalent in man is 20% recombination fraction from HLA, one might begin to wonder if that is significant. These arguments are much more difficult to substantiate than the arguments about what happens within a cluster.

Brenner: You omitted to mention a very important difference between prokaryotes and eukaryotes which may account for some of the complex genes. In eukaryotes ribosomes cannot reinitiate on the same messenger; there is only one entry site. Prokaryotes can reinitiate on the same message, which is why operons can work.

Bodmer: This is connected with the whole complexity of processing of HLA.

Crick: That explains the tryptophan synthetase too.

Brenner: And it explains the results with the rudimentary locus in *Drosophila*.

Bodmer: That is part of the argument of the transition from prokaryotes to eukaryotes.

Brenner: I am just saying that many of the elements that you think may be together for other reasons may have to be together for this reason.

References

BODMER, W. F. (1975) in *Chromosome Variation in Human Evolution (Symp. Study Hum. Biol. 14)* (Boyce, A. J., ed.), pp. 53-61, Halsted Press, New York

MOLE, L. E., JACKSON, S. A., PORTER, R. R. & WILKINSON, J. M. (1971) Allotypically related sequences in the Fd fragment of rabbit immunoglobulin heavy chains. *Biochem. J. 124,* 301-318

OHNO, S. (1973) Ancient linkage groups and frozen accidents. *Nature (Lond.) 244,* 259-262

The genetics of the complement system

P. J. LACHMANN and M. J. HOBART

MRC Group on Mechanisms in Tumour Immunity, The Medical School, Cambridge

Abstract It is suggested that the classical and alternative pathways of C3 activation and the terminal complement reaction pathway represent three replicates formed by gene duplication. It seems likely that the precursor system was what is now the alternative pathway and that its original function was a positive feedback loop for amplifying inflammatory reactions. These views are supported by physicochemical data on the complement components; by genetic data; and by similarities in the reaction patterns.

Study of subjects with isolated complement component deficiencies provides information on the function of complement *in vivo*. It has been found that complement component deficiency is associated with bacterial infection, particularly with *Neisseria;* and even more so with immune complex disease. This second, rather surprising, association is likely to be the result of a requirement for complement in the catabolism of immune complexes on the one hand, and the need for an intact complement system to eliminate low virulence infectious organisms on the other.

Complement is familiar as the plasma constituent that lyses antibody-coated red cells and is consumed by antigen–antibody reactions. It is also fairly well known as a triggered enzyme cascade with forbidding nomenclature. Something is now known of its genetics and an attempt is to be made here to interpret this from an 'evolutionary' point of view. Evolutionary in this context is a theoretical concept since no convincing biochemical fossils with an incomplete complement system have been discovered. The most primitive group that has been looked at in any detail is the shark (Ross & Jensen 1973*a*, *b*) which seems to have a complement system quite analogous to that found in mammals.

The current terminology of the complement system is shown in Table 1. The central, most abundant and biologically most important of the complement components is C3. It is activated by the classical and alternative pathways and

TABLE 1

Complement terminology

C1 C4 C2 } C3 } C5 C6 C7 C8 C9	Classical pathway
	Terminal pathway
FD C3 FB FP }	Alternative pathway
C$\bar{1}$-inhibitor C3b-inactivator (KAF) β1H-globulin (and others)	Control proteins
Fragments e.g. C3a C3b	
Activated components e.g. C$\bar{1}$ C$\overline{42}$	
Inactivated components e.g. C3bi	
Short-lived bindings sites e.g. C3b* (nascent C3b)	

is the link to the terminal pathway. The components of the classical pathway are given numbers, prefixed with a C and are listed (but not numbered) in the order in which they interact. The components of the terminal pathway which produces cellular lysis are similarly numbered. Although it was recognized a long time ago that there were alternative methods of activating C3 that did not use the classical complement components, it was not until the 1950s that Pillemer and his group (1954) recognized that there were further and distinct components involved in this alternative pathway of C3 activation. These are now designated as factors, abbreviated as F followed by a capital letter. Pillemer and his group discovered a protein, called properdin (Factor P) about which little will be said since nothing is known of its genetics. Pillemer and colleagues then discovered that two other factors were also involved. One of these, called Factor A, resembled C4 and to everyone's surprise turned out to be C3 itself. The second, called Factor B, was very similar to C2 and was indeed a new factor. Subsequently, further factors involved in the alternative pathway were discovered. Of these, only one survived critical scrutiny as a separate component. This is Factor D which has some similarities with but also considerable differences from C$\bar{1}$s. The fourth group of proteins listed in Table 1 comprises the so-called control proteins. The C$\bar{1}$ inhibitor is a competitive substrate which inhibits C$\bar{1}$ in a stoichiometric fashion. The C3b inactivator (KAF) is an enzyme which destroys C3b (and C4b). For this purpose it requires a co-factor which is still described by its immuno-electrophoretic designation, β1H globulin.

When complement components become fragmented during the complement activation sequence the fragments are given small letter suffixes, the smaller fragment being given the suffix 'a' and the larger suffix 'b' in most cases. Where complement components acquire new biological activities (usually enzymic) as a result of activation this is designated by a bar over the activated component. When an activity is lost as a result of further fragmentation this may be designated by the suffix 'i'.

We wish to suggest that the three pathways, the classical pathway, the alternative pathway and the terminal pathway, are likely to represent replicates formed by gene duplication; that they are variations on a common theme; and that the 'archaeo-theme' is the alternative pathway. Fig. 1 shows complement components divided into families showing recognizable similarities to each other. The first of these comprises C3 and C4 and C5. These are all proteins which show characteristic susceptibilities to chemical inactivation. They are all destroyed by chaotropic ions and by primary amines and they are all activated during the complement sequence by proteolytic cleavage to a small fragment (which with C3 and C5 have very similar biological activities) and a large fragment which (with C4 and C3 in particular) has a short-lived strong binding site. In serum, C3 and C5 have markedly similar chain structures whereas C4 has an extra third chain. It is, however, now believed that the components of this family are all likely to be synthesized within the cell as single chains (Brade *et al.* 1977; Hall & Colten 1977) and split post-synthetically rather like insulin. The second family comprises C2 and Factor B, two proteins which are similar physicochemically, and possibly two other proteins, C6 and C7. These are again similar to each other and show a less close similarity to C2 and Factor B. All these are single-chain proteins with varying degrees of heat susceptibility. Factor B is the most heat-labile followed by C2, then C6 and finally C7, which is rather heat-resistant. They show similarities of reaction patterns, Factor B reacting with C3b, C2 with C4b and C6 with C5b. However, while C3bB and C4b2 give rise to proteolytic enzymes, C5b6 is the initiating step of the multimolecular self-associating complex which eventually produces cell lysis. The genetic evidence for the interrelatedness of the members of these families will be shown below.

Table 2 lists the distinct loci coding for complement components which have so far been described. There are three loci known for human complement components, identified partly by studies of charge polymorphisms and partly by the study of isolated complement component deficiencies. The first of these loci is extremely close to the B locus of HLA. It is

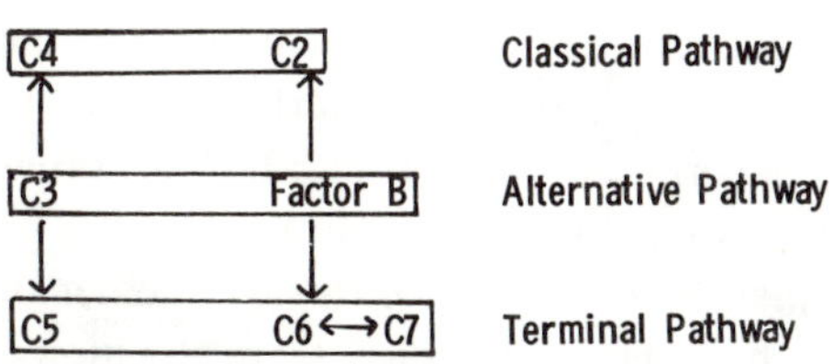

FIG. 1. 'Families' of complement components. Arrows connect components which may be related and which may have evolved in the directions shown.

TABLE 2

Complement loci

Components	Linkage data
C4, C2, Factor B	Very close to B locus of HLA
C6, C7	Unassigned
C3	Loose linkage to Lewis blood groups (N.B. In mice loose linkage to *H-2*)
C1 C5 C8 Factor D C$\bar{1}$ inhibitor C3b inactivator	Unassigned and show no linkage to any of the three loci above. (Insufficient data to exclude linkage within this group of components)

doubtful whether any convincing recombinants between any of the complement components coded at this locus—factor B, C2 and C4—and the B locus of HLA have so far been described. On the other hand there are recombinants reported between the D locus and these complement components. The three components in the HLA region include two that are homologues in the alternative pathway (Factor B) and classical pathway (C2) and two (C4 and C2) which act sequentially in the classical pathway. However, the protein that is the homologue of C4 in the alternative pathway (C3) is, in man, not within mapping distance of HLA. It is therefore of considerable interest that da Silva *et al.* (1978), using Swiss-Webster mice, have recently published data showing that murine C3 can be mapped at about 12 centimorgans from *H-2*. This supports the idea that there is a difference between the major histocompatibility complex in the mouse and that in other species. The human C3 locus has been claimed to show loose linkage to the Lewis blood groups (Weitkamp *et al.* 1974) and to certain genes controlling the serum cholesterol level (Ott *et al.* 1974; Berg & Heiberg 1976; Elston *et al.* 1976) but is not otherwise assigned. The third locus is that which codes for C6 and C7. These two components are coded very close to each other and since a very unusual complement deficiency involving both these components has been discovered (Lachmann *et al.* 1978) it is possible that they are in fact contiguous in the genome. However, no convincing linkage with any other marker has so far been established. There is genetic information on a number of other components as shown in Table 3. These do not show close linkage to any of the three established loci but there is insufficient evidence to say whether or not they show any linkage with each other. However, data from the mouse show that C5, the homologue of C3 in the terminal pathway, is not coded within mapping distance of *H-2* and it has been assigned to chromosome 5 of

TABLE 3

Unassigned complement loci

Component	HLA linkage excluded by	Gene frequencies (allotypes)			
			Commonest	Second	Others
C1r	Deficiency	No data			
C3	Allotypes	Caucasoids	0.77	0.22	Rare
	Deficiency	Mongoloids	0.99	<0.01	
		Negroids	0.93	0.06	
C5	Deficiency	Caucasoids and negroids—very low or absent (0/2262)			
C6	Allotypes Deficiency	All races	0.6	0.35	0.5
C7	Allotypes Deficiency	Caucasoids	0.99	<0.01	<0.01
C8	Allotypes Deficiency	Negroids	0.99	6 rare	
Factor D	No data	Negroids	0.98	0.02	

Data from: Vaz Guedes *et al.* (1978); Alper *et al.* (1975); Mittal *et al.* (1976); Rittner *et al.* (1976).

the mouse. The genetic arrangements of complement components are, therefore, somewhat complicated. If they did all arise by gene duplication within their 'family groups', some have stayed together while others have drifted apart. It would seem a little difficult to accept that C4 and C2 and Factor B arose from the same nucleotide sequence that also gave rise to the other HLA products (Bodmer, this symposium) whereas C3 and C5 on the one hand and C6 and C7 on the other hand do not arise from this sequence. C4 is very much more like C3 and C5 than it is like C2; and C2 and Factor B show much greater similarity to C6 and C7 than they do to C4.

We would like to suggest that the original complement system, the 'archaeo-complement', may have had nothing particularly to do with immunology, that it is essentially what is now the alternative pathway, and that it functioned originally as an amplification system for inflammatory reactions (Fig. 2). This function would have needed only two components, C3 and Factor B. Proteolytic enzymes derived from inflammatory cells can split C3, as can plasmin and thrombin, to generate C3a and C3b. The C3b then combines with Factor B in the presence of magnesium ions and the complex of C3b and B is itself a C3-splitting enzyme (albeit a rather weak one) so that positive feedback occurs. The Factor B in the complex can be split to Bb and Ba, the latter being released by a variety of proteolytic enzymes. In serum, where there are many proteolytic inhibitors, there is a specific enzyme (Factor

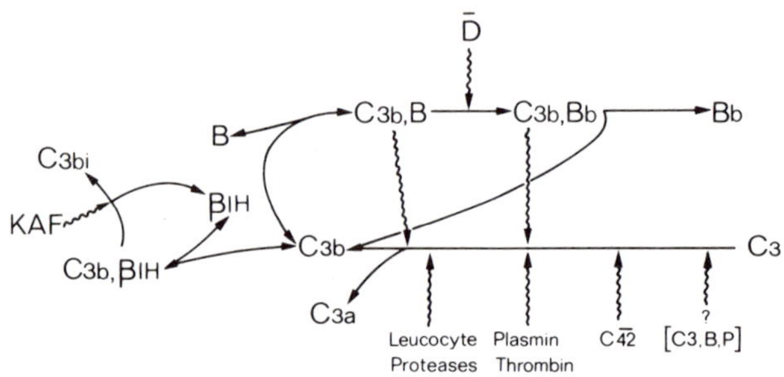

FIG. 2. The alternative pathway. C3 is activated by many proteases, including those produced by inflammatory cells. C3a is a mediator of inflammation and C3b is an opsonin. Bound C3b also plays the triggering role in the feedback cycle which leads to the formation of a C3 convertase (C3b,Bb). Wavy arrows denote catalysis; straight arrows denote conversion.

D) which cleaves C3b, B. This cleavage gives rise to a much more powerful C3 splitting enzyme, C$\overline{3bBb}$, which is the principal 'feedback C3 convertase'. This then splits more C3, giving more C3b, allowing the formation of more enzyme and providing a powerful amplification mechanism for fixing more C3b on bacteria or for promoting inflammation. A number of the factors produced in the C3b-feedback cycle have biological activity. For example, C3a releases mediators from mast cells and will attract leucocytes to the site; and Bb is a 'macrophage spreading factor' (Götze *et al.* 1978). To prevent this cycle going to exhaustion, control processes have evolved. These, also, are more complex than was once thought. Essentially, though, there is an enzyme in plasma that splits C3b to an inactive product (C3bi) which will no longer combine with Factor B. The extent to which the feedback cycle works depends on the rate of C3b-generation on the one hand and the rate of its destruction on the other. There are, for example, rare humans with isolated deficiencies of the C3b-inactivator. In consequence, without the need for any other activation, the whole cycle runs to exhaustion when no Factor B or no C3 are left.

Gene duplication seems to have produced an enzyme, the classical pathway C3 convertase, that is highly efficient at producing rapid conversion from C3 to C3b in serum. It provides the 'interface' with antibody-mediated reactions and can also trigger the amplification cycle via C3b. Fig. 3 shows how closely analogous the classical pathway is to the alternative pathway. The significant difference is that the same enzyme that splits C4 to C4b in the classical pathway (C$\overline{1}$) also splits C4b,2 to the C3 convertase (C4b2a), whereas this does not occur in the alternative pathway (i.e. $\overline{\text{D}}$ does not split C3) and if it did all the control mechanisms would fail. It has been suggested that D may be a

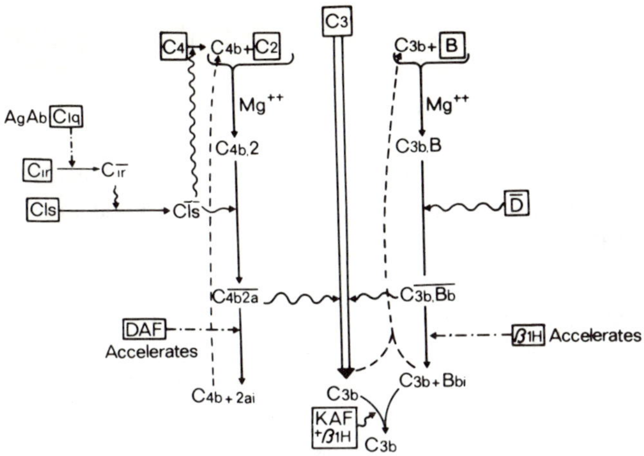

FIG. 3. The classical and alternative pathways are drawn to emphasize their similarity. Native serum components are shown in boxes.

thrombin fragment (Davis *et al.* 1978) but since we have found that levels of $\bar{D}$ are not affected by prolonged warfarin treatment and the enzyme is not inhibited by hirudin this seems unlikely. There is not enough genetic information or enough structural data to show whether C$\bar{1}$ and $\bar{D}$ are in any way related.

There is thus good evidence that the classical pathway and the alternative pathway are evolutionarily related and probably duplicates, and reason to believe that the alternative pathway is the older.

The evidence for the terminal pathway (shown in Fig. 4) is much less direct. The terminal pathway is not an enzyme cascade but the formation of a self-associating complex which lyses cells. However, the initial stages are very

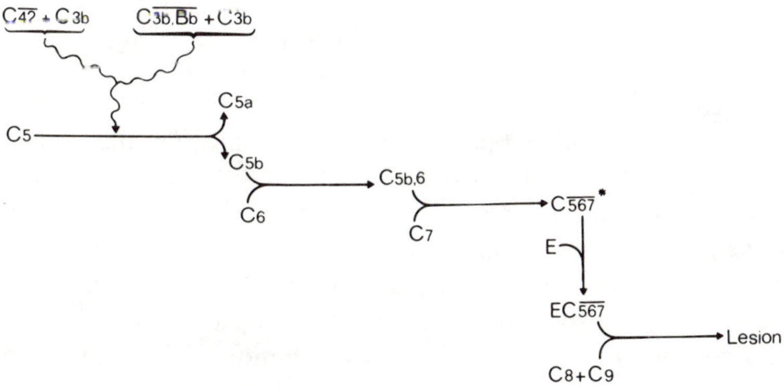

FIG. 4. The terminal pathway.

similar to the other pathways. C5 is split into C5b and C5a and, just as C2 reacts with C4b and Factor B with C3b, C6 reacts with C5b. However, in this instance there is a further protein, C7, to be added to the complex before it binds stably to the cell membrane. Too little is known about C8 and C9 to conjecture whether they are analogous to anything else in the complement system but C6 and C7 are closely similar to each other and bear a family resemblance to C2 and Factor B.

POLYMORPHISM OF COMPLEMENT COMPONENTS

Charge polymorphisms have been described for a number of complement components. These include the three components, C4, C2 and Factor B, that are linked to HLA and that have already been discussed in this symposium (Bodmer). O'Neill *et al.* (1978) have reported that the two common C4 variants, C4F and C4S, are isotypes and that the haplotype C4F has a 'null' or silent allele at the C4S locus, and vice versa. The simplest explanation for this situation is that duplication of the C4 locus arose by unequal crossing-over between two different C4 alleles and that this duplication is itself distributed polymorphically in the population (cf. haptoglobin). In this case, the reduplicated haplotype C4FS will make twice as much C4 as either of the single haplotypes, C4F and C4S, and this is indeed what O'Neill and his colleagues have found. The haplotype frequency for the C4FS is about 0.2, taking the data of Teisberg *et al.* (1977). It is not yet known whether the duplication affecting the C4 locus may have affected the C2 and Factor B loci, and this is now being studied.

Of the complement polymorphisms not linked to HLA (Table 3) two are well studied and at high frequency. The first of these is C3, which is sufficiently common to be useful in genetic mapping although the C3F gene (the second most common allele) is found at high frequency only among Caucasians. By contrast the gene frequencies for the other common polymorphism, C6, are remarkably constant through all the racial groups so far tested (Table 3). This suggested to us that polymorphism must at least be old enough for the variant gene to have disseminated throughout all human population groups and also raises the question of whether there is selective pressure maintaining the polymorphism. This would be surprising since there are no functional differences known between the two C6 alleles and even deficiency of C6 is not incompatible with health. The closely linked locus for C7 shows in man a polymorphism where the second most common allele is distinctly infrequent. The polymorphisms of these two closely linked components have now been looked at in a number of different species by P. Eldridge & M. J.

Hobart (unpublished observations 1978) to see whether there is any suggestion of a selective pressure maintaining polymorphism at one or other of these loci. The results are shown in Table 4, which is restricted to mammalian species. It can be seen that occasionally the assay fails because of incompatibilities among the various complement components but, of the species which could be typed for both components, four species showed common polymorphisms for both C6 and C7; five species for C7 only; eight species for C6 only; and eight for neither C6 nor C7. This 'random' distribution of C6 and C7 polymorphisms throughout different mammalian species, some of which are quite closely related (for example, five of these are different species of cat), suggest to us that there can be no major selective pressure maintaining these polymorphisms and that these are likely to be neutral mutations.

TABLE 4

C6 and C7 polymorphism in various mammalian species (numbers of species). (Data from P. Eldridge & M. J. Hobart, unpublished observations 1978)

		C6 polymorphism		
		+	–	Assay failed
	+	4	5	0
C7 polymorphism	–	8	8	1
	Assay failed	4	3	

'Polymorphic' means at least two band patterns observed by isoelectric focusing. 'Non-polymorphic' means only a single band pattern seen in at least three individuals. Some polymorphisms have been confirmed by family studies.

ISOLATED COMPLEMENT COMPONENT DEFICIENCIES

In recent years there has been a great increase in the measurement of complement components in hospital laboratories and one result of this has been the discovery of an increasing number of subjects with what appear to be genetic isolated complement component deficiencies. These have been described for all the components of the classical and terminal pathways except for C1q and for C9 and, even here, it has been rumoured that such patients have recently been encountered. Isolated deficiencies are also known for two of the control proteins, $C\bar{1}$ inhibitor and C3b inactivator but, interestingly enough, none have so far been described for Factor B, Factor D or Factor P of the alternative pathway. The failure to find deficiencies of these alternative pathway components may reflect their great importance but it is more likely to be the result of inadequacy of techniques used for measuring complement in the clinic. These would not normally pick up deficiencies of alternative pathway factors and so the number of patients screened for such deficiencies is very much smaller than for those of the other components. Table 5 lists the number of affected subjects known to the authors. This was updated recently but even so is unlikely to represent the present state of knowledge.

By far the commonest isolated complement deficiency in man is that of C2, which is almost as common as all the rest put together. About half the C2-deficient patients are healthy and about half of them suffer from immune complex diseases, of which systemic lupus erythematosus is the most common. Systemic lupus erythematosus is a widespread vasculitic disorder where immune complexes involving various auto-antigens, particularly derived from cell nuclei, are deposited in blood vessels in the kidney and in many other sites, giving rise to a serious disease which occurs predominantly in young women. Rather similar diseases have been described in mice and dogs and in these species there has been some suggestion that the disease is associated with oncornavirus infection. This association of immune complex disease with complement deficiency is surprising and the mechanisms that may possibly be involved are discussed below.

Another group of diseases which one would anticipate finding among

TABLE 5

Clinical associations of complement deficiency

			Immune complex disease			*Bacterial defects*	
Component	*Total no. of affected subjects*	*Healthy*	*SLE or SLE-like syndrome*	*GN*	*Other*	*Multiple (pyoder mal etc.)*	*Neisserial*
C3	5	0	0	0	2	4	0
C3b inactivator	4	1	0	0	0	1	2
C5	4	1	1	0	0	0	2
C6	5	1	0	0	0	0	4
C6 + C7	1	1	0	0	0	0	0
C7	9	2	0	1	2[a]	0	4
C8	9	4	3	0	0	0	2
C1r	3	0	2	1	0	2	0
C4	2	0	2	0	0	0	0
C2	>40	~20	~10	~5	6[b]	0	0
C1 inhibitor	many	—	7	11	0	0	0

[a]1 Raynaud's disease; 1 rheumatoid arthritis
[b]3 Henoch-Schönlein purpura; 1 rheumatoid arthritis; 1 chronic vasculitis; 1 polymyositis
SLE: systemic lupus erythematosus; GN: glomerulonephritis

patients with complement deficiency is bacterial infection. This is indeed found from time to time. In a number of patients who have deficiencies of C3, whether these are primary or secondary to deficiencies of C3b inactivator, multiple bacterial infections have been encountered. It is of interest, however, that occasionally patients with apparently identical serum findings appear to have a relatively normal immunity to bacterial infections and to suffer only from the third main clinical concomitant of complement deficiency, which is infection with *Neisseria*. This is manifested either as repeated attacks of meningococcal meningitis or of disseminated gonococcal sepsis. Groups of patients with these two forms of neisserial infection have proved a very rewarding hunting ground for finding complement-deficient subjects, and deficiencies of many complement components, not only of the terminal pathway, but also, for example, C3b inactivator, have now been found that show this syndrome. The association of these three syndromes with complement component deficiencies is shown in Table 6.

One other deficiency that must be mentioned is that of the $C\bar{1}$ inhibitor. Heterozygous deficient individuals tend to suffer from the disease known as hereditary angio-oedema, which is due to the production of a kinin-like fragment from C2 by the combined action of plasmin and $C\bar{1}$. It is, however, interesting that patients with $C\bar{1}$ deficiency also tend to have very low levels of C2 and C4 throughout their lives, this being due to hypercatabolism rather than lack of synthesis. Nevertheless, these patients, just like subjects with genetic deficiencies of C2 and C4, appear to have an increased incidence of immune complex diseases such as glomerulonephritis and systemic lupus erythematosus (see Lachmann & Rosen 1978).

The association of complement deficiency with bacterial infections is not surprising and even the particular association with neisserial infections makes sense, in as much as these are organisms that are lysed in the circulation, but

TABLE 6

Clinical associations of complement deficiency

Bacterial infection	
General (especially pyococcal)	C1r (2:3); C3 (4:5); KAF (1:4)
Neisserial	KAF (2:4); C5 (2:4); C6 (4:5); C7 (4:9); C8 (2:9)
Immune complex disease	C1r (3:3); C1s (1:1); C4 (2:2); C2 (~50%); C5 (1:4); C8 (3:9); $C\bar{1}$-inhibitor (raised incidence, about 18 described cases).

Numbers in parentheses show the number of subjects affected compared with the total number described with that deficiency.

the association with immune complex disease is puzzling. Since complement is known to be a major mediator of immune complex damage one might have expected the contrary finding that patients with complement deficiencies should show a freedom from immune complex disease. For this reason the nature of the association of immune complex disease with complement deficiency has received some attention. It has, for example, been suggested that the association is an ascertainment artifact due to the phenomenon that it is patients with putatively immunological diseases who tend to have their complement levels measured and that it is therefore among this group that complement deficiencies are most commonly discovered. This is a plausible point and difficult to exclude entirely because the amount of data on the true incidence of complement deficiency is rather small. F. S. Stratton (personal communication) found one C2-deficient subject among 10 000 or more donors on the Manchester Blood Transfusion Panel (a group who are entirely healthy). In a study of the Cambridge outpatient population—who are all sick—we found one genetic complement component deficiency among some 3500 outpatients. There are older studies where it is not at all clear that any of the deficiencies found were in fact genetic. However, taking as an approximate figure the incidence of homozygous complement component deficiency in the normal population as around 1 in 10 000, and a figure for homozygous complement deficiency among patients with systemic lupus erythematosus of 1 in 100 to 1 in 200 (Glass *et al.* 1976), an average relative risk of between 25 and 50 can be calculated. This makes it unlikely, particularly for C2 deficiency, that the association with systemic lupus is an ascertainment artifact. For hereditary angio-oedema the ascertainment is very easy and here it is also clear that the incidence of lupus is at least 10 times that expected in the control population.

The second possibility that has been discussed is that the complement deficiency is indeed associated with immune complex diseases but the association is not causal, linkage disequilibrium with a true susceptibility gene being responsible. Such suggestions are made particularly for C2 and C4 deficiency, components that are linked to HLA. Here the suggestion that there may be an immune response gene in linkage disequilibrium with the complement deficiency is quite plausible. It would, however, be surprising that if there were a particular immune response gene in linkage disequilibrium with C2 deficiency it should also show the same linkage disequilibrium with C4 deficiency. Nor would such an explanation operate for the same diseases occurring in C1r deficiency (which does not show linkage to HLA) and above all in hereditary angio-oedema where the gene is certainly not linked to HLA and where the complement component deficiency is in fact due to hypercatabolism. The data on heterozygotes, which should be informative in this respect, are, in fact, highly confusing. There are a number of diseases that have been claimed to be commoner in 'apparent heterozygotes' for C2 deficiency than in the control population. These include some such as the atopic state (Turner *et al.* 1978) and multiple sclerosis (Trouillas & Betuel 1977) which have never so far been described in homozygous C2 deficiency, as well as an increased incidence of rheumatic diseases which are commoner among the homozygotes (Glass *et al.* 1976). Since the ascertainment of heterozygote complement component deficiencies in families where there are no homozygous deficiency states is very difficult, these data have to be taken with some reserve. The third possibility is that there is a causal relationship between complement component deficiency and immune complex disease. This is the explanation which we favour. Two types of mechanism may be involved. The first stems from the findings of Miller & Nussenzweig (1974, 1975) that an intact complement system is needed to catabolize immune complexes properly and that, in the absence of an efficiently working complement system, immune complexes may persist bound to cell receptors for abnormally long periods and perhaps be particularly capable of generating inflammatory reactions. Another class of explanation, to which perhaps more weight should be given, is the growing evidence that an intact complement system is required to deal with certain infections, particularly with viruses which may not themselves be pathogenic but which may be capable of giving rise to immune complex disease under conditions where their elimination is impaired. Although there is no direct evidence that such mechanisms are involved in human disease, a number of models of this sort have been described. For example, oncornaviruses are lysed in the absence of antibody by normal human serum (Welsh *et al.* 1976). Vesicular stomatitis virus is similarly lysed by normal human serum, although curiously enough low density lipoproteins are required as well as an intact complement system (Cooper *et al.* 1976). Cells infected with measles virus can be shown to fix complement by the alternative pathway in the presence of large amounts of antibody. Joseph *et al.* (1975) have, however, found that if such cells are exposed to antibody in the absence of an intact complement system, the measles virus antigens are modulated from the cell surface by capping and extrusion. A cell results which still contains measles virions but which, having no surface bound antigens, is protected from immunological attack. This is a model of a mechanism by which an inoperative complement system may enable virus infection to become sequestered in such a way that it can continue to shed antigens without being eliminated, and this is an attractive model for a disease such as subacute sclerosing panencephalitis. We have ourselves been involved in studies of the antibody-independent lysis by the alternative pathway of cells transformed by the EB virus (Budzko *et al.* 1976; McConnell *et al.* 1978), and this may again be a mechanism by which early transformed cells are normally sequestered. It is possible to envisage situations where a complement system compromised by other infections, e.g. malaria, may be incapable of eliminating EB-virus-transformed cells during a primary infection and that this may have pathogenic sequelae later on.

If explanations of this kind are right it seems likely that the selective pressures that maintain an intact complement system are the same as those that may maintain the polymorphism of the major histocompatibility locus, that is, the requirement to have optimal resistance to infectious disease.

References

ALPER, C. A., HOBART, M. J. & LACHMANN, P. J. (1975) Polymorphism of the sixth component of complement, in *Isoelectric Focusing* (Arbuthnott, J. P. & Beeley, J. A., eds.), pp. 306-312, Butterworths, London

BERG, K. & HEIBERG, A. (1976) Linkage studies on familial hyperlipoproteinaemia with xanthomatosis: normal lipoprotein markers and the C3 polymorphism. *Birth Defects 12,* 266-270

BODMER, W. F. (1979) Gene clusters and the HLA system, this symposium, pp. 205-223

BRADE, V., HALL, R. E. & COLTEN, H. R. (1977) Biosynthesis of pro-C3, a precursor of the third component of complement. *J. Exp. Med. 146,* 759-765

BUDZKO, D. B., LACHMANN, P. J. & MCCONNELL, I. (1976) Activation of the alternative complement pathway by lymphoblastoid cell lines derived from patients with Burkitt's lymphoma and infectious mononucleosis. *Cell. Immunol. 22,* 98-109

COOPER, N. R., JENSEN, F. C., WELSH, R. M. & OLDSTONE, M. B. A. (1976) Lysis of RNA tumour viruses by human serum: direct antibody independent triggering of the classical complement pathway. *J. Exp. Med. 144,* 970-984

DA SILVA, P., HOECKER, G. F., VIENNE, K. & RUBINSTEIN, P. (1978) Murine complement component C3: genetic variation and linkage to H-2. *Proc. Natl. Acad. Sci. U.S.A. 75,* 963-965

DAVIS, A. E., ROSENBERG, R. D., FENTON, J. W., BING, D. H., ROSEN, F. S. & ALPER, C. A. (1978) Is Factor D of the alternative pathway a fragment of thrombin? *J. Immunol. 120,* 1771 (abstr.)

ELSTON, R. C., NAMBOODIRI, K. K., GO, R. C., SIERVOGEL, R. M. & GLUECK, C. J. (1976) Probable linkage between essential hypercholesterolaemia and the third complement component (C3). *Birth Defects 12,* 294-297

GLASS, D., RAUM, D., GIBSON, D., STILLMAN, J. S. & SCHUR, P. H. (1976) Inherited deficiency of the second componenent of complement. Rheumatic disease association. *J. Clin. Invest. 58,* 853-861

GÖTZE, O., BIANCO, C. & COHN, Z. A. (1978) The induction of macrophage spreading factor by factor B of the properdin system. *J. Immunol. 120,* 1776 (abstr.)

HALL, R. E. & COLTEN, H. R. (1977) Cell-free synthesis of the fourth component of guinea pig complement (C4): identification of a precursor of serum C4 (pro-C4). *Proc. Natl. Acad. Sci. U.S.A. 74,* 1707-1710

JOSEPH, B. S., COOPER, N. R. & OLDSTONE, M. B. A. (1975) Immunologic injury of cultured cells infected with measles virus. 1. Role of IgG antibody and the alternative complement pathway. *J. Exp. Med. 141,* 761-774

LACHMANN, P. J. & ROSEN, F. S. (1978) Genetic defects of complement in man. *Semin. Immunopathol. 1,* 339-353

LACHMANN, P. J., HOBART, M. J. & WOO, P. M. (1978) Combined genetic deficiency of C6 and C7 in man. *Clin. Exp. Immunol. 33,* 193-203

McCONNELL, I., KLEIN, G., LINT, T. F. & LACHMANN, P. J. (1978) Activation of the alternative complement pathway by human B cell lymphoma lines is associated with Epstein Barr virus (EBV) transformation of the cells. *Eur. J. Immunol. 8,* 453-458

MILLER, G. W. & NUSSENZWEIG, V. (1974) Complement as a regulator of interactions between immune complexes and cell membranes. *J. Immunol. 113,* 464-469

MILLER, G. W. & NUSSENZWEIG, V. (1975) A new complement function: solubilisation of antigen-antibody aggregates. *Proc. Natl. Acad. Sci. U.S.A. 72,* 418-422

MITTAL, K. K., WOLSKI, K. P., LIM, D., GEWURZ, A., GEWURZ, H. & SCHMIDT, F. R. (1976) Genetic independence between the HLA system and deficits in the first and sixth components of complement. *Tissue Antigens 7,* 97-104

O'NEILL, G. J., YOUNG-YANG, S., TEGOLI, J., BERGER, R. & DUPONT, B. (1978) Chido and Rodgers blood groups are distinct antigenic components of human complement. *Nature (Lond.) 273,* 668-670

OTT, J., SCHROTT, H. G., GOLDSTEIN, H. C., HAZZARD, W. R., ALLEN, F. H., FALK, C. T. & MOTULSKY, A. G. (1974) Linkage studies in a large kindred with hypercholesterolaemia. *Am. J. Hum. Genet. 26,* 598-603

PILLEMER, L., BLUM, L., LEPOW, I. H., ROSS, O. A., TODD, E. W. & WARDLAW, A. C. (1954) The properdin system and immunity. 1. Demonstration and isolation of a new serum protein, properdin, and its role in immune phenomena. *Science (Wash. D.C.) 120,* 279-285

RITTNER, CH., OPFERKUCH, W., WELLEK, B., GROSSE-WILDE, H. & WERNET, P. (1976) Lack of linkage between gene(s) controlling the synthesis of the seventh component of complement and the HLA region on chromosome No. 6 in man. *Hum. Genet. 34,* 137-142

ROSS, G. D. & JENSEN, J. A. (1973*a*) The first component (C1n) of the complement system of the nurse shark (*Ginglymostoma cirratum*). I. Hemolytic characteristics of partially purified C1n. *J. Immunol. 110,* 175-182

ROSS, G. D. & JENSEN, J. A. (1973*b*) The first component (C1n) of the complement system of the nurse shark (*Ginglymostoma cirratum*). II. Purification of the first component by ultracentrifugation and studies of its physicochemical properties. *J. Immunol. 110,* 911-918

TEISBERG, P., OLAISEN, B., JONASSEN, R., GEDDE-DAHL, T. Jr. & THORSBY, E. (1977) The genetic polymorphism of the fourth component of human complement. Methodological aspects of and a presentation of linkage and association data relevant to its localization in the HLA region. *J. Exp. Med. 146,* 1380-1389

TROUILLAS, P. & BETUEL, H. (1977) Hypocomplementaemic and normocomplementaemic multiple sclerosis. Genetic determinism and association with specific HLA determinants (B18 and B7). *J. Neurol. Sci. 32,* 425-435

TURNER, M. W., MOWBRAY, J. F., HARVEY, B. A. M., BROSTOFF, J., WELLS, R. S. & SOOTHILL, J. F. (1978) Defective opsonisation and C2 deficiency in atopic patients. *Clin. Exp. Immunol. 34,* 253-259

VAZ GUEDES, M. A., HOBART, M. J. & LACHMANN, P. J. (1978) Absence of variation in human C5. *J. Immunogenet. 5,* 279-282

WEITKAMP, L. R., JOHNSTON, E. & GUTTORMSEN, S. A. (1974) Probable genetic linkage between the loci for the Lewis blood groups and complement C3. *Cytogenet. Cell Genet. 13,* 183-184 (abstr.)

WELSH, R. M., JENSEN, F. C., COOPER, N. R. & OLDSTONE, M. B. A. (1976) Inactivation and lysis of oncornaviruses by human serum. *Virology 74,* 432-440

Discussion

Nossal: Has any evidence accrued that cancer is more common in humans with complement deficiency affecting either the classical or the alternative pathway than in other people?

Weissmann: Perhaps we should first ask whether people come down with leukaemia, say, because of oncornavirus.

Nossal: It is a circular argument, of course.

Lachmann: I know of no report of complement-deficient subjects with tumours. We recently lost a complement-deficient patient with cancer but he was 69 and that is probably not unusual. An attractive speculation about why oncornaviruses cannot be found in humans is that they are immediately lysed and therefore one should look only for proviral forms. There is however no evidence that subjects with complement deficiency have complete oncornaviruses or even raised antibody titres to them.

Henry Harris: How long do these people live?

Lachmann: They vary very much. About half the C2-deficient subjects live a normal lifespan. It is not entirely coincidental that the two subjects were immunologists.

Koprowski: I was never very satisfied with the explanation of the dengue shock syndrome. If you were to study the south-east Asian population, do you think you would find complement deficiencies that could explain why that particular population is more susceptible to dengue shock syndrome than other populations?

Lachmann: In south-east Asia epidemics of dengue have occurred in recent decades. Children who had a second attack of dengue with a different serotype of the virus were found sometimes to suffer what seems to be a severe immune complex disease instead of getting a straightforward viral disease. These children had marked complement activation and, in severe cases, a syndrome resembling endotoxic shock with disseminated intravascular coagulation, which could cause death. There are of course no prospective studies but if a pre-existing complement deficiency was responsible it would have to be extremely common. In the big epidemics in south-east Asia a large percentage of children who had a second attack of dengue within two years appear to have been afflicted.

Weatherall: This has happened in other areas as well, such as the Caribbean.

Cavalli-Sforza: I disagree with two of your evolutionary interpretations. For C6 you found frequency values which are basically the same all over the world, so you thought it must be a very old polymorphism of neutral mutants.

But that doesn't really follow because there is likely to have been enough drift to differentiate these frequencies, if the polymorphism is really neutral. In fact this might be an interesting example of natural selection of stabilizing type, which is almost always due to the advantage of the heterozygote. This is not terribly new; sickle cell anaemia is a perfect case. But sickle cell anaemia varies a lot in different parts of the world, simply because the selective conditions vary a lot. So one also has to postulate in the case of C6 that the same selective conditions exist around the world. This may be strange but is worth considering as an important alternative to the neutral argument.

The other point referred to the independence between C6 and C7 in the frequency with which the polymorphism appeared or didn't appear in different species. Again I don't see how that can follow from a neutral argument. All it means is that the two are independent in their polymorphisms.

Lachmann: If polymorphisms in two tandemly duplicated genes whose products have intimately related functions are maintained by heterozygote advantage, it would surely be reasonable to expect the selective pressure giving the advantage to be seen similarly in closely related species. That the occurrence of either or both C6 and C7 polymorphisms appears to occur at random in various species therefore suggests to us that they are not maintained by selective pressures.

Siniscalco: I would also like to know about the relationship between C3 and the Lewis substances.

Lachmann: That is a loose linkage and not everyone is convinced of it (Weitkamp *et al.* 1975). One trouble is that Lewis typing requires saliva which is frequently not available.

Siniscalco: Is there any chance that C3 is a Lewis substance?

Lachmann: Presumably not or the linkage would be very tight.

Bodmer: Lewis substances are glycolipids, as far as we know. C3 is a protein that is well characterized so we would have to talk about sugars on it.

Loose linkage may well exist between *H-2* and mouse C3 but a 12 centimorgan average could be anywhere from 5 to 50, so it is dangerous to say anything about the relationship to the *H-2* region. It is quite clear that *H-2* and HLA, within which there is a recombination factor of 1–2%, is already about the maximum one would expect for an interrelated system of genes with linkage disequilibrium. The loose linkage of C3 to *H-2* may have nothing to do with that being a part of *H-2*.

Siniscalco: O'Neill *et al.* (1978) have shown a correlation between the absence of C4 fast bands and of Rodgers blood group antigen as well as between the absence of C4 slow bands and of the Chido blood group. I understand that the latter two gene products have already been proved to be under the control of different loci. Couldn't Chido and Rodgers antigens simply be the result of adsorption of C4 slow and C4 fast components on the red cells?

Lachmann: That is so. It has been shown that these antigens are picked up from plasma and that C4F is Rodgers and C4S is Chido.

Siniscalco: Nevertheless, I seem to remember from O'Neill *et al.* (1978) that a C4-deficient individual was found who was also deficient in both Chido and Rodgers blood groups.

Lachmann: That's right; the totally C4-deficient subject is said to be the only case of absence of both Rodgers and Chido.

Bodmer: An obvious suggestion I would make on your C4F and C4S is that there are two genes present in all individuals and the C4F individual has one gene active and not the other, and vice versa. Both seem equally plausible and it is a matter of establishing whether that duplication occurs in individuals who only have one allele.

Your conceptual homologies are very interesting in relation to the other discussions we have had. But is there really good biochemical evidence for structural homology between any part of C3 and any part of C4, and between any parts of C6 and C7 and C2 or Bf, or is that just by analogy with the activity?

Lachmann: There are no sequence homologous data because there are so few sequences on these proteins. There is some information on the γ chain of C4 (Bolotin *et al.* 1977; Gigli *et al.* 1977) and the C3a fragment of C3 has been fully sequenced (Hugli 1975). There is some homology between C3a and C5a (Fernandez & Hugli 1977). There are not enough data yet on C2 and Factor B.

Bodmer: R.R. Porter's group is beginning to get sequence data but the properties of the molecules are so similar that it is impossible to say. One has to be a little cautious in assuming that there are structural homologies between C3 and C4, C6 and C7, and C2 and Factor B.

Lachmann: Yes, though the functional homologies between C3 and C4, C2 and Factor B, and C6 and C7 are overwhelmingly strong.

Neel: How close is the linkage of C4F and C4S?

Lachmann: No cross-overs have yet been described. The homozygotes for null alleles at both C4 loci are extremely rare. There are only three known cases of what appears to be total C4 deficiency in the world. If there were a lot of homozygotes for nulls at one or other of the two loci, half levels of C4 should commonly be found.

Henry Harris: In haemorrhagic dengue there is, I think, an increased yield of virus from cells *in vitro* in the presence of antibody to the virus.

Lachmann: The idea is that if there are antibodies to a different viral serotype this does not give immunity to the new serotype, but enhances its growth.

Henry Harris: Has anybody got any ideas about how that happens?

Lachmann: Halstead and his colleagues (1973) had the very attractive hypothesis (though it has been superseded) that the consequence of first infection with the virus was to generate a population of sensitized lymphocytes. Infection with the second virus then caused these cells to divide, producing a large number of blast cells which could then act as a 'selective culture medium' for the second strain of virus.

Henry Harris: But if you put antibody into an *in vitro* system you can get an increased yield too, can't you?

Lachmann: Yes; it has since been shown (Halstead & O'Rourke 1977; Halstead *et al.* 1977) that the virus grows only in monocytes and not in lymphocytes and that non-neutralizing antibody enhances viral growth. This is ascribed to immune complexes obtaining access to the monocytes by way of the Fc receptor.

Nossal: It is of course exceedingly common for us to have immune complexes briefly in our circulation, for example in any chronic infection, in some acute infections or after immunization. The complement system is heavily involved in the opsonization that allows us to get rid of those immune complexes. It is also true that the complement system is one of the factors that render immune complexes potentially injurious to tissues, but there are other factors, for example platelet aggregation and the chemotactic events which follow immune complex formation. So in this very dynamic and complex equilibrium, the opsonization role of complement might be the most important, with the dynamic equilibrium shifting a bit more towards persistence of complexes and damage in these complement-deficient individuals.

Lachmann: This is the Miller & Nussenzweig argument (1974, 1975): that intact complement systems are needed to solubilize immune complexes and to remove them from cellular receptors. That is quite possible. It is worth pointing out that in experimental situations complement deprivation prevents immmune complex damage and doesn't potentiate it. For example, attempts have been made to induce acute serum sickness in rabbits which had been de-complemented with cobra venom factor. Although complement deprivation doesn't always prevent the renal disease completely, it does not produce a potentiated form. In extra-renal sites complement depletion tends to prevent the disease altogether. But such models are blunt tools and in real life it may all be much more complicated!

The Miller and Nussenzweig effect probably acts together with the effect of persisting antigen.

References

BOLOTIN, C., MORRIS, S., TACK, B. & PRAHL, J. (1977) Purification and structural analysis of the fourth component of human complement. *Biochemistry 16,* 2008-2015

FERNANDEZ, H.N. & HUGLI, T.E. (1977) Chemical evidence for common genetic ancestry of complement components C3 and C5. *J. Biol. Chem. 252,* 1826-1828

GIGLI, I., VON ZABERN, I. & PORTER, R.R. (1977) The isolation and structure of C4, the fourth component of human complement. *Biochem. J. 165,* 439-444

HALSTEAD, S.B. & O'ROURKE E.J. (1977) Dengue viruses and mononuclear phagocytes. 1. Infection enhancement by non-neutralising antibody. *J. Exp. Med. 146,* 201-207

HALSTEAD, S.B., MARCHETTE, N.J. & CHOW, J.S.S. (1973) Enhancement of dengue virus replication in immune leukocytes as a mechanism in the immunopathogenesis of dengue shock syndrome. *Adv. Biosci. 12,* 401

HALSTEAD, S.B., O'ROURKE, E.J. & ALLISON, A.C. (1977) Dengue viruses and mononuclear phagocytes. II. Identity of blood and tissue leukocytes supporting in-vitro infection. *J. Exp. Med. 146,* 218-229

HUGLI, T.E. (1975) Human anaphylatoxin (C3a) from the third component of complement primary structure. *J. Biol. Chem. 250,* 8293-8301

MILLER, G.W. & NUSSENZWEIG, V. (1974) Complement as a regulator of interactions between immune complexes and cell membranes. *J. Immunol. 113,* 464-469

MILLER, G.W. & NUSSENZWEIG, V. (1975) A new complement function: solubilisation of antigen-antibody aggregates. *Proc. Natl. Acad. Sci. U.S.A. 72,* 418-422

O'NEILL, G.J., YANG, S.Y., TIGOLI, J., BERGER, R. & DUPONT, B. (1978) Chido and Rodgers blood groups are distinct antigenic components of human complement C4. *Nature (Lond.) 273,* 668-670

WEITKAMP, L.R., LOWIEN, E.W., OLAISEN, B., FEUGER, K., GEDDE-DAHL, T. JR., SORENSEN, S.A., CONNEALLY, P.M., BIAS,W.B. & OTT, J. (1975) in *Human Gene Mapping 2 (Rotterdam Conference 1974)* (Bergsma, D., ed.) *(Cytogen. Cell Genet. Ser. 14),* p. 276, Karger, Basel

Monoclonal antibodies and cell surface antigens

C.MILSTEIN, G. GALFRE, D.S. SECHER and T. SPRINGER*

MRC Laboratory of Molecular Biology, Cambridge

Abstract Antibody chains are encoded in three gene clusters containing genes for the variable and constant regions. V and C genes are separated in the germ line and during differentiation a rearrangement takes place. But even after this rearrangement the V and C coding sequences are not contiguous. A final splicing must take place in committed cells between the transcription of a discontinuous V- and C-region DNA and the expression of a continuous mRNA coding for an antibody chain. Analysis by cell fusion indicates that the splicing is *cis*.

When two antibody-producing cell lines are fused, the resulting hybrids express the two antibodies that characterize the parental lines. Permanent cell lines producing antibody of predefined specificity have now been derived in this way. Spleen cells from hyperimmunized donors are fused with myeloma cells and a proportion of the hybrids that are established synthesize and secrete antibodies directed against the immunogen. The heterogeneous cell population can be cloned and propagated. This is a potent way of producing monospecific antibodies to complex antigens such as cell membranes and transplantation antigens.

Monoclonal xenogeneic antibodies to rat cell-surface membranes have proved very valuable for characterizing and separating rat lymphocyte subpopulations. In more recent experiments, monoclonal xenogeneic antibodies to mouse and human cell-surface antigens have also been produced which permit the characterization of the hitherto undescribed differentiation antigens.

When an animal is injected with cells from another animal of a different species or a different strain, an antibody response is initiated. This response involves the production of circulating antibodies directed against the injected cells. Cells contain a large number of surface antigens, i.e. structures which are recognized as non-self by the immune system of the responding animal.

*Department of Pathology, Harvard Medical School, 25 Shattuck Street, Boston, Mass. 02115, USA

Each antigen is recognized by different antibodies directed to different regions (i.e. independent antigenic determinants), but each antigenic region can be recognized by a number of antibodies directed to overlapping determinants, and even the same determinant is usually recognized by several antibody structures. In this way an enormously complex and heterogeneous response is initiated which is the hallmark of the antibody system. According to the clonal selection theory each antibody-producing clone is committed to the production of a single antibody structure which is selected by the antigen for further proliferation (Burnet 1969). The heterogeneity is therefore built up by an enormous complexity of clones of responding cells.

Tumours of immunoglobulin-producing cells either occur spontaneously, like the myelomas in man, or can be induced experimentally. These tumours produce and secrete large amounts of myeloma protein, which is an immunoglobulin with an unknown antibody activity. Myeloma proteins, therefore, are a random collection of antibodies directed against unknown antigens. Myeloma proteins, however, were for a long time the only source of pure Ig preparations and were for that reason an invaluable tool in structural studies.

Antibodies are made up of light and heavy chains. Each of these chains consists of two segments, one referred to as the variable and the other as the constant region (Hilschmann & Craig 1965). The antibody combining site is defined by the amino acid sequence of the variable region which determines the unique diversity in protein structure that is characteristic of the immune system. The realization that V regions were coded by multiple V genes while C regions could be coded by single genes (Milstein 1967) made inescapable the assumption that V and C genes should occur separately in the genome (Dreyer & Bennett 1965). This has now been supported in a more direct way by the difference in restriction fragments of DNA derived from myelomas and embryonal or other tissues. At some stage in differentiation, therefore, rearrangement of DNA involved in antibody synthesis takes place. But it appears that even in myeloma cells, after this rearrangement has taken place, V- and C-region DNA sequences are not contiguous (Tonegawa *et al.* 1976; Rabbitts & Forster 1978; Brack & Tonegawa 1977). On the other hand, sequence analysis of mRNA indicates that the V and C sequences are uninterrupted and that the protein is translated from a single continuous mRNA (Milstein *et al.* 1974). So a final rearrangement or splicing takes places somewhere between the transcription of a discontinuous V- and C- myeloma DNA and the expression of a continuous mRNA.

HYBRID MYELOMA CELLS AND THE SPECIFICITY OF V–C INTEGRATION

We further know that the splicing or rearrangement is *cis,* i.e. that it does not involve the production of a pool of V- and C-region precursor RNA or DNA molecules. This information comes from an experiment done by Cotton & Milstein (1973) in which hybrid cells expressing two different sets of V- and C-region kappa light chains were produced. The C-region difference was ensured by using cells of mouse and rat origin (Fig. 1). The experiment

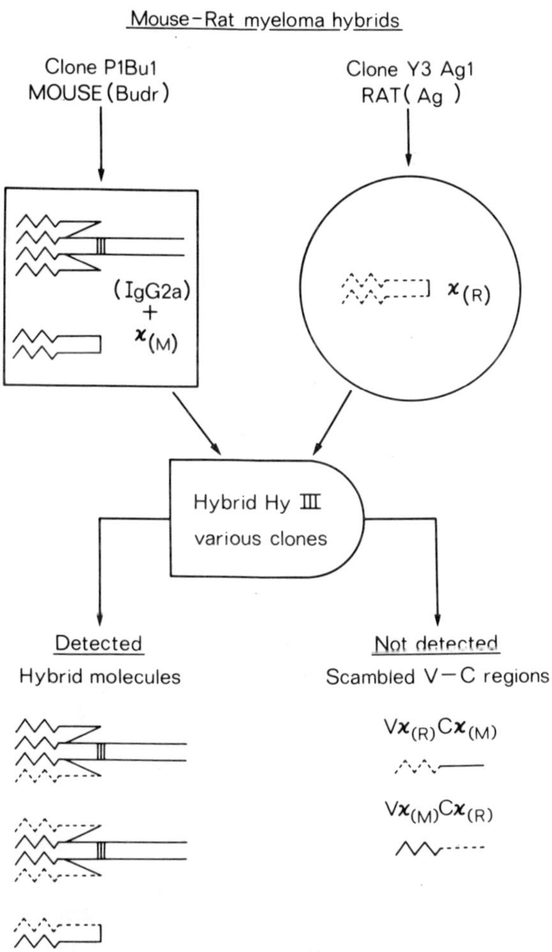

Fig. 1. Codominant *cis* expression of antibody genes in hybrids of myeloma cells. The data are presented in Cotton & Milstein (1973). Diagram reproduced from Milstein & Köhler (1977).

described in Fig. 1 was by necessity between mouse and rat cells because the C regions of kappa light chains of mice have no known genetic markers and therefore no distinction between two different C_κ regions of mice was possible.

Markers for the heavy chains are much more simple because the VH region is shared by all classes and subclasses of heavy chains. The random expression of different combinations of V_H and C_H on cells synthesizing two different heavy chains can therefore be checked, even within the same strain of mice. Experiments along these lines confirmed the general conclusions drawn from the experiments of Fig. 1 (Köhler & Milstein 1975; Milstein *et al.* 1977; Margulis *et al.* 1977).

This type of observation has been extended more recently to a large number of other examples with other myelomas and antibody-producing cells, including different types and classes of heavy chains. In some experiments efforts have been made to induce trans-integration but with no success. In other words, hybrid cells which express two immunoglobulins, one of the μ class, capable of lysing sheep red blood cells (SRBC), and another, an IgG1 immunoglobulin with no antibody activity, retain the separate expression. Out of 10^5 clones, none was found which expressed the anti-SRBC activity in IgG1 molecules. A similar result was obtained with an anti-TNP (trinitrophenyl)-producing hybrid (Milstein *et al.* 1977, and unpublished experiments).

DERIVATION OF ANTIBODY-PRODUCING LINES BY CELL FUSION

Fusion of two immunoglobulin-producing cells therefore results in the codominant expression of both immunoglobulins. Mixed molecules expressing different combinations of H and L chains are produced by hybrid cells but not by a mixed population of cells, indicating that the mixed molecules are formed intracellularly (Cotton & Milstein 1973; Köhler & Milstein 1975). These experiments suggested the production of hybrids between a myeloma cell producing an immunoglobulin with no known antibody activity but with the ability to grow permanently in culture, and another cell capable of producing a well-defined antibody but with a limited lifespan, as it is an antibody-producing cell of an animal (Köhler & Milstein 1975). Such fusions (Fig. 2) produced permanent hybrid lines capable of permanent growth in tissue culture and as transplantable tumours, which have the capacity to produce the specific antibody expressed by the normal primed parental cell. The first antibodies produced in this way were anti-SRBC and anti-TNP hapten antibodies (Köhler & Milstein 1975, 1976; Milstein & Köhler 1977).

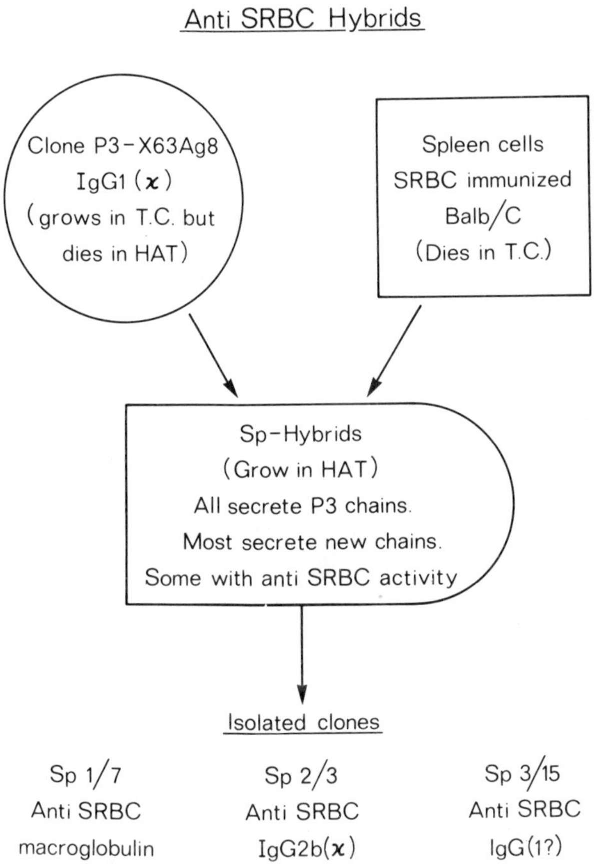

Fig. 2. Fixation of specific antibody production by a transient spleen cell in a permanent tissue culture line.

Technical modifications in the fusion technique, screening and purification of active clones were introduced for the derivation of hybrid myelomas secreting monoclonal antibodies to the cell surface antigens of the major histocompatibility complex of the rat (Galfre *et al.* 1977). Since then a large number of other antibodies have been produced (Table 1), and the results obtained formed the subject of a recent workshop (Melchers *et al.* 1978). The experience accumulated so far tends to indicate that the procedure is quite general and that any antibody an animal can produce can also be prepared as a monoclonal antibody through myeloma hybrid lines.

The major interest of this technique derives from two fundamental points. First, the monoclonal antibody produced by an isolated clone is a well-defined

TABLE 1

Examples of monoclonal antibodies produced (by us and others) using the hybrid-myeloma technique

Antibodies against:
red blood cells (sheep, human)
haptens (TNP, NP)
proteins (γ-globulins)
carbohydrate (bacterial cell wall, Forssman antigen)
enzymes (peroxidase, lysozyme)
viruses (influenza)
major histocompatibility antigens
other cell surface components (differentiation antigens)

chemical and not an undefined heterogeneous mixture which changes with each immunized animal and even with each bleed of the same animal. The permanent cultures are capable of producing an unlimited supply of exactly the same chemical structure. Second, the technique is ideally suited for the preparation of pure antibodies using non-purified antigens.

PURE ANTIBODIES FROM IMPURE ANTIGENS

As pointed out in the introduction, the antibody response of an animal to the challenge of a given antigen is highly complex. This is particularly so when cells from a given species are injected into an animal of a different species. The serum of the immunized animal contains a large number of antibodies directed against a large number of cell surface components. Multiple antibodies are likely to be produced against a single antigen and even against a single antigenic determinant (Fig. 3).

There is a more interesting complication. The immunizing cells are often themselves a complex set of different cell types; the antigenic structures shown in the hypothetical cell of Fig. 3 used as immunogen are therefore somewhat different in different cell types. These antigens, present in some cell types but not in others, are referred to as differentiation antigens. They allow us to recognize and separate one cell type from another. An extreme example of cell heterogeneity which can only be distinguished by the antigenic character of the cell surface are those B cells which differ only in the structure of the (antibody) receptor they synthesize.

The consequence of such complex antigenic stimulation is an even more complex antibody response. The use of such complex antibody mixtures is severely restricted and requires extensive purification. However, each antibody-producing cell is committed to the synthesis of a single antibody.

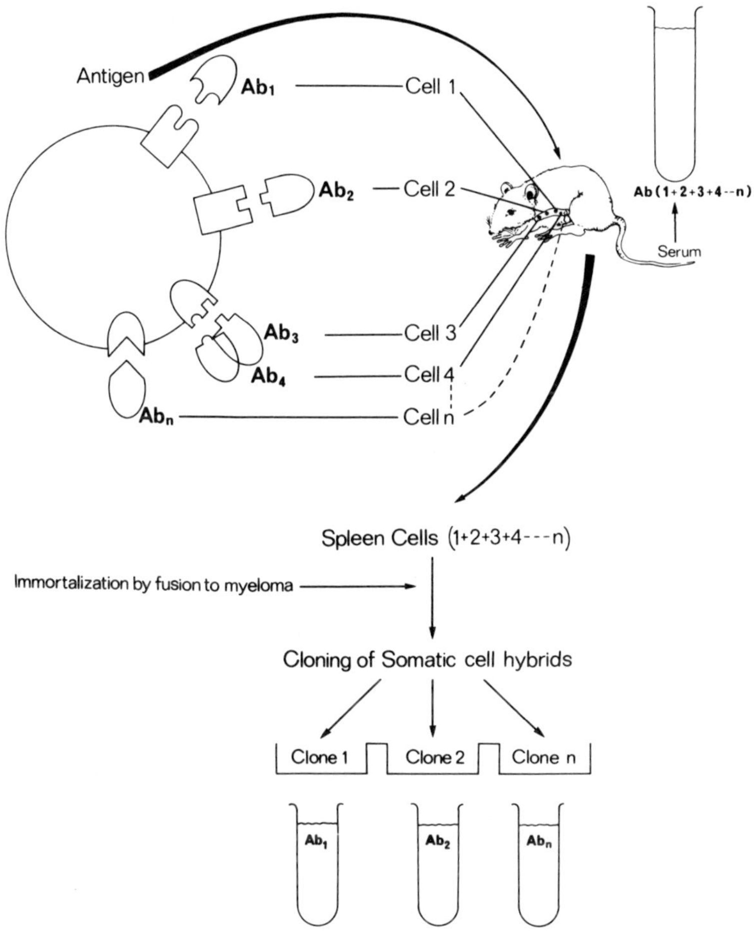

Fig. 3. Antibody production by animals and by hybrid myelomas. Isolated clones from hybrid myeloma fusions permit the preparation of pure monospecific antibodies directed to single antigenic determinants, starting from a complex mixture of antigens.

By fusion with a myeloma it is possible to fix the transient expression of individual cells in the form of permanent hybrid cultures. Such cultures can be subjected to the usual techniques of cell culturing and can therefore be purified and stabilized by subcloning (Fig. 3). By segregation analysis, subclones can be derived so that the myeloma components which do not contribute to the antibody activity can be eliminated (Köhler & Milstein 1976; Milstein *et al.* 1977). The end-product is a collection of permanent lines, each of them producing a different monoclonal antibody.

DIFFERENTIATION ANTIGENS FROM RATS, HUMANS AND MICE

The above type of approach to the detection of differentiation antigens was first used to derive mouse antibodies to rat cell membranes (Williams *et al.* 1977) and to human tonsil cells (Barnstable *et al.* 1978). A number of clones have been established and their characteristics are listed in Table 2; full details can be found in the references just cited.

In recent experiments monoclonal antibodies to mouse differentiation antigens have been produced. About 20 differentiation antigens in the mouse have been described by conventional procedures.

TABLE 2

Monoclonal antibodies to cell surface antigens in rat and man

Clone	*Antigenic target*	*Tissue*	*Cellular recognition*	*Antibody*
		Mouse anti-rat[a]		
W3/13		Lymphoid, bone marrow, brain	T cells; myeloid (?)	IgG1
W3/25 HL		Lymphoid	T cell subset (including[b] helper and GVH functions)	IgG1
W3/15 HLK		Blood, bone marrow, thymus	Red cells, erythroid (?) precursors	IgG
W3/4 HLK		Lymphoid	B cells	IgM
		Mouse anti-human[c]		
W6/1 HLK	Blood group A	Blood	Erythrocytes	IgM
W6/32	HLA	Widespread	Widespread	IgG2
W6/34 HLK	Controlled by chromosome 11 (carbohydrate ?)	Widespread	Widespread	IgG2

[a] Williams *et al.* (1977).
[b] White *et al.* (1978).
[c] Barnstable *et al.* (1978).
GVH: graft versus host.

It was interesting to obtain information on the frequency at which new antigens could be found by random collection of monoclonal antibodies prepared by xenogeneic immunization. Mouse spleen cells were fused to the mouse myeloma NSI/1-Ag4-1. Immediately after fusion, the cell population was divided into a large number of cultures. After a couple of weeks hybrid lines grew in all the cultures and the presence of antibody in the supernatants was tested by mixing the spent culture medium with target cells (mouse spleen cells in this case). The presence of bound rat antibodies was detected by a ^{125}I-labelled anti-rat Ig in almost all the cultures (Fig. 4). Since there were so many positive cultures we grew them further while we were freezing stocks for later studies. The most vigorous clones were in this way allowed to dominate each culture and 10 different hybrid clones were randomly purified and studied. Some of the cloned antibodies gave very low binding to spleen cells. It was

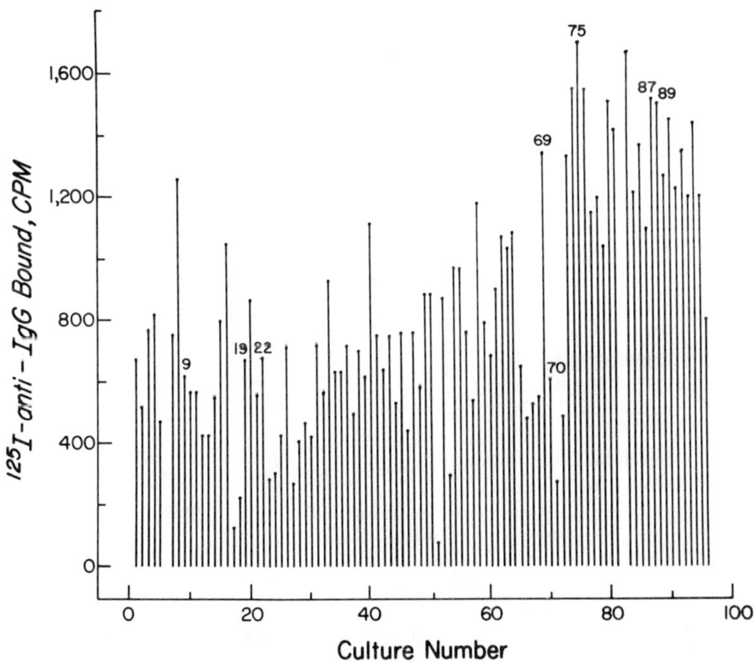

Fig. 4. Spleen cell binding activity of rat anti-mouse hybrid myeloma supernatants. Numbers refer to cultures from which hybrid clones were derived in this study.

only when we started to use panels of tumour cells that their true interest began to emerge (Stern *et al.* 1978; Springer *et al.* 1978*a*).

Table 3 summarizes the properties of the isolated antibodies and their antigenic targets. The table shows that the 10 monoclonal antibodies can be classified into four groups. Within each group, differences between the antibodies themselves can be demonstrated by several properties, especially by the electrophoretic properties of the isolated chains (Springer *et al.* 1978*a*). All of the antigens are differentiation antigens. None had been previously recognized by allogeneic immunization methods although one (reacting with M1/9.3 and M1/89.18) may be the same as a major iodinated membrane component observed by chemical and immunological methods (Trowbridge & Mazauskas 1976). The Forssman antigen is a heterophile antigen whose existence has been known for a long time (Humphrey & White 1963). The

TABLE 3

Rat monoclonal antibodies to mouse differentiation antigens

Clone	*Antibody*	*Cellular recognition*	*Antigen*	*Stability of antigen at 120°C*
M1/9.3 M1/89.18	IgG IgG	White cells	210 000 mol. wt.[a]	Unstable
M1/70	IgG	Macrophages and precursors	190 000 mol. wt.[a] 105 000 mol. wt.	Unstable
M1/75	IgG	Mouse RBC. Not on thymocytes	No iodinated component[a,b]	Stable
M1/69 M1/22.54 M1/89.1 M1/9.47	IgG IgG IgG IgG	Mouse RBC and most leucocytes. Thymocytes but not peripheral T cells.		
M1/22.25 M1/87	IgM IgM	Sheep RBC but not mouse RBC. Mouse teratocarcinomas. Minor cell subpopulations and early embryos[c]	Forssman	Stable

[a] The immuno precipitates were analysed by sodium dodecylsulphate—polyacrylamide gel electrophoresis.

[b] This group of antibodies compete between themselves for binding to mouse red blood cells (mouse RBC).

[c] Stern *et al.* (1978); Willison & Stern (1978).

other two have a very interesting cellular distribution (Springer *et al.* 1978*b* and unpublished). One of them—M1/70.15—appears to be present in macrophages and their precursors. Interesting relationships between the two can be deduced, in that they are expressed in a quantitatively inverse manner in the macrophage and its precursors. Thus differentiation from monocyte to peritoneal exudate macrophage is accompanied by a dramatic increase in the antigen recognized by M1/70, whereas the reverse appears true for M1/69.16. It appears, therefore, that the random derivation of monoclonal xenogeneic antibodies is a profitable approach to the detection of new antigenic surface components. A comparison of the ^{125}I-labelled membrane components precipitated by the antiserum from the rat which contributed the spleen for the fusion revealed many more bands than the composite of bands detected by the monoclonal antibodies (Fig. 5). Thus, the hybrid clones so far isolated in the experiment described represent only a few out of many clones responding to the immunization.

The use of monoclonal antibodies introduces a more reliable tool for studies of the quantitative expression of differentiation antigens. Instruments such as the fluorescent activated cell sorter (Loken & Herzenberg 1975) can measure the fluorescence intensity of large numbers of individual cells and permit a quantitative comparison of the antigenic expression in each cell. At the same time cell size can be measured by light scattering.

A simultaneous display of these two parameters introduces a new dimension to the analysis of cell subpopulations. This is well illustrated by Fig. 6 where distinct subpopulations are clearly defined by the monoclonal antibodies in a two-dimensional map or cellular fingerprint. Such cellular fingerprints may be of general use in the characterization of antigenic targets and antigenic variation as well as in defining new cell subpopulations.

W3/13 was found on thoracic duct T cells but not B cells, in a subpopulation (probably myeloid) of bone marrow cells and in brain (Table 2). M1/22.25 is expressed in a small subpopulation of spleen cells (apparently normoblasts), very strongly on teratocarcinomas and in four-day mouse embryos but not at earlier stages (Table 3). M1/69 is expressed in most leucocytes and thymocytes but not in closely related peripheral T cells. These antigens are therefore detected by monoclonal antibodies in different subpopulations and often unrelated populations of cells. The cell surface antigens of an animal may therefore be commonly expressed in different cell populations as a mosaic, the pattern of the mosaic being more characteristic of the individual cell type than the individual antigens themselves. This type of expression emphasizes the importance of the hybrid myeloma approach and the limitations of the absorption of antisera with tissues or cells different from the immunizing one.

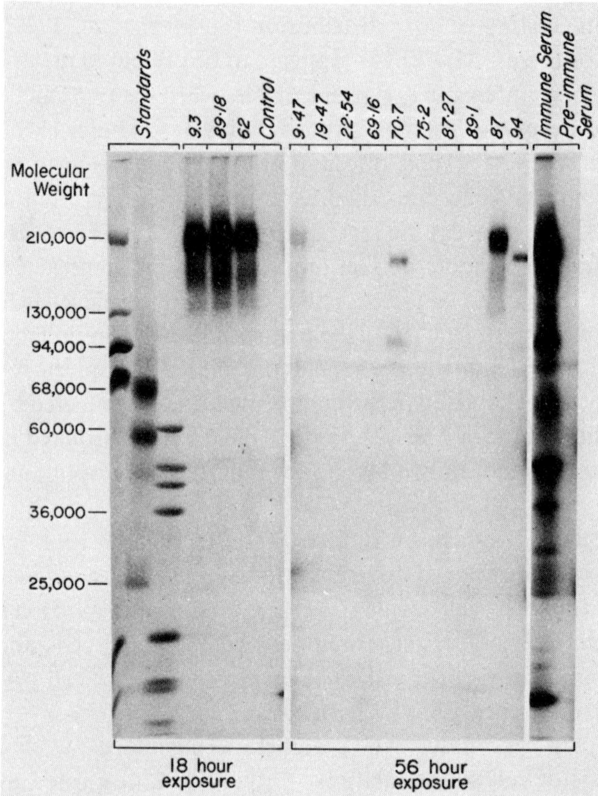

Fig. 5. Surface molecules precipitated by antibodies in the serum of the spleen (used for fusion) donor animal, in culture supernatants before cloning and in monoclonal supernatants. Concanavalin A-stimulated spleen cells were surface-labelled by lactoperoxidase and ^{125}I, solubilized with Triton X-100 and precleared by precipitation with anti-rat IgG. The subsequent specific immuno-precipitates were analysed by sodium dodecyl sulphate—polyacrylamide gel electrophoresis after reduction. Numbers refer to culture supernatants from Fig. 4, and individual clones are indicated by numbers after the dot.

FUSION PARTNERS AND THE EXPRESSION OF DIFFERENTIATED FUNCTIONS

One aspect of the hybridization experiments puzzled us from the very beginning. The desired hybrid myelomas were much easier to derive than we had thought. It soon became obvious that this was due to some form of selection or enrichment of the hybrids which expressed the antibody-secretion phenotype (Köhler & Milstein 1975, 1976). While the proportion of cells that secrete immunoglobulin is about 5% of the spleen cells used for hybridization, the proportion of hybrid clones derived that secrete immunoglobulin is well

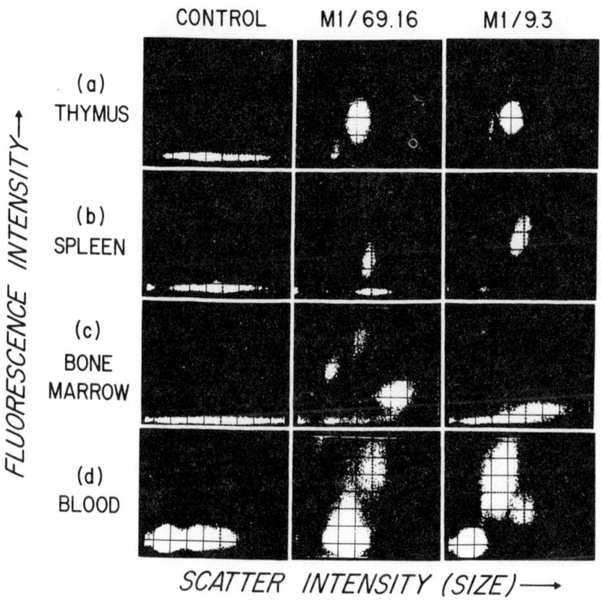

Fig. 6. Cellular fingerprints. Cell suspensions were depleted of red cells by Isopaque-Ficoll sedimentation and labelled with monoclonal culture supernatants followed by fluorescent anti-rat IgG.

over 50% (Table 4). Furthermore, if the mouse was immunized with SRBC, less than 1% of the spleen cells would secrete anti-SRBC, while the proportion of hybrid clones secreting anti-SRBC was 10%. In both cases the enrichment factor was about 10%.

The parental cell used for hybridization was a myeloma. However, when other parental cell types were used, the results were dramatically different. In other words, spleen cell hybridization to either a myeloma line or a T lymphoma line results in the selective expression of the spleen antibody in the first case and no expression of such a phenotype in the second. Conversely, the expression of the T cell antigen Thy-1 is selectively recovered from the spleen in the form of a permanent hybrid line when a T lymphoma, but not a myeloma, is used as the fusion partner (Table 4). This selective recovery of function may in part be due to an incompatibility of phenotypes since a fusion between a T lymphoma and a myeloma resulted in the suppression of the Thy-1 in the resulting hybrids. But that this is not the only explanation is suggested by the fact that the myeloma function (Ig secretion) was retained (Köhler *et al.* 1977*a*).

TABLE 4

Selective fixation of differentiated functions in established hybrids

Parental and hybrid lines	*Cell phenotype*			
	Ig secreted		*Thy-1 surface antigen*	
	Parental myeloma	*Other Ig*	*Thy-1, 1*	*Thy-1, 2*
X63 (myeloma)	>95%	0	0	0
(X63 × spleen) hybrids[a]	>90%	~65%	0	0
Spleen	0	~ 5%	0	~40%
(BW × spleen) hybrids[b]	0	0	>90%	~70%
BW (T lymphoma)	0	0	>95%	0

0 means none detected so far.
The spleens used were of an appropriate mouse strain. Further details in original references.
[a] Köhler *et al.* (1977*a*)
[b] Goldsby *et al.* (1977); Hämmerling (1977); Köhler *et al.* (1977*b*).

FUTURE PROSPECTS

The technique and the results described here seem to offer new vistas and some promising prospects in fundamental as well as applied aspects of human biology and human genetics (Table 5). In particular it appears worth while to explore the general use of cell fusion techniques for the recovery of transient differentiated functions in the form of permanent tissue culture lines. The results so far obtained strongly support our suggestion (Köhler *et al.* 1977*a*) that for the recovery of a transient differentiated function it is important to use cell lines phenotypically similar to the ones carrying the property in question.

TABLE 5

Some future prospects

1.	Anti-viral antibodies (for diagnosis and therapy)
2.	World standard reagents for organ transplantation
3.	Diagnostic reagents for clinical biochemistry
4.	Biochemistry and genetics of cell surfaces
5.	Affinity chromatography
6.	Fixation of other differentiated functions in permanent cell lines (T cells, hormone production, etc.)

References

BARNSTABLE, C.J., BODMER, W.F., BROWN, G., GALFRÈ, G., MILSTEIN, C., WILLIAMS A.F. & ZIEGLER, A. (1978) Production of monoclonal antibodies to group A erythrocytes, HLA and other human cell surface antigens–new tools for genetic analysis. *Cell 14,* 9-20

BRACK, C. & TONEGAWA, S. (1977) Variable and constant parts of the immunoglobulin light chain gene of a mouse myeloma cell are 1250 non-translated bases apart. *Proc. Natl. Acad. Sci. U.S.A. 74,* 5652-5656

BURNET, F.M. (1969) *Cellular Immunology,* p. 453, Melbourne University Press, Carlton, Victoria

COTTON, R.G.H. & MILSTEIN, C. (1973) Fusion of two Ig producing myeloma cells. *Nature (Lond.) 244,* 42-43

DREYER, W.J. & BENNETT, C.J. (1965) The molecular basis of antibody formation: a paradox. *Proc. Natl. Acad. Sci. U.S.A. 54,* 864-869

GALFRÈ, G., HOWE, S.C., MILSTEIN, C., BUTCHER, C.W. & HOWARD, J.C. (1977) Antibodies to major histocompatibility antigens produced by hybrid cell lines. *Nature (Lond.) 266,* 550-552

GOLDSBY, R.A., OSBORNE, B.A., SIMPSON, E. & HERZENBERG, L.A. (1977) Hybrid cell lines with T-cell characteristics. *Nature (Lond.) 267,* 707-708

HÄMMERLING, G.J. (1977) T lymphocyte tissue culture lines produced by cell hybridisation. *Eur. J. Immunol. 7,* 743-746

HILSCHMANN, N. & CRAIG, L.C. (1965) Amino acid sequence studies with Bence-Jones proteins. *Proc. Natl. Acad. Sci. U.S.A. 53,* 1403-1409

HUMPHREY, J. & WHITE, R.G. (1963) *Immunology for Students of Medicine,* p. 175, Blackwell, Oxford

KÖHLER, G. & MILSTEIN, C. (1975) Continuous cultures of fused cells secreting antibody of predefined specificity. *Nature (Lond.) 256,* 495-497

KÖHLER, G. & MILSTEIN, C. (1976) Derivation of specific antibody-producing tissue culture and tumour lines by cell fusion. *Eur. J. Immunol. 6,* 511-519

KÖHLER, G., PEARSON, T. & MILSTEIN, C. (1977*a*) Fusion of T and B cells. *Somatic Cell Genet. 3,* 303-312

KÖHLER, G., LEFKOVITS, I., ELLIOT, B. & CONTINHO, A. (1977*b*) Derivation of hybrids between a thymoma line and spleen cells activated in a mixed leukocyte reaction. *Eur. J. Immunol. 7,* 758-760

LOKEN, M.R. & HERZENBERG, L.A. (1975) Analysis of cell populations with a fluorescence activated cell sorter. *Ann. N.Y. Acad. Sci. 254,* 163-171

MARGULIS, D.H., CIEPLINSKI, W., DHARMGRONGARTAMA, B., GEFTER, M.L., MORRISON, S.L., KELLY, T. & SCHARFF, M.S. (1977) Regulation of immunoglobulin expression in mouse myeloma cells. *Cold Spring Harbor Symp. Quant. Biol. 41,* 781-791

MELCHERS, F., POTTER, M. & WARNER, N. (eds.) (1978) *Lymphocyte Hybridomas (2nd Workshop on Functional Properties of Tumors of T and B Lymphocytes, Bethesda, Md.,* April 1978) *(Current Topics in Microbiology and Immunology,* vol. 81), Springer, Berlin

MILSTEIN, C. (1967) Linked group of residues in immunoglobulin $\varkappa$-chains. *Nature (Lond.) 216,* 330-332

MILSTEIN, C. & KÖHLER, G. (1977) Cell fusion and the derivation of cell lines producing specific antibody, in *Antibodies in Human Diagnosis and Therapy* (Haber, E. & Krause, R.M., eds.), pp. 271-284, Raven Press, New York

MILSTEIN, C., BROWNLEE, G.G., CARTWRIGHT, E.M., JARVIS, J.M. & PROUDFOOT, N.J. (1974) Sequence analysis of immunoglobulin light chain mRNA. *Nature (Lond.) 252,* 354-359

MILSTEIN, C., ADETUGBO, K., COWAN, N.J., KÖHLER, G., SECHER, D.S. & WILDE, D.C. (1977) Somatic cell genetics of antibody-secreting cells: studies of clonal diversification and analysis by cell fusion. *Cold Spring Harbor Symp. Quant. Biol. 41,* 793-803

RABBITTS, T.H. & FORSTER, A. (1978) Evidence for non-contiguous variable and constant region genes in both germ line and myeloma DNA. *Cell 13,* 319-327

STERN, P., WILLISON, K., LENNOX, E., GALFRÈ, G., MILSTEIN, C., SECHER, D.S. & ZIEGLER, A. (1978) Monoclonal antibodies as probes for differentiation and tumour associated antigens: a Forssman specificity on teratocarcinoma stem cells. *Cell 14,* 775-783

SPRINGER, T., GALFRÈ, G., SECHER, D.S. & MILSTEIN, C. (1978*a*) Monoclonal xenogeneic antibodies to murine cell surface antigens: identification of novel leukocyte differentiation antigens. *Eur. J. Immunol. 8,* 539-551

SPRINGER, T., GALFRÈ, G., SECHER, D.S. & MILSTEIN, C. (1978*b*) Mac-1: a macrophage differentiation antigen identified by monoclonal antibody. *Eur. J. Immunol.,* in press

TONEGAWA, S., HOZUMI, N., MATTHYSSENS, G. & SCHULLER, R. (1976) Somatic changes in the content and context of immunoglobulin genes. *Cold Spring Harbor Symp. Quant. Biol. 41,* 877-889

TROWBRIDGE, I.S. & MAZAUSKAS, C. (1976) Immunological properties of murine thymus-dependent lymphocyte surface glycoproteins. *Eur. J. Immunol. 6,* 557-562

WHITE, R.A.H., MASON, D.W., WILLIAMS, A.F., GALFRÈ, G. & MILSTEIN, C. (1978) T lymphocyte heterogeneity in the rat: separation of functional subpopulations using a monoclonal antibody. *J. Exp. Med., 148,* 644-673

WILLIAMS, A.F., GALFRÈ, G. & MILSTEIN, C. (1977) Analysis of cell surfaces by xenogeneic myeloma-hybrid antibodies: differentiation antigens of rat lymphocytes. *Cell 12,* 663-673

WILLISON, K.R. & STERN, P.L. (1978) Expression of a Forssman specificity in the preimplantation mouse embryo. *Cell 14,* 785-793

Discussion

Koprowski: We were able to produce intraspecies hybridomas by fusing P3 × 63Ag8 mouse myeloma cells (Köhler & Milstein 1975) with splenocytes from mice immunized against viral antigens or human tumour cells grown in culture (Koprowski *et al.* 1978*a*) (Table 1). In addition, we were successful in producing interspecies hybridomas between P3 × 63Ag8 cells and human lymphocytes obtained from either the blood or cerebrospinal fluid of patients suffering from a CNS disease (Koprowski *et al.* 1978*a*). The human × mouse hybridomas were found to segregate into clones that produce human IgM and clones that do not. It will thus be possible to assign the gene(s) for human immunoglobulin production to specific chromosomes.

Fusion of influenza (Gerhard *et al.* 1978), rabies (Wiktor & Koprowski 1978) or parainfluenza virus (Gerhard *et al.* 1978)-primed mouse spleen cells with P3 × 63Ag8 cells frequently and reproducibly resulted in the production of somatic cell hybrids. When 1.6×10^6 splenocytes of mice immunized with the PR8 strain of influenza A virus were fused with mouse myeloma cells, 100% of the hybrids secreted anti-influenza antibody. Through a gradual decrease in the number of influenza-primed splenocytes in the fusion experiments, it was possible to determine that antigenically stimulated B cells have a distinctive advantage over non-stimulated B cells in forming hybrids with mouse myeloma cells.

Antibody-producing hybridomas could be maintained indefinitely either in tissue culture or in ascitic form in BALB/c mice (Koprowski *et al.* 1977), and in the course of *in vitro* or *in vivo* passage neither the secretion nor the reactivity of the antibody changed significantly (Gerhard *et al.* 1978). Since in several hybridomas the IgG 3 fraction exhibited anti-influenza activity, it was possible to purify the hybridoma antibody for production of anti-idiotype antiserum in order to compare the idiotypic determinants of antibodies produced by hybridomas with those produced *in vivo* (Gerhard *et al.* 1978). The antigenic drifts characteristic for influenza A viruses were investigated through the study of variants of the virus selected after passage of the parental virus in the presence of monoclonal hybridoma antibody (Gerhard & Webster 1978). The results suggested that a single amino acid substitution may account for the antigenic modifications on the haemagglutinin molecule. Finally, interaction between monoclonal anti-influenza antibodies and a panel of influenza viruses provided a tool for rapid determination of binding constants for antibody–antigen reactions (Frankel & Gerhard 1978).

Until now only minor antigenic differences have been described for rabies virus strains. The interaction of monoclonal antibodies produced by hybridomas with various strains of rabies virus allowed us to detect marked differences in antigenic determinants within strains of fixed and street rabies virus (Wiktor & Koprowski 1978). Furthermore, since it was possible to produce monoclonal antibodies reacting only with nucleocapsids or only with envelope proteins of the virus, this technique greatly facilitates biochemical characterization of isolated viral components. Hybridoma antibodies also protected mice against the lethal effect of the virus.

Until now, transforming proteins of cells transformed by oncogenic viruses such as SV40 or polyoma could not be identified or biochemically characterized because of their cross-reactivities. Monoclonal antibodies produced by a hybridoma against an SV40 tumour (T) antigen showed specificities for the SV40 T antigen only and opened the way for its isolation in a pure form and its biochemical analysis (Martinis & Croce 1978).

Fusion of P3 × 63Ag8 cells with splenocytes from mice immunized against human melanoma produced hybridomas secreting antibodies that reacted only with melanoma tumours grown in culture or obtained from patients (Koprowski *et al.* 1978*a, b*). The hybridoma antibodies did not cross-react with normal cells (skin fibroblasts, liver cells, erythrocytes) obtained from the same patients who donated the melanoma cells (Steplewski *et al.* 1979). Antibody secreted by hybridoma cells maintained in diffusion chambers in the peritoneal cavity of nude mice suppressed growth of human melanoma implanted in these mice (Koprowski *et al.* 1978*b*).

TABLE 1 (Koprowski)

Profiles of monoclonal antibodies produced by hybridomas

Hybridomas obtained by fusion of P3 × 63Ag8 mouse myeloma with:													
Splenocytes of mice immunized against:													
Viral antigens								*Tumour antigens*				*Human lymphocytes obtained from:*	
Influenza		Parainfluenza	Rabies	Herpes		Measles	SV40	SV40 trans-formed cells	Polyoma trans-formed cells	Human mela-noma	Human colo-rectal Ca	Blood[a]	CSF[b]
A	B	1		Herpes simplex	Herpes canis								

CSF = cerebrospinal fluid from patients with CNS disease
[a] Producing human IgM
[b] Producing small amounts of human IgG

Fusion of P3 × 63Ag8 cells with splenocytes of mice immunized *twice* with human colorectal carcinoma cells resulted in the production of hybridomas (Koprowski *et al.* 1978*a*) that secreted antibodies cross-reacting with a number of antigenic determinants present on human cells derived from neoplastic or from normal tissues. However, when splenocytes were obtained from mice after only one immunization with colorectal carcinoma cells, it was possible to produce hybridoma antibodies that reacted only with colorectal carcinoma cells.

This brief survey of antibody profiles produced by hybridomas grown in tissue culture and studied at the Wistar Institute should end on a note of congratulation to Cesar Milstein for opening new vistas in immunology, cell biology and genetics.

Nossal: Anybody who starts using hybridoma antibodies may be a little disappointed with their strength in an immunofluorescent assay, for example. This requires a bit of thought. The molar concentration of antibodies can be enormously high in tissue culture hybridoma supernatants or in the sera of mice bearing the hybridoma tumours *in vivo.* At 1 in 10 000 or 1 in 100 000 dilution they can still react. However, at saturation they may appear to be rather weak reagents, giving less intense immunofluorescence than polyvalent antisera. If one is, for example, making an anti-human thymocyte serum in a rabbit there are two points to note if that is done in the conventional way, just through an immunization. First of all, the rabbit will recognize the foreignness of a number of different proteins and carbohydrates on the human cell surface. For that conceptually quite trivial reason, the antiserum that is made, which is a very heterogeneous population of antibody molecules, will have more molecules bound to the human cell surface at saturation.

But there is a more subtle point. Supposing you are talking about all of the antibodies made to a single molecule. Any protein is a mosaic of a very large number of antigenic determinants. In a conventional immunization, the sole factor limiting how many antibody molecules can be squeezed onto that single protein molecule will be steric hindrance. The protein could have 20 different antigenic determinants. But with monoclonal antibody, which is a pure protein directed against just one antigenic determinant, at saturation just one antibody molecule binds to each protein molecule on the cell surface.

There is a two-fold lesson to this. First of all, don't be disappointed if the fluorescent glow on your cell is rather weaker than you expected. Secondly, be sure to do your experiments at saturation and you will have a very beautiful quantitative technology for determining how many of the molecules being recognized are present on your cell.

Dr Milstein, your human T cell antigen looks extremely like TL in the mouse, which is an MHC-linked antigen that appears on thymus lymphocytes but not peripheral T cells, and also is expressed in the thymic-type leukaemias. One would predict then that your marker would be present on human T cell leukaemic lymphocytes.

Milstein: In collaboration with A. McMichael and his colleagues in Oxford we have studied monoclonal antibodies produced by hybrids made with spleens of mice immunized with human thymocytes (McMichael *et al.* 1979). One of the hybrids has been cloned and the pure line is now well established, producing an antibody which is highly specific for human thymocytes. The antigen detected by this antibody, which we have named HTA.1 (human thymocyte antigen) is expressed in T cell leukaemias but not in all of them, and is not expressed in B cell leukaemias. (These tests were performed in collaboration with Dr G. Janossy from the Royal Free Hospital, London.) The expression of this antigen in normal thymocytes appears to be reciprocal to HLA in that thymocytes positive for one are negative for the other and vice versa. The antigen contains a component of about 45 000 mol.wt. In these properties it resembles the TL antigen of mice. However, no β_2-microglobulin component associated with it was detected, which seems to depart from the mice analogy, but there are some doubts about these results. In particular Dr A. Ziegler in my laboratory has strong reservations and wants to reinvestigate the problem.

I fully agree with Dr Nossal's comment on the use of monoclonal antibodies to quantitate the number of molecules of antigen per cell. This is something we have been doing with our reagents. In the case I have just been talking about, HTA.1 seems to have about 150 000 molecules per cell and is therefore a major surface component.

Bodmer: The published evidence for the β_2 association is in fact fairly weak. It may turn out that you are right and the others are wrong.

In the work we did together, two of the antibodies against the chromosome 11 determinants, W6.34 and W6.46, were very similar. Subsequently we found that sort of thing in other fusions. Sometimes perhaps one could get the same antibody in fusions because different cells from the same antibody clone in the spleen are being fused, and presumably in that case one would expect them, for example, to have the same isoelectric points.

I think there is a major distinction between looking at the carbohydrates, whether they are on the lipid or the protein, and looking at protein determinants. As you and others including ourselves find, a lot of these antibodies may be against sugar determinants. A lot of the things that give the cellular fingerprints may be those carbohydrates. One may expect to find that relatively minor perturbations of the biosynthesis of the sugar side chains may produce these differences, and may have questionable biological significance. We don't know anything about the role of these sugars anyway, and of course there are well defined examples like the loss of ABO specificity on fibroblasts in tissue culture. Perhaps that sort of thing should be clearly distinguished from the expression of a given protein, which tends to be more definite: either it is in a particular class of cells or it isn't.

Harry Harris: Are you saying that if monoclonal antibodies against a glycoprotein are made, most of the antibodies would be against carbohydrates?

Bodmer: No. When one immunizes with whole cells or with membranes in xenogeneic immunizations, a substantial proportion of the antibodies obtained are apparently against sugar determinants, probably on glycolipids. An important class of these determinants, as Don Marcus and others have shown, have quite distinctive tissue distributions.

When we immunize with purified glycoproteins, which we have done for example with HLA, we don't find antibodies to the sugars on them. We have a whole collection of antibodies to HLA determinants, including a number which are polymorphic, by immunizing with purified glycoprotein. It is probably the sugars on the lipids that are the most antigenic. That is also the classical blood group experience.

Milstein: The response to carbohydrates in hetero immunization, from our experience so far, seems to be quite substantial. This is true not only when we immunize with membranes but also when we immunize with whole cells. In the case I described in my talk whole mouse cells were used as immunogen. Of 10 hybrid clones, seven seem to recognize what appears to be carbohydrate or lipopolysaccharide. We can't find protein bands in the gels, and they are thermostable, which is a classical property of carbohydrate antigens. Perhaps anti-carbohydrate differentiation antigens have been largely absent from previous studies using heterologous antisera because they require absorption with different cell types which may also include the antigen in minor subpopulations. If one was doing intraspecies immunizations one wouldn't find them because they are not polymorphic. The cloning of hybrid myelomas eliminates the problem.

Bodmer: We have found them immunizing with hybrids.

Klein: Is the class of antibody made by the hybridoma frozen or can it be switched?

Milstein: The class of antibody seems to be quite frozen. We could not switch from one to the other, even when we tried it in different ways (Milstein *et al.* 1977). One attempt was with a tumour which expressed both the specific anti-SRBC μ chain and the γ_1 chain from the parental myeloma, to see whether we could switch from μ to γ antibodies. This was tested by suppressing the direct plaques (μ) with antimacroglobulin and developing indirect plaques to detect γ anti-SRBC activity. We tested 10^6 cells in this way (with appropriate controls) and they were negative. *In vitro* cellular switch must therefore occur at a frequency of, say, less than 1 in 10^6. If there is an error in switching in individual cells of the order of a few per cent we wouldn't know, but the cell itself doesn't switch in that way. We also injected the cells back into the animal but we couldn't switch them in that way either. Nor did ultraviolet irradiation work.

Gardner: Could this system be used to look at whether gene inactivation or repression is a particularly stable phenomenon in relation to determination in general? I don't find that the rarity of transdetermination detected in imaginal disks is very compelling. One could say that it is a problem not of the frequency of the event but of the frequency of detectability. Your system, with a very well-defined specificity in terms of the molecule that is being produced, and the possibility for enormous clonal expansion, might be ideal for looking at that particular problem.

Milstein: We don't know whether a translocation event or some DNA rearrangement is needed for the switch but it is definitely needed for the V-C combination. Whatever the mechanism of the switch, it appears to be extremely stable.

Gardner: If a hybrid is producing an antibody of a particular specificity and that clone is built up, is the specificity retained?

Milstein: The quick answer is yes. There is of course a certain loss of activity. We studied that earlier by looking at the clonal diversification of the myeloma line MOPC 21 and a certain frequency of irreversible loss of antibody gene expression was described. It is not excluded that this is due to chromosomal losses. Also there is a certain frequency of spontaneous mutations and that is where the mutants that I was talking about earlier came from.

Ruddle: So the stability of the hybrids is affected by the trivial loss of chromosomes?

Milstein: Absolutely; especially at the beginning. That is why multiple cloning is important in stabilizing the hybrids.

Peto: In the list of possible future uses why didn't you include tumour therapy? An incredible amount of very non-specific immunotherapy research is being done.

Milstein: I included diagnostic reagents. If I had put tumour therapy on the list I would have to deal with the press tomorrow!

Gurdon: How effectively does this procedure work with solid tissues? Most of the cells you described were single cells. If you cut frozen sections does the increased specificity make it easier to detect markers?

Milstein: Yes, we have examples of that. It works well with solid tissues.

Jacob: I would like to return to the homology between the chromosomes of man and mouse and discuss the possibility of having something in the human like the *T* locus in the mouse. In the mouse, the H-2 complex is located on chromosome 17. On the same chromosome there is also a segment on which a number of mutations are known. These complex mutations have been detected because they affect the tail and for this reason are called *T*. Some are recessive *(t)* and some are dominant *(T)* for tail length. Most of them are recessive lethals. They also affect sperm formation in males. Many of the +/*t* heterozygote males transfer the abnormal allele with a high frequency to the progeny: some 90% of the progeny inherit the *t* mutation. They also affect recombination on a chromosome segment from the centromere to further than the H-2 locus (for details, see Bennett 1975).

In humans, a disease which might recall some of the lesions due to the *t* anomalies in the mouse is spina bifida, a disease of the vertebral column in which the vertebrae do not close over the spinal cord. Some are widely open and are called 'aperta'; others are barely open and are called 'occulta'. There are many genetic and environmental causes of spina bifida. The work I am going to report was done in Paris by Marc Fellous, Joëlle Boué and Jacques Hors, who studied a family from Charleville, in northern France. In this family, a high incidence of disease of the sacrum is found, going from a complete absence of sacrum to a spina bifida occulta which can be detected only by X-rays. The disease appeared in a man who was the father of four children. These four children were affected and among 29 of his progeny, 19 were affected to various degrees.

In the mouse, the *T* locus is linked to the major histocompatibility complex, H-2. So the possibility of a linkage between the disease and HLA was investigated in detail (see Fig. 1). The two haplotypes of the man who had the original disease are unknown. It is unlikely, however, that this man was homozygous for the HLA region, the frequency of homozygotes in France being less than 1%. It can be seen on the figure that haplotype *d* is very frequently transmitted with the disease. Among 23 informative patients, 19 have an HLA haplotype *d*. Another marker which is known to be located on the same segment of chromosome 6, phosphoglucomutase-3 *(PGm3)*, was also analysed. Among 12 informative patients, the allele $PGm3^2$ was found in 11 cases, again suggesting a linkage. In contrast, no linkage was observed between the disease and other markers such as ABO blood groups or Rh.

To interpret these data, one could propose the hypothesis that in this family

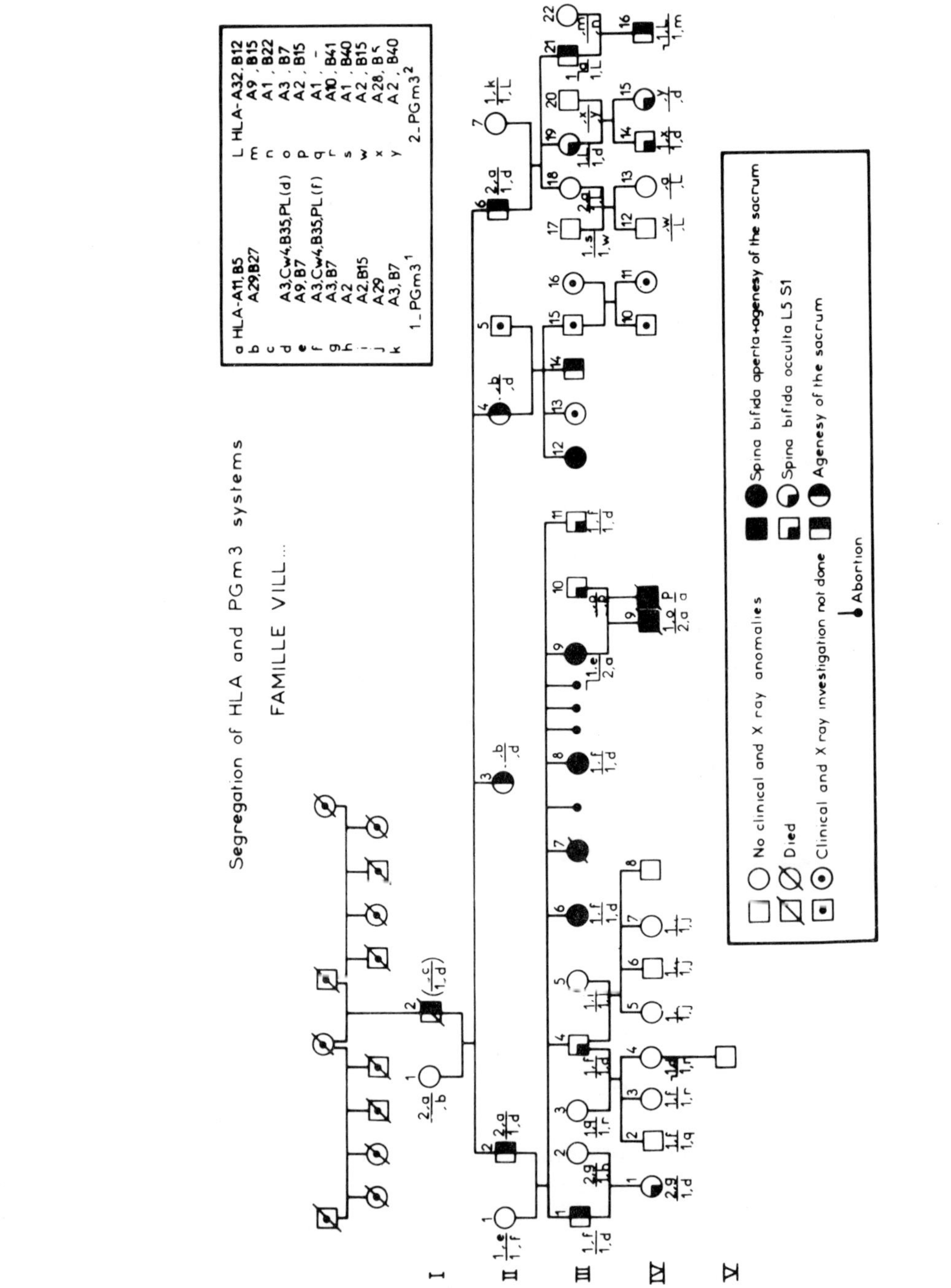

Fig. 1 (Jacob). Segregation of HLA and *PGm3* systems in a family from northern France.

the disease is due to a dominant gene, *Sp* (for spina bifida). This gene would be located on chromosome 6 and linked to HLA and *PGm3*. Although the figures are small, the order could be determined through the structure of recombinants, and especially children III_9 and III_{21}. III_9 is recombinant for the disease and HLA as well as for the disease and *PGm3*. III_{21} is recombinant between the disease and HLA but not *PGm3*.

The most likely order is therefore:

centromere—Sp—PGm3—HLA

Brenner: Is there any evidence for segregation distortion in man?

Jacob: Yes, there is an excess of disease in the family over random transmission. There is correlatively an excess of transmission of HLA haplotype *d*.

Brenner: Is there any HLA haplotype which shows segregation distortion?

Bodmer: Not as far as we know, but this would be an example of a rare mutant which is associated with segregation distortion. In general one would not expect HLA haplotypes to show that association. This family is extremely interesting.

Pontecorvo: Is it a damp area?

Jacob: Not especially. In this family the men were plasterers for three generations and then they changed their occupation.

Klein: What about the dominance relationships in the mouse *T* system?

Jacob: There are some mutations which are dominant and others which are recessive for the tail anomalies. They are recessive lethals.

Polani: This interesting family with sacral aplasia/hypoplasia reminded me of an observation by Degos and his colleagues (1974) on HLA in a tribe of Tuaregs in the Sahara Desert. Evidence from this tribe is pertinent to the question of the existence in humans of a situation analogous to that in the mouse, in respect to the *T* locus. In this highly inbred Tuareg isolate, the number of homozygotes for HLA haplotypes was significantly deficient. A possible explanation for this is that a locus on chromosome 6 of man, linked to HLA, may prevent homozygosity for HLA because homozygosity for one or more alleles at this linked locus could result in embryonic malformations such as those with some *T*-locus alleles in the mouse, and thus be embryo-lethal. Indirectly this would prevent homozygosis for the linked HLA alleles.

References

BENNETT, D. (1975) The T-locus of the mouse. *Cell 6,* 441-454

DEGOS, L., COLOMBANI, J., CHAVENTRE, A., BENGTSON, B. & JACQUARD, A. (1974) Selective pressure on HL-A polymorphism. *Nature (Lond.) 249,* 62-63

FRANKEL, M.E. & GERHARD, W. (1978) The rapid determination of binding constants for antiviral antibodies by a radioimmune assay. An analysis of the interaction between hybridoma proteins and influenza virus. *Molecular Immunology 16,* 101-106

GERHARD, W. & WEBSTER, R.G. (1978) Antigenic drift in influenza A viruses. I. Selection and characterization of antigenic variants of A/PR/8/34 (HON1) influenza virus with monoclonal antibodies. *J. Exp. Med. 148,* 383-392

GERHARD, W., CROCE, C.M., LOPES, D. & KOPROWSKI, H. (1978) Repertoire of antiviral antibodies expressed by somatic cell hybrids. *Proc. Natl. Acad. Sci. U.S.A. 75,* 1510-1514

KÖHLER, G. & MILSTEIN, C. (1975) Continuous cultures of fused cells secreting antibody of predefined specificity. *Nature (Lond.) 256,* 495-497

KOPROWSKI, H., GERHARD, W. & CROCE, C.M. (1977) Production of antibodies against influenza virus by somatic cell hybrids between mouse myeloma and primed spleen cells. *Proc. Natl. Acad. Sci. U.S.A. 74,* 2985-2988

KOPROWSKI, H., GERHARD, W., WIKTOR, T., MARTINIS, J., SHANDER, M. & CROCE, C.M. (1978*a*) Somatic cell hybrids producing antibodies for viral or tumor antigens. *Curr. Top. Microbiol. Immunol. 81,* 8-19

KOPROWSKI, H., STEPLEWSKI, Z., HERLYN, D. & HERLYN, M. (1978*b*) Study of antibodies against human melanoma produced by somatic cell hybrids. *Proc. Natl. Acad. Sci. U.S.A. 75,* 3405-3409

MARTINIS, J. & CROCE, C.M. (1978) Somatic cell hybrids producing antibodies specific for the tumor antigen of simian virus 40. *Proc. Natl. Acad. Sci. U.S.A. 75,* 2320-2323

MCMICHAEL, A.J., PILCH, J.R., GALFRÈ, G., MASON, D.Y., FABRE, J.W. & MILSTEIN, C. (1979) A human thymocyte antigen defined by a hybrid myeloma monoclonal antibody. *Eur. J. Immunol.,* in press

MILSTEIN, C., ADETUGBO, K., COWAN, N.J., KÖHLER, G., SECHER, D.S. & WILDE, C.D. (1977) Somatic cell genetics of antibody-secreting cells: studies of clonal diversification and analysis by cell fusion. *Cold Spring Harbor Symp. Quant. Biol. 41,* 793-803

STEPLEWSKI, Z., HERLYN, M., HERLYN, D., CLARK, W.H. Jr. & KOPROWSKI, H. (1979) Reactivity of monoclonal antibodies with melanoma cells freshly isolated from primary and metastatic melanoma. *Eur. J. Immunol.*

WIKTOR, T.J. & KOPROWSKI, H. (1978) Monoclonal antibodies against rabies virus produced by somatic cell hybridization: detection of antigenic variants. *Proc. Natl. Acad. Sci. U.S.A. 75,* 3938-3942

General discussion II

Neel: A reported relationship that is causing quite a bit of controversy is that between the DNA content per haploid genome and the probability of mutation per rad per locus. Abrahamson *et al.* (1973) pointed out that there was a linear relationship from *E.coli* with the least DNA and lowest probability of mutation, up to barley with the highest content and the highest probability. Another way of saying this is that the dose of radiation needed to double the spontaneous mutation rate goes down as the DNA content goes up.

One interpretation given to this relationship is that the number of genes is more or less constant but that the size of the sensitive target is increasing as the amount of DNA increases, and that a hit anywhere in that target is translated into a genetic effect. This observation is support of a sort for the 'super-gene' concept.

Crick: I am baffled by it.

Brenner: Is it possible to scale lethal doses for this variety of organisms?

Bodmer: Are these meiotic or mitotic data?

Neel: These are germinal mutations.

Brenner: They are probably aberrations, aren't they?

Neel: With radiation we would think they probably were, although in the mouse and in *Drosophila* the rates are got by the 'specific locus' method. It turns out that most of Bill Russell's so-called point mutations are probably small deletions (Russell 1971; Erickson *et al.* 1974) but nobody has gone back to re-examine the *Drosophila* data. I am not sure about yeast but certainly the lethal dose for *Drosophila* is an order of magnitude greater than for the mouse. In that sense the relationship with total DNA holds.

Fincham: In *Neurospora* loss of function of a particular gene in a heterokaryon was being measured. That would include both viable point mutants and haploid-lethal deletions.

Neel: We now think that for the mouse this is measuring both true point mutations and deletions.

Fincham: If tests of different organisms are measuring different spectra of mutants, that makes it an even more astonishing coincidence.

Brenner: One explanation of this boils down to the following. If we assume that the mutations are aberrations, in the higher organisms, then we only have to consider the abundance of break and join points. I happen to believe that it is the latter that determine the results.

Neel: That implies that there is not much silent DNA.

Fincham: I don't really understand this. Even assuming that there is a limited and perhaps fairly constant number of 'hook' sequences for rejoining with DNA, why should a rejoining involving that sequence rescue a break made somewhere else?

Brenner: My argument is that breaks occur anywhere. Such breaks would then produce DNA with free ends which would probably be degraded, possibly to give stretches of single strands. These can then recombine through the homologous repeats, and lead to intact structures. Depending on the orientation and location of the repeats with respect to the original breaks, so one generates inversions, deletions, duplications and so on. Even if the DNA involved in rejoining has no function, this does not mean that there will be no mutations. Clearly if there is a deletion there is a mutation.

One should contrast this with a chemical mutagen such as ethane methane sulphonate. It produces largely single base-pair transitions. We know that the absolute induced mutation rate per base pair in the nematode is the same as that for *E.coli* using the same dose of mutagen. The organism seems to be transparent to the chemical.

However the interesting question is the replication error rate in different organisms. In a small RNA bacteriophage the mutation rate per nucleotide is about 10^{-4}, which is the reciprocal, more or less, of the number of nucleotides. Here we know that replication has no editing function. This is also probably true for the reverse transcriptase of the RNA viruses. It seems that 10^{-4} is the best you get for specificity using one site, and that must set the basic limit to the size of the genome. In bacteria the spontaneous frequency is about 10^{-8} per base pair, which is about the square of the basic error rate. These organisms contain a replication system with an editing function, so that, roughly speaking, everything is done twice. In theory, this basic error rate would allow for genomes of the order of 10^7 to 10^8 base pairs in size. Bacteria have genomes at the lower end of this range. What happens in organisms with much larger DNA contents, such as man, with more than 10^9 base pairs? Is there a superediting function to keep down spontaneous replication errors?

Weissmann: But there you get diploidy coming along to help.

Brenner: That would be one argument but it really only helps you in somatic cells. J. Drake thought many years ago that mutation rates would have to be specially reduced in higher organisms.

Bodmer: Isn't that in the other direction?

Brenner: Yes, but those were radiation experiments; what I would like to know is the spontaneous mutation frequency per base pair in man. Let me express my prejudice. I believe it will be the same as that in bacteria—about 10^{-7} per base pair. I don't think there will be superediting.

Bodmer: That is two or three orders of magnitude too high.

Brenner: Nobody knows.

Milstein: In the heavy chain the mutation frequency is about 10^{-5}.

Brenner: You must be careful in defining what kind of mutants you are studying. My statements apply to *replication error mutations,* and I am asserting that the frequency will be the same in all organisms above and including bacteria. Let me just add that this is consistent with the thought that perhaps all organisms have about the same size of essential genome.

Neel: Most of the mouse mutations are small deletions, not chromosome eliminations.

Crick: How small is small?

Neel: They embrace several loci. In the best-studied situation, involving induced lethal alleles at the albino locus, complementation studies suggest that the defects involve one to two centimorgans (Gluecksohn-Waelsch *et al.* 1974). Of course, by molecular standards these are really rather large deletions.

Crick: There is something called a 'domain' by people who have studied chromosome structure. It is rather ill-characterized. It can mean a region of supercoiling, where one nick will un-supercoil a certain length. It is also defined by the length that needs two double-stranded cuts to take it out. In a *Drosophila* cell line the average size is of the order of 50 000 base pairs. I don't think it is known how much bigger it is in the frog than in *Drosophila.* One shouldn't lose sight of the fact that there are *some* structural units of that type, although characterized only in a preliminary way, which might be the units that are knocked out by radiation. One suspects that a domain is a big loop held at the bottom by DNA, by a four-stranded structure or something like that.

Henry Harris: That is essentially Peter Cook's model. His estimate of the size of the structural unit in HeLa cells is 220 000 base pairs (Cook & Brazell 1978). My point is simply that in animal cells the structural unit may be much larger than in simpler organisms.

Crick: What I was trying to get at is exactly what you are saying: maybe the unit is increased in size as DNA increases and that is the unit that in some ways is a unitary thing for damage by radiation.

The question would be, can you measure the average size of these units with a little more precision, and could you do it for a series of things like *Drosophila*, mouse, barley and so on?

Henry Harris: It depends on how much faith you put in target theory. Peter Cook gives some comparative figures but he makes certain assumptions.

Crick: All we want at this stage is to see whether the unit is about the same size in *Drosophila* and man or whether it differs by a factor of 30, but the measurements must be done by the same method.

Henry Harris: The units seem to be about the same size in man and *Drosophila* (Cook & Brazell 1976).

Fincham: I don't see how these radiation targets can have anything to do with Walter Bodmer's gene clusters. Even if higher organisms have gene clusters, I don't see why knocking out one component of a cluster should kill all the others. If anything, it should work the other way round.

The presence of a cluster (in so far as this implies repetition of homologous genes) should provide a fail-safe mechanism. If you knock out the active one you still have the possibility of reactivating one of the others.

Bodmer: It depends on the mechanism of knocking out.

Peto: If a lesion is produced by radiation somewhere on the DNA, could the repair induced from that have some non-local effect? By really bad luck one might find a peculiar class of side effects.

Brenner: What I said earlier was exactly that. Radiation only initiates breaks; the final consequences depend on the distribution of special sequences.

Has anybody measured the average size of the replication units in different organisms? It is known that they vary in the same organism.

Polani: Callan (1972) has looked at this and more data are in Davidson's book (1976).

Fincham: In comparing *Triturus* and *Xenopus* there is some correspondence between the amount of DNA and the size of the replication unit (Callan 1973).

Walker: It is not exact. The rate of replication also varies.

Weissmann: The mutation rate measured in *E.coli* is partly due to insertion sequences, so the estimate of the so-called spontaneous mutations really goes back to the insertion of these sequences in the gene. If you assume that there is something like an insertion sequence in higher organisms and that the number of insertion sequences is proportional to the amount of DNA, with the number of genes remaining constant, then if you activated them with X-rays you would also get this kind of effect.

Brenner: Can jumping genes in maize be activated with X-rays?

Fincham: No, not as far as I know.

References

ABRAHAMSON, S., BENDER, M.A., CONGER, A.D. & WOLFF, S. (1973) Uniformity of radiation-induced mutation rates among different species. *Nature (Lond.) 245,* 460-462

CALLAN, H.G. (1972) Review lecture. Replication of DNA in the chromosomes of eukaryotes. *Proc. R. Soc. Lond. B Biol. Sci. 181,* 19-41

CALLAN, H.G. (1973) DNA replication in eukaryotic organisms. *Heredity 30* (2), 256

COOK, P.R. & BRAZELL, I.A. (1976) Conformational constraints in nuclear DNA. *J. Cell Sci. 22,* 287-302

COOK, P.R. & BRAZELL, I.A. (1978) Spectrofluorometric measurement of the binding of ethidium to superhelical DNA from cell nuclei. *Eur. J. Biochem. 84,* 465-477

DAVIDSON, E.H. (1976) *Gene Activity in Early Development,* 2nd edn., pp. 126-128, Academic Press, New York

ERICKSON, R.P., EICHER, E.M. & GLUECKSOHN-WAELSCH, S. (1974) Demonstration in mouse of X-ray induced deletions for a known enzyme structural locus. *Nature (Lond.) 248,* 416-418

GLUECKSOHN-WAELSCH, S., SCHIFFMAN, M.B., THORNDIKE, J. & CORI, C.F. (1974) Complementation studies of lethal alleles in the mouse causing deficiencies of glucose-6-phoshate, tyrosine-aminotransferase, and serine dehydratase. *Proc. Natl. Acad. Sci. U.S.A. 71,* 825-829

RUSSELL, L.B. (1971) Definition of functional units in a small chromosomal segment of the mouse and its use in interpreting the nature of radiation-induced mutations. *Mutat. Res. 11,* 107-123

Human gene mapping and cancer biology

M. SINISCALCO

Sloan Kettering Institute for Cancer Research, New York

Abstract The topics covered in this paper include: (i) a somatic cell genetics approach for measuring the individual variability in the susceptibility to DNA damage/repair at the level of specific chromosomal sites; (ii) a rationale for a selective chemotherapy and/or immunotherapy of chromosomally unbalanced tumours; and (iii) studies on complementation for sister chromatid exchange (SCE) in rodent–human hybrids.

Preliminary studies indicate that the 'radiation co-transfer method' for gene mapping can be simplified and used in screening for differences in individual susceptibility to radiation induced chromosomal damage.

The second topic is dealt with only speculatively, with the explicit aim of emphasizing a practical application in clinical medicine which may potentially derive from the admittedly esoteric activity of gene mappers.

The third topic summarizes a somatic cell genetic approach to the study of SCE in mammalian cells. The high rate of SCE observed in an established rodent cell line can be fully suppressed after hybridization with SCE-normal human cells. However, this suppression can be fully removed after extensive loss of the human chromosome complement. Correspondingly, the high SCE rate of Bloom syndrome cells is fully corrected after hybridization with a Chinese hamster line, though the chromosomes of the latter parental cells continue to exhibit in the hybrid cells the moderately high rate of SCE which is typical of this animal line.

These complementation experiments indicate that more than one lesion can upset the normal chromatid replication of mammalian cells and lead to high SCE. The experimental studies described promise to be of significant help for studies on the biology of SCE in general and offer a suitable way of screening for possible genetical heterogeneity among different Bloom syndrome patients.

THE STATE OF THE ART

During the last ten years, the harvest of data on human gene mapping has proceeded at an unprecedented pace which shows no sign of abating. The last edition of Victor McKusick's *Newsletter* (21 November 1977) lists the chromosomal assignments of 183 genes, 61 more than in the previous

Newsletter circulated only seven months earlier. This means that new assignments are nowadays made at an average rate of two per week. To appreciate this exceptional situation, it is enough to remember that after the rediscovery of Mendel's laws it took about 50 years to identify the first human autosomal linkage, and that in 1967, at the time of the first gene assignment with somatic cell hybrids, the total number of human autosomal linkages known could still be counted on the fingers of one hand. The situation was of course different for the X-linked traits, whose existence had actually been acknowledged 2000 years before Mendel, as witnessed by the remarkable precision with which the inheritance of bleeding at circumcision (today's haemophilia) is described in the Talmud (Stern 1960). However, all of the 156 human X-linked markers, classified as such on the basis of their typical criss-cross mode of inheritance, had remained 'unsited' with respect to their subregional mapping until very recently, when the cytological location of about a tenth of them was directly or indirectly established (Miller *et al.* 1978).

As we know, the major event which made this mapping possible was the discovery of Weiss & Green (1967) that murine–human somatic cell hybrids undergo the gradual and preferential loss of human chromosomes; the association between the retention or loss of one or more genetic markers and that of a specific human chromosome could thus be screened for. The main advantage of studying human linkage with interspecific somatic cell hybrids is that, owing to the evolutionary distance between the parental species, these hybrids can be considered as 'multiple heterozygotes' at all loci whose products are expressed in cultured cells. Thus, one can now investigate linkage relationships between genes independently of whether they are polymorphic within the species, which is the *sine qua non* condition for the detection and estimation of linkage within pedigrees and populations. But the greatest attraction for newcomers to this field of study is the possibility of being able to investigate man's inheritance experimentally. Those who attended the Ciba Foundation symposium on the *Biochemistry of Human Genetics* in Naples in 1959 may remember the first sentence of Professor Penrose's introduction: 'Human genetics is an applied science' (Penrose 1959). I am sure that if Penrose had been here to open the present gathering his first sentence would have been: 'Human genetics is no longer an applied science'. That such a change would eventually take place was everybody's feeling at the Naples meeting, but Pontecorvo was the only one to venture an estimate of the time required for this to happen, and we know now that he turned out to be correct to the nearest year (Pontecorvo 1959). Indeed, thanks to these new methodologies, human gene mapping is now an experimental endeavour. However, it would be wrong to assume that the dramatic success of the cell hybrid approach has rendered all other approaches obsolete. On the contrary, experience has shown that the best-mapped chromosomes are those for which data gathered with different approaches could be collated together, as happened for autosomes 1,2,6,9 and for the X-chromosome (McKusick & Ruddle 1977).

Fig. 1 illustrates the status of the art as on 1 February 1977. Of course this map is already outdated. There is no question in my mind that during the next few years the rate of accumulation of new mapping data will increase even further. This I expect to be largely the result of the powerful immunological tools that can now be prepared following the rationales devised by Buck & Bodmer (1975) and by Köhler & Milstein (1975). I refer in particular to the current possibility that mice can be immunized with highly reduced man–mouse hybrid cells which will trigger the production of splenocytes secreting monoclonal antibodies against chromosome-specific human surface antigens and secure their permanent supply by means of the so-called 'hybridomas' derived from the fusion of the splenocytes with mouse myeloma cells. Drs Bodmer and Milstein have already referred to the latest developments in this exciting new chapter of immunogenetics (Bodmer 1979; Milstein 1979).

For the speculations that I am going to make, I need to point out that in consequence of these developments our knowledge of human genes coding for surface antigenic proteins will increase exponentially, so it is not unrealistic to expect that the human gene map will soon include one or more such markers per chromosome. This means that there might soon be available a series of immunoreagents each of which will react specifically with one or more cell surface proteins coded by genes carried by an individual human chromosome. A glance at the current list of mapped human genes (Table 1) shows that this expectation is now being rapidly fulfilled, since surface antigen markers have already been identified for the human autosomes 6,7,10,11,12,14 and for the two sex chromosomes (SAX-1, SAX-2, SAX-3 and H-Y).

I expect that the above-mentioned immunoreagents, coupled with the use of techniques which permit the isolation of cells on the basis of their surface antigen moieties, will dramatically simplify the experimental strategies for the production of somatic cell hybrids, and in particular will make possible the isolation of mouse–man hybrids with the wanted human chromosome(s) or chromosomal fragment.

Apart from being an addition to the basic knowledge of the human gene map, the discovery of each new chromosome-specific surface antigen (with the corresponding monoclonal antibody) provides an additional precious tool for investigating the expression of specific gene products at the level of the

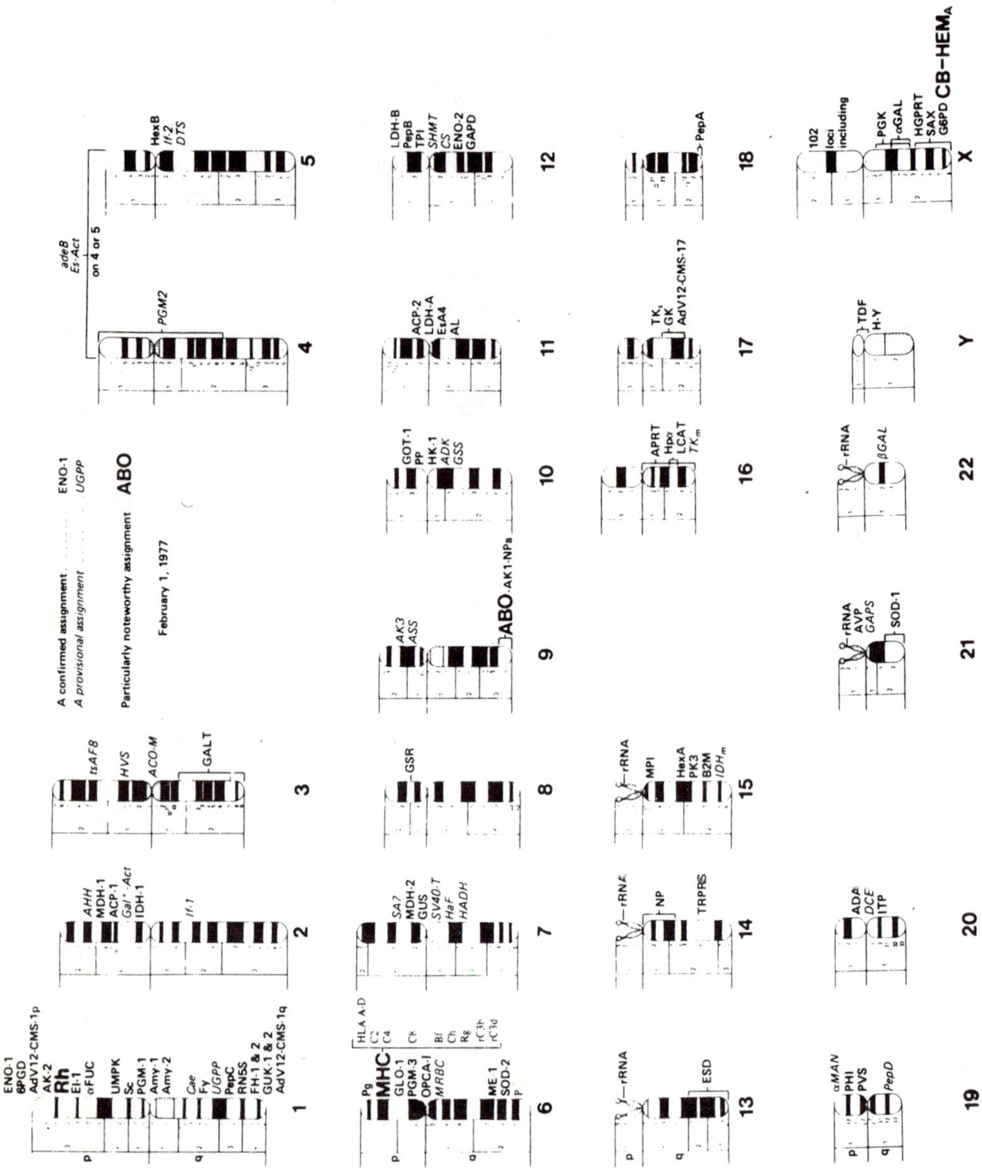

FIG. 1. Map of the human genome as per 1 February 1977 (from McKusick & Ruddle 1977). The key to the abbreviated symbols is given in Table 1 where the list of the human genes mapped is updated to 21 November 1977.

TABLE 1

List of human genes assigned to individual chromosomes as on 21 November 1977 (McKusick 1977). The abbreviated symbols are the same as those reproduced in Fig. 1. The asterisks denote the new assignments not reported in Fig. 1

	ABO	= ABO blood group (chr. 9)
	ACO-M	= Aconitase, mitochondrial (chr. 22)
	ACO-S	= Aconitase, soluble (chr. 3 or 9)
	ACP-1	= Acid phosphatase-1 (chr. 2)
	ACP-2	= Acid phosphatase-2 (chr. 11)
	adeB	= FGAR amidotransferase (chr. 4 or 5)
	ADK	= Adenosine kinase (chr. 10)
	AdV12-CMS-1p	= Adenovirus-12 chromosome modification site-1p (chr. 1)
	AdV12-CMS-1q	= Adenovirus-12 chromosome modification site-1q (chr. 1)
	AdV12-CMS-17	= Adenovirus-12 chromosome modification site-17 (chr. 17)
	AHH	= Aryl hydrocarbon hydroxylase (chr. 2)
	AK-1	= Adenylate kinase-1 (chr. 9)
	AK-2	= Adenylate kinase-2 (chr. 1)
	AK-3	= Adenylate kinase-3 (chr. 9)
	AL	= Lethal antigen: 3 loci (a1, a2, a3) (chr. 11)
*	Alb	= Albumin (chr. 4)
	Amy-1	= Amylase, salivary (chr. 1)
	Amy-2	= Amylase, pancreatic (chr. 1)
*	An-2	= Aniridia, type II Baltimore (chr. 1)
*	Ant-III	= Antithrombin III (chr. 1)
*	ARS-A	= Arylsulphatase A (chr. 22)
*	ARS-B	= Arylsulphatase B (chr. 5)
	APRT	= Adenine phosphoribosyltransferase (chr. 16)
*	ASD-2	= Atrial septal defect, secundum type (chr. 6)
	ASS	= Argininosuccinate synthetase (chr. 9)
	AVP	= Antiviral protein (chr. 21)
*	AVr	= Antiviral state regulator (chr. 5)
	Bf	= Properdin factor B (chr. 6)
	β2M	= β_2-microglobulin (chr. 15)
*	BVR	= Baboon virus replication (chr. 6 or 19)
	C2	= Complement component-2 (chr. 6)
	C4	= Complement component-4 (chr. 6)
	C6	= Complement component-6 (chr. 6)
	C8	= Complement component-8 (chr. 6)
	Cae	= Cataract, zonular pulverulent (chr. 1)
	CB	= Colour blindness (deutan and protan) (X chr.)
*	CF7	= Clotting factor VII (chr. 8)
	Ch	= Chido blood group (chr. 6) (same as C4S)
*	CML	= Chronic myeloid leukaemia (chr. 22)
*	Co	= Colton blood group (chr. 7)
*	Col-1	= Collagen 1 (1 and 2) (chr. 7 and 17)
	CS	= Citrate synthase, mitochondrial (chr. 12)
*	Dia-1	= NADH-diaphorase (chr. 6 or 22)
*	DMJ	= Juvenile diabetes mellitus (chr. 6)

TABLE 1 *(continued)*

*	Do	= Dombrock blood group (chr.1)
	DCE	= Desmosterol-to-cholesterol enzyme (chr. 20)
	DTS	= Diphtheria toxin sensitivity (chr. 5)
*	EBS-1	= Epidermolysis bullosa, Ogna type (chr. 10)
*	E-1	= Pseudocholinesterase-1 (chr. 1)
*	E-2	= Pseudocholinesterase-2 (chr. 16)
	E11S	= Echo 11 sensitivity (chr. 19)
	El-1	= Elliptocytosis-1 (chr. 1)
*	EMP-130 or SA 10	= External membrane protein-130 (chr. 10)
*	EMP-195 or SA 14	= External membrane protein-195 (chr. 14)
	ENO-1	= Enolase-1 (chr. 1)
	ENO-2	= Enolase-2 (chr. 12)
	Es-Act	= Esterase activator (chr. 4 or 5)
	EsA4	= Esterase A4 (chr. 11)
	EsD	= Esterase D (chr. 13)
	FH-M	= Fumarate hydratase, mitochondrial (chr. 1)
	FH-S	= Fumarate hydratase, soluble (chr. 1)
	αFUC	= Alpha-L-fucosidase (chr. 1)
*	FUSE	= Polykaryocytosis inducer (chr. 10)
	Fy	= Duffy blood group (chr. 1)
	Gal+-Act	= Galactose + activator (chr. 2)
	αGAL A	= α-galactosidase A (Fabry disease) (X chr.)
	βGAL	= β-galactosidase (chr. 3 or 22)
	GALT	= Galactose-1-phosphate uridyltransferase (chr. 2, 3 or 9)
	GAPD	= Glyceraldehyde-3-phosphate dehydrogenase (chr. 12)
	GAPS	= Phosphoribosyl glycineamide synthetase (chr. 21)
	Gc	= Group-specific component (chr. 4)
*	GDH	= Glucose dehydrogenase (chr. 1)
	GK	= Galactokinase (chr. 17)
	GLO-1	= Glyoxylase I (chr. 6)
*	Gm	= Immunoglobulin heavy chain (chr. 8)
*	GOT-M	= Glutamate oxaloacetate transaminase, mitochondrial (chr. 6)
	GOT-S	= Glutamate oxaloacetate transaminase, soluble (chr. 10)
*	GPI	= Glucosephosphate isomerase (chr. 11)
*	GPT-1	= Glutamate pyruvate transaminase, soluble (chr. 10)
*	GPx-1	= Gluthathione peroxidase-1 (chr. 3)
	G6PD	= Glucose-6-phosphate dehydrogenase (X chr.)
	GSR	= Glutathione reductase (chr. 8)
	GSS	= Glutamate-gamma-semialdehyde synthetase (chr. 10)
	GUK-1 & 2	= Guanylate kinase-1 & 2 (chr. 1)
	GUS	= Beta-glucuronidase (chr. 7)
*	H4	= Histone H4 (chr. 7)
	HADH	= Hydroxyacyl-CoA dehydrogenase (chr. 7)
	HaF	= Hageman factor (chr. 7)
*	Hbα	= Haemoglobin alpha chain (chr. 2, 4, 5 or 16)
*	Hbβ	= Haemoglobin beta chain (chr. 2, 4 or 5)
*	hCG	= Human chorionic gonadotropin (chr. 18)

TABLE 1 *(continued)*

*	Hch	= Haemochromatosis (chr. 6)
	HEM-A	= Classic haemophilia (X chr.)
	HexA	= Hexosaminidase A (chr. 15)
	HexB	= Hexosaminidase B (chr. 5)
	HGPRT	= Hypoxanthine-guanine phosphoribosyltransferase (X chr.)
	HK-1	= Hexokinase-1 (chr. 10)
	HLA	= Human leucocyte antigen (chr. 6)
	Hpα	= Haptoglobin, alpha (chr. 16)
	HVS	= Herpes virus sensitivity (chr. 3)
	H-Y	= Y histocompatibility antigen (Y chr.)
*	Ia-1	= Immune response associated-1 (chr. 6)
*	Ia-2	= Immune response associated-2 (chr. 6)
*	IgAS	= Immunoglobulin heavy chains attachment site (chr. 2)
	If-1	= Interferon-1 (chr. 2)
	If-2	= Interferon-2 (chr. 5)
	IDH-M	= Isocitrate dehydrogenase, mitochondrial (chr. 15)
	IDH-S	= Isocitrate dehydrogenase, soluble (chr. 2)
*	Ir	= Immune response (chr. 6)
	ITP	= Inosine triphosphatase (chr. 20)
*	Jk	= Kidd blood group (chr. 7)
*	Km	= Kappa immunoglobulin light chains, Inv (chr. 7)
*	LCAT	= Lecithin-cholesterol acyltransferase (chr. 16)
	LDH-A	= Lactate dehydrogenase A (chr. 11)
	LDH-B	= Lactate dehydrogenase B (chr. 12)
*	Leu-RS	= Leucyl-tRNA synthetase (chr. 5)
*	αMAN-A	= Cytoplasmic α-D-mannosidase (chr. 15)
	αMAN-B	= Lysosomal α-D-mannosidase (chr. 19)
	MDH-M	= Malate dehydrogenase, mitochondrial (chr. 7)
	MDH-S	= Malate dehydrogenase, soluble (chr. 2)
	ME-1	= Malic enzyme-1 (chr. 6)
	MHC	= Major histocompatibility complex (chr. 6)
*	MLC-W	= Mixed lymphocyte culture, weak (chr. 6)
	MPI	= Mannosephosphate isomerase (chr. 15)
	MRBC	= Monkey red blood cell receptor (chr. 6)
*	NCR	= Neutrophil chemotactic response (chr. 7)
*	NDF	= Neutrophil differentiation factor (chr. 6)
	NP	= Nucleoside phosphorylase (chr. 14)
	NPa	= Nail-patella syndrome (chr. 9)
	OPCA-1	= Olivopontocerebellar atrophy I (chr. 6)
	P	= P blood group (chr. 6)
*	PA	= Plasminogen activator (chr. 6)
*	PDB	= Paget disease of bone (chr. 6)
	PepA	= Peptidase A (chr. 18)

TABLE 1 *(continued)*

	PepB	= Peptidase B (chr. 12)
	PepC	= Peptidase C (chr. 1)
	PepD	= Peptidase D (chr. 19)
*	PepS	= Peptidase S (chr. 4)
	Pg	= Pepsinogen (chr. 6)
	PGK	= Phosphoglycerate kinase (X chr.)
	PGM-1	= Phosphoglucomutase-1 (chr. 1)
	PGM-2	= Phosphoglucomutase-2 (chr. 4)
	PGM-3	= Phosphoglucomutase-3 (chr. 6)
	6PGD	= 6-phosphogluconate dehydrogenase (chr. 1)
*	PRPP-AT	= Phosphoribosylpyrophosphate amidotransferase (chr. 4)
	PK3	= Pyruvate kinase-3 (chr. 15)
	PP	= Inorganic pyrophosphatase (chr. 10)
	PVS	= Polio sensitivity (chr. 19)
*	RB-1	= Retinoblastoma-1 (chr. 13)
	rC3b	= Receptor for C3b (chr. 6)
	rC3d	= Receptor for C3b (chr. 6)
	Rg	= Rodgers blood group (chr. 6) (same as C4F)
	Rh	= Rhesus blood group (chr. 1)
	RN5S	= 5S RNA gene(s) (chr. 1)
*	RP-1	= Retinitis pigmentosa-1 (chr. 1)
	rRNA	= Ribosomal RNA (chr. 13, 14, 15, 21, 22)
*	RwS	= Ragweed sensitivity (chr. 6)
*	SA6	= Surface antigen 6 (chr. 6)
	SA7	= Species antigen 7 (chr. 7)
*	SA11	= Surface antigen 11 (chr. 11)
*	SA12	= Surface antigen 12 (chr. 12)
	SAX1, SAX2, SAX3	= X-linked species (or surface) antigen (X chr.)
	Sc	= Scianna blood group (chr. 1)
	SHMT	= Serine hydroxymethyltransferase (chr. 12)
	SOD-1	= Superoxide dismutase-1 mitochondrial (chr. 21)
	SOD-2	= Superoxide dismutase-2 soluble (chr. 6)
*	Sph-1	= Spherocytosis, Denver type (chr. 8 or 12)
	SV40-7	= SV40-integration site-7 (chr. 7)
*	SV40-17	= SV40-integration site-17 (chr. 17)
	TDF	= Testis determining factor (Y chr.)
	Tf	= Transferrin (chr. 1)
	TK-M	= Thymidine kinase, mitochondrial (chr. 16)
	TK-S	= Thymidine kinase, soluble (chr. 17)
	TPI-1 & 2	= Triosephosphate isomerase-1 and -2 (chr. 12)
	Trp-RS	= Tryptophanyl-tRNA synthetase (chr. 14)
	tsAF8	= Temperature sensitive (AF8) complementing (chr. 3)
	UGPP-1	= Uridyl diphosphate glucose pyrophosphorylase-1 (chr. 1)
*	UGPP-2	= Uridyl diphosphate glucose pyrophosphorylase-2 (chr. 2)
	UMPK	= Uridine monophosphate kinase (chr. 1)
*	UP	= Uridine phosphorylase (chr. 7)

TABLE 1 *(continued)*

*	WS-1	= Waardenburg syndrome-1 (chr. 9)
*	W-AGR	= Wilms tumour—aniridia/ambiguous genitalia/mental retardation (chr. 11)
*	XP-E	= Xeroderma pigmentosum, Egyptian (chr. 9)

individual cell surfaces. This means that in due course cell–cell interactions, which are believed to be at the root of differentiation, can be investigated in terms of specific genetic signals appearing on the cell surfaces.

Is normal differentiation directly dependent on given sequential patterns with which specific gene products are expressed on the cell surfaces? What patterns are these, and what is the relative role of nature and nurture in determining them? Are 'tumour-specific antigens' altogether new aberrant products or the unbalanced immunological phenotype resulting from the overproduction of one or more of the regular chromosome-specific surface antigens?

These are some of the questions which may become the obvious targets of experimentation for somatic cell geneticists if the harvest of mapping data on chromosome-specific surface antigens proceeds during the next decade in the way that the mapping of human biochemical markers has gone during the past ten years.

It is almost a rule for students of human linkage to justify their addiction by emphasizing the potential application of this type of specialized knowledge to medical biology. From the title of my paper it is clear that I am no exception to this rule and, since I work in a cancer research centre, I am now going to speculate on how human gene mapping could help cancer biology.

DISRUPTION OF LINKAGE GROUPS IN CELL HYBRIDS AS A MEASURE FOR DNA DAMAGE REPAIR

In the early days of human somatic cell genetics my colleagues and I showed that the murine–human cell hybrids, besides their unique potential for mapping human genes with the classical 'synteny test', could be used to screen for the 'mitotic separation' of already known linked genes, and hence to map their sequential order in a statistical fashion (Siniscalco 1970; Miller *et al.* 1971). This approach has recently been further developed and improved by Goss and Harris. They have demonstrated that if human cells are irradiated with increasing doses of γ rays before fusion with murine cells, the probability of co-transference of linked genes into the cell hybrids is inversely proportional to the distance between them (Goss & Harris 1975). With this elegant method (which I will refer to as the 'radiation co-transfer test'), Goss and Harris have obtained estimates of genetic distances between six human X-linked markers (one of which, phosphoribosyl-pyrophosphate synthetase, has actually been assigned to the X chromosome through this very type of study) (J.E. Seegmiller, personal communication, 1978) and of eight markers of autosome number 1 (Goss & Harris 1977*a,b*). One of the most interesting messages stemming from these experiments is that at each radiation dose, the percentage of co-transference for a given pair of linked genes is remarkably constant in different experiments. To secure this concordance, which is the *sine qua non* for their mapping studies, Goss and Harris were careful to use human lymphocytes from the same donor (S.J. Goss, personal communication, 1977) in all their experiments.

It is indeed to be expected that lymphocytes from individuals highly susceptible to radiation-induced chromosomal damage may require a lower radiation dose for the splitting of a given pair of linked genes. If this is the case, the 'radiation co-transfer test' could obviously become a powerful tool for measuring the individual variation to DNA damage repair at the level of the functional chromosomal DNA. Moreover, it would allow such measurements to be made in diploid cells freshly derived from the living propositi. However, it is clear that to be of practical use for cancer biologists and/or therapists, this test should be significantly simplified in its experimental design and shortened in its duration. For the last year, Ms Angelina Petrovic and I have been trying to do just this, with some encouraging results. Though still incomplete, these results are worth mentioning here. Firstly, we have restricted the screening for radiation-induced chromosomal damage to the small target site of the human X chromosome long arm between Xq26 and Xqter which includes the genes for hypoxanthine-guanine phosphoribosyltransferase (HGPRT) and glucose-6-phosphate dehydrogenase (G6PD) (Fig. 1). Secondly, we have chosen as murine parental partner a Chinese hamster line (CHO-YH21) whose cells are deficient in both these enzymic markers (Chasin 1975). This choice has considerably simplified the screening for the co-transfer of these two human markers, which is sequentially performed in two steps, by isolating the HGPRT-positive hybrid cells in the HAT selective medium and then testing a random sample of individual hybrid cells for G6PD activity with a cytochemical assay. Fig. 2 shows the typical appearance of G6PD(+) and G6PD(-) hybrid cells in a hybrid HAT clone derived from a PEG-induced fusion of CHO-YH21 cells with irradiated normal human lymphocytes. Table 2 and Fig. 3 summarize the over-all results of an experiment of this type in

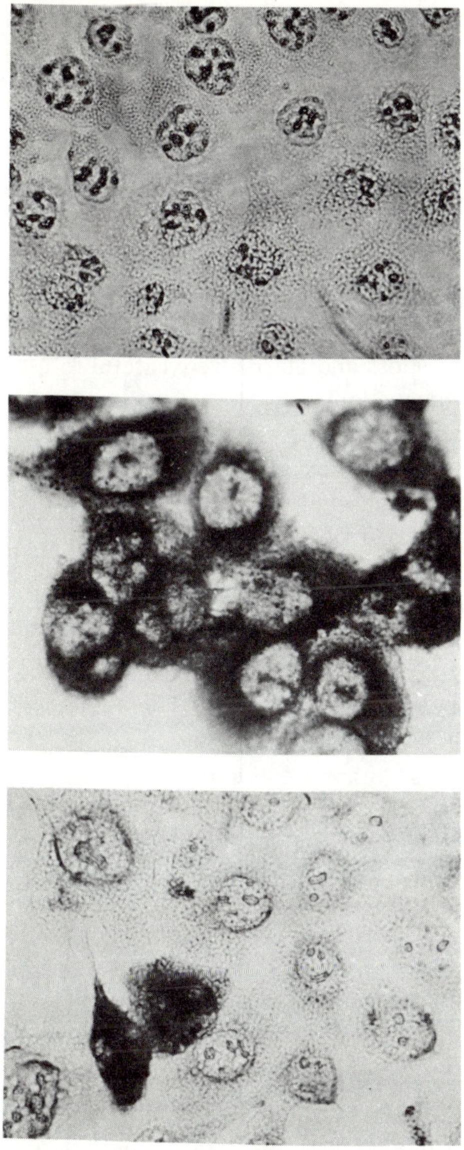

A Chinese hamster cells CHO-YH21 : all G6PD (-) and HGPRT (-)

B Somatic cell hybrid between CHO-YH21 and human lymphoblasts : all G6PD (+) and HGPRT (+)

C Same type of hybrid as above but with human parental cells irradiated prior to fusion : only two G6PD (+) and all HGPRT (+)

FIG. 2. Classification of the G6PD phenotype at the level of individual cells with the cytochemical staining of Chasin (1975). The unstained cells are G6PD negative and those stained dark are G6PD positive. The rodent parental cells (CHO-YH21) are all G6PD negative (Panel A); they are also HGPRT negative and are therefore killed by HAT selection within one week after the fusion event. The HAT-resistant hybrid clones are G6PD positive in all of their cells (Panel B) when the relevant portion of the human X chromosome long arm is retained unbroken by the hybrid genome. The hybrid clone reproduced in Panel C derives from fusion with human lymphoblasts irradiated with 1000 rad. Further discussion in the text.

TABLE 2

Co-transfer of human X-linked markers in HAT-selected hybrids between CHO-YH21 cells and irradiated human lymphoid cells

Dose of X-rays (kilorads)	*Clone identification number*	*Human X-linked markers retained by HAT-selected hybrid clones*			*% of co-transfer of G6PD*		
		G6PD	*PGK*	*α-gal*	*(i) at the clonal level*	*(ii) at the cell level*	
0	1	+ (199/310)	+	+		64	
	4	+ (310/450)	+	+		69	
	5	+ (330/500)	+	+	100	66	67
	8	+ (220/300)	+	+		73	
	10	+ (189/300)	+	+		63	
1	1	+ (60/300)	–	–		20	
	2	+ (59/280)	+	+		21	
	5	+ (87/430)	+	+		20	
	6	+ (150/390)	+	+		38	
	8	+ (60/275)	–	+	67	22	21
	11	+ (132/400)	+	+		33	
	12	– (22/320)	–	–		7	
	19	– (17/290)	–	–		6	
	20	– (33/410)	–	–		8	
2	1	– (8/517)	–	–		2	
	4	– (9/290)	+	+		3	
	5	– (12/300)	+	+		4	
	8	+ (53/405)	–	–	50	13	10
	10	+ (86/390)	–	–		22	
	12	+ (42/270)	–	–		16	
	15	+ (50/350)	–	+		14	
	20	– (6/280)	+	+		2	
4	3	– (0/290)	–	–		0	
	9	– (4/250)	–	–		2	
	10	– (5/300)	–	–	33	2	5
	11	+ (38/310)	–	–		12	
	17	+ (33/290)	–			11	
	19	– (4/150)	–	–		3	
5	1	+ (51/390)	–	–		13	
	6	– (0/240)	–	–		0	
	17	+ (16/500)	–	–	20	13	5
	18	– (0/190)	–	–		0	
	19	– (9/210)	–	–		4	
	20	– (0/360)	–	–		0	

This table describes the percentage of co-transfer of X-linked genes into cell hybrids after fusion of CHO-YH21 cells (deficient in HGPRT and G6PD) with human lymphoid cells irradiated with 0, 1, 2, 4, 5 kilorads before fusion. All HAT-selected clones reported were found to be HGPRT-positive and their hybrid nature was established independently by Giemsa-11 differential staining

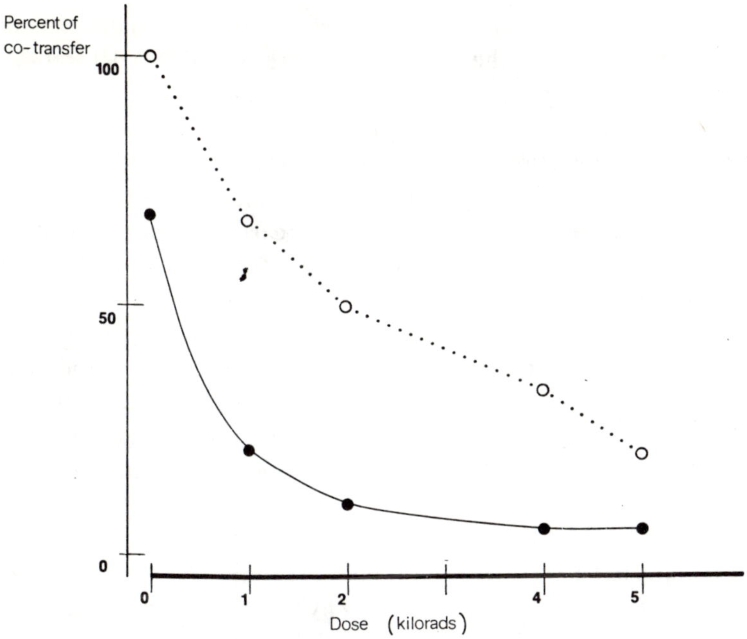

Co-transfer of human HGPRT and G6PD measured at the clonal level (○) and at the individual cell level (●)

FIG. 3. This figure illustrates graphically the discordance in the estimates of the percentage of HGPRT-G6PD co-transfer obtained at the clonal level (empty circles) and at the individual cell level (full circles). The data are the same as those reported in Table 2. Further discussion in the text.

which CHO-YH21 cells were fused to normal diploid human lymphoblasts irradiated with 0, 1, 2, 4 or 5 kilorads. From these data it is clear that the frequency with which human HGPRT and G6PD are co-transferred is grossly overestimated when the presence of human G6PD is screened at the clonal level rather than on the individual hybrid cells. In particular the percentage of

---

←

according to Bobrow & Cross (1974). The co-transfer of human G6PD was classified either electrophoretically on lysate of clonal cell populations (Miller *et al.* 1971) or cytochemically (Chasin 1975) on random cell samples from each clone. The number of cells found positive for G6PD and the total number of cells screened are given in parentheses. The cytochemical assay has never been positive on the parental CHO-YH21 cells. From these data it is clear that the percentage of co-transfer of human G6PD at the clonal level has been grossly overestimated. This point is further illustrated in Fig. 3 and discussed in the text. The pattern of co-transfer of phosphoglycerate kinase and α-gal (classified only electrophoretically at the clonal level), at different doses of radiation, agrees in general with that reported by Goss & Harris (1975, 1977*a*), but the situation might be considerably changed if these markers were also studied at the individual cell level.

G6PD-deficient cells in the hybrid clones derived from non-irradiated human parental cells is a measure of the spontaneous rate of 'mitotic separation' between the two loci under the given experimental conditions. However, the discordance between the two estimates of the percentage of co-transfer (i.e. at the cellular and at the clonal level) is clearly larger when the hybrid clones are derived from irradiated human parental cells. This indicates that irradiated chromosomes may be more prone to *de novo* rearrangements in culture and/or that in some hybrid clones the human loci for HGPRT and G6PD must have been co-transferred on separate fragments of the X chromosome with the consequent obligatory retention of the HGPRT and the random loss of G6PD. Finally, the sharp fall in the proportion of G6PD hybrid cells observed at 1000 rads tells us that a much smaller range of irradiation must be applied to the human parental cells before fusion if this test is meant to detect differences in the individual susceptibility to radiation-induced chromosomal breakage.

Accordingly we now treat the human test cells with 0, 50, 100, 300 or 500 rads and, to minimize the risk of chromosomal rearrangement in culture, we screen for G6PD activity in the HAT-surviving hybrid cells when they are still at a very early stage of their clonal growth (i.e. about 10^2 to 10^3 cells). This latest version of the radiation co-transfer test, which has the additional advantage that it can be completed within two to three weeks, is being extensively used in our laboratory to estimate the gamma radiation dose required to split the above-mentioned X-linked genes in 50% of the hybrid cells derived from lymphoid cells of normal individuals as well as from patients with a known predisposition to chromosomal breakage and cancer. Though the data accumulated so far are too scanty and incomplete to warrant a detailed report, we have been encouraged by the finding that the percentage of HGPRT-G6PD co-transfer has been consistently higher at all radiation doses when lymphoblasts of normal individuals were used as opposed to lymphoblasts from patients with ataxia telangectasia, a recessive condition with well-documented defective excision repair of γ-ray-damaged DNA (Paterson *et al.* 1976).

HUMAN GENE MAPPING AND CANCER THERAPY: A RATIONALE FOR SELECTIVE CHEMOTHERAPY AND IMMUNOTHERAPY

A variable degree of heteroploidy is a common aspecific connotation of malignant cells but, since the banding era began, the examples of tumours which are associated with specific numerical or structural aberrations have been steadily increasing. Recently Levan & Mitelman (1977) have reviewed the over-all karyological data of 706 cases of human neoplasms exhibiting chromosomal aberrations. From their analysis it appears that, apart from the regular association of the 9–22 translocation with chronic myeloid leukaemia, the aberrations found so far tend to cluster to several specific chromosomes, namely 1, 7, 8, 9, 14, 17, 21 and 22. I am glad to leave to Dr Klein the difficult task of attempting to interpret the nature of these associations. What I need to emphasize, as introduction to my own speculations, is the well-established fact that *in vivo* and *in vitro* growth of tumour somatic cells is associated with genomic unbalance which leads to redundancy or loss of individual chromosomes or chromosomal fragments. This means that most tumour cells turn out to be monosomic or trisomic (though, more rarely, also nullisomic or multisomic) for part of their genome, with a consequent dosage effect at the level of production of the relevant gene products. Addicts of studies of human linkage know well that these gene dosage variations are used for mapping the human chromosomes with the so-called 'gene dosage method' (McKusick & Ruddle 1977). For instance, the first evidence in favour of the linkage between the loci for 6-phosphogluconate-dehydrogenase (6PGD) and the Rh complex was obtained by Fialkow and associates from a patient with chronic myeloid leukaemia whose tumoral peripheral blood cells happened to be monosomic for part of autosome 1 in addition to being carriers of the Philadelphia chromosome (Fialkow *et al.* 1972). My question is whether these situations could be exploited in a selective approach to the chemotherapy and/or immunotherapy of tumours when the variation in gene dosage brought about by the aneuploid mutation makes the cells in question amenable to some already available selective procedure.

If, for instance, a tumour clonal line happens to lose the functional human gene for HGPRT, it is to be expected, at least in theory, that its cells would be killed *in vivo,* as they are *in vitro,* by doses of amethopterin (methotrexate) which are not toxic for the normal cells. Likewise, monosomy for autosome 10 or 16 would lead to a 50% decrease in the amount of adenosine kinase or adenine phosphoribosyltransferase, respectively—two enzymes whose absence is known to be associated with sensitivity to the antibiotic alanosine (Kusano *et al.* 1971; Chan *et al.* 1978). A few years ago, Dr Old, Dr Boyse and their colleagues reported an interesting case of a dog lymphosarcoma which was killed *in vivo* by treatment with asparaginase and could not grow *in vivo* or *in vitro* in the absence of asparagine. One of the possible explanations of these findings was that an early somatic mutation had rendered the tumour cells auxotrophic for the synthesis of asparagine (Old *et al.* 1968). Similar situations could also arise, and probably have, in humans. In fact the work of

Puck and associates has shown that auxotrophic mutations can arise in cultured mammalian somatic cells (Puck 1973) and that some human chromosomes, such as chromosomes 4 or 5 (Kao & Puck 1972) and 12 (Jones *et al.* 1972), carry genes which complement for these types of mutations. Dr Puck maintains, and I entirely agree with him, that each human chromosome may carry gene(s) for auxotrophic mutants. Thus monosomy for these human genes in tumour cells may result in partial auxotrophy and render the tumour line susceptible to selective killing *in vivo* by amino acid deprivation in the diet or by treatment with the specific enzyme which further depletes the organism of the gene product that the auxotrophic cells are unable to synthesize.

Trisomic alterations of tumour cells are also expected to result in a quantitative variation of specific gene products, which, in this case, should be increased by 50%. In particular, if it is true that every human chromosome carries genes for surface antigens, the trisomy of any one chromosome could result in an excessive accumulation of the relevant chromosome-specific antigen on the cell surface. Such a situation could again be exploited for selective killing of the tumour cells provided that the appropriate specific antibody is available and that this can be used at a threshold level which is lethal only for the trisomic cells. For this to be possible, however, the antibody should be highly specific, cytotoxic and available in large quantities. These circumstances are perhaps not far from being fulfilled, in view of the dramatic progress which is being made in the production of monoclonal antibodies, as we have heard from Dr Milstein and from Dr Koprowski (see p. 251 *et seq.* and p. 266 *et seq.*).

The selection against trisomic or multisomic cells can also be achieved in culture biochemically when the appropriate chemical reagent is available. It is known, for instance, that the sensitivity of mammalian cells to guanine analogues (6-thioguanine, 8-azaguanine etc.: Szybalski & Szybalska 1962) or to 5-bromodeoxyuridine (Kit *et al.* 1963) is based on the normal level of activity of the enzymes HGPRT and thymidine kinase, respectively. Likewise, sensitivity to cytosine-arabinoside and to 5-bromodeoxycytidine is based on normal enzyme activity for 2-deoxycytidine kinase and deoxycytidine deaminase, respectively (Siminovitch 1976). Thus, a 50% increase in the activity for any of these enzymes, expected in a tumour cell that happens to be trisomic for the corresponding structural gene, may again offer us the possibility of finding a threshold concentration of the drug which is lethal for the tumour and not for the normal cells.

The two following points summarize the essence of my reasoning. On the one hand, it is becoming clearer than ever that malignant cells undergo numerical and/or structural chromosomal mutations which most probably occur at random but may all the same become specific for a given type of tumour (at least at a subclonal level) as a result of somatic selection. On the other hand, there can be little doubt that alongside the rapidly expanding knowledge on human gene mapping there will soon be available an array of chemical and immunological reagents which can selectively kill—at least *in vitro*—only those cells bearing a specific type of somatic mutation at the genic, chromosomal or genomic level. Given these two sets of circumstances, it seems to me not too unrealistic to start thinking about the feasibility of a selective chemotherapy or immunotherapy of tumour cells based on the strategies described above.

I am, of course, fully aware of the very speculative nature of these considerations at this time, and of the very slim chance that the several expectations presented will prove to be correct. Perhaps the only virtue of these speculations is that they are amenable to experimental verification. This could be achieved by testing the appropriate tumour lines and their controls for differential rate of growth in athymic mice treated with the specific chemical or immunological selective reagents. With this programme in mind we have isolated two clonal sublines, with or without monosomy at the locus for thymidine kinase from a human quasi-diploid tumour line (HUTU-80, from the cell repository of the Sloan-Kettering Institute), and we intend to use them in due course to test the possibility of selective chemotherapy with amethopterin. Likewise, clonal sublines of Burkitt lymphoma, with or without the chromosome-14 trisomy, could be isolated and used to test the feasibility of a selective immunotherapy with the already-reported specific antiserum raised against a surface antigen coded by genes of this human chromosome (McKusick 1977).

If selective killing of tumour cells, based on the above rationale, proves to be possible also *in vivo*, the cancer therapists of tomorrow will have to devise *ad hoc* protocols (chemical or immunological as the case may be) for each individual tumour, based on the results of *in vitro* screening tests which, by analogy with the bacteriologist's antibiogram, could be called 'tumour chemogram' or 'immunogram'.

COMPLEMENTATION FOR SISTER CHROMATID EXCHANGE (SCE) IN MURINE–HUMAN HYBRID CELLS

I would now like to turn to a less speculative area which illustrates another, as yet unexplored, application of somatic cell hybridization to cancer biology. I shall refer to some studies in my laboratory during the past year of the genetics of sister chromatid exchange (SCE), a well known cancer-related phenomenon, by means of complementation analysis with murine–human somatic cell hybrids. I am grateful to Dr B. Alhadeff, who is the main author of this research endeavour, for allowing me to report some results of this study, which will soon be published in full.

As is well known, a microfluorometric method is now being used (Latt 1973; Perry & Wolff 1974) for detecting SCE during the replication of chromosomes in cultured cells. Cells allowed to grow in 5-bromodeoxyuridine (BUdR) for two cycles and then treated with the fluorescent dye Hoechst 33258 show differential staining of their sister chromatids. The chromatid with only one BUdR-substituted DNA strand stains much darker than the one with both DNA strands substituted. Consequently, exchanges between chromatids can be readily recognized by examination of the pattern of differential staining within each chromatid. This technique has proved to be a valuable device for evaluating the cytological effect of mutagenic carcinogenic agents (Perry & Evans 1975) as well as for detecting malignant and premalignant conditions (Wolff 1977). One of the most dramatic contributions of this technology has been the discovery by German and his colleagues (Chaganti *et al.* 1974) of a 12-fold increase of SCE in cultured lymphocytes from patients with Bloom syndrome, an autosomal recessive disorder of growth (Bloom 1966).

Our interest in complementation studies for SCE with somatic cell hybrids developed out of serendipity. Over a year ago, we started to investigate the pattern of chromatid replication of interspecific somatic hybrids derived from parental cells with marked differences in rate of growth. The main motivation of these studies was to test a working hypothesis that the chromosomal evolution of the hybrid genome could be largely dependent on the degree of asynchrony of chromatid replication among the two parental sets of chromosomes. In order to investigate this issue we devised a technique which summarizes the properties of the so-called Giemsa-11 differential staining of Bobrow & Cross (1974) and the FPG method of Perry & Wolff (1974), and which thus allows us to distinguish the human from the mouse chromosomes and simultaneously to analyse in the same metaphase the pattern of chromatid replication for both sets of parental chromosomes (Alhadeff *et al.* 1977). The first series of observations with this technique were made on freshly produced as well as on long-established murine–human hybrids derived from the fusion of the rapidly growing L-A9 cells (deficient in HGPRT and therefore unable to grow in the HAT selective medium) with normal human diploid cells. Besides their relevance to the hypothesis under test (on which I will not elaborate further here), the results of these preliminary studies were of interest especially in view of the unexpected finding that that L-A9 cells exhibited an

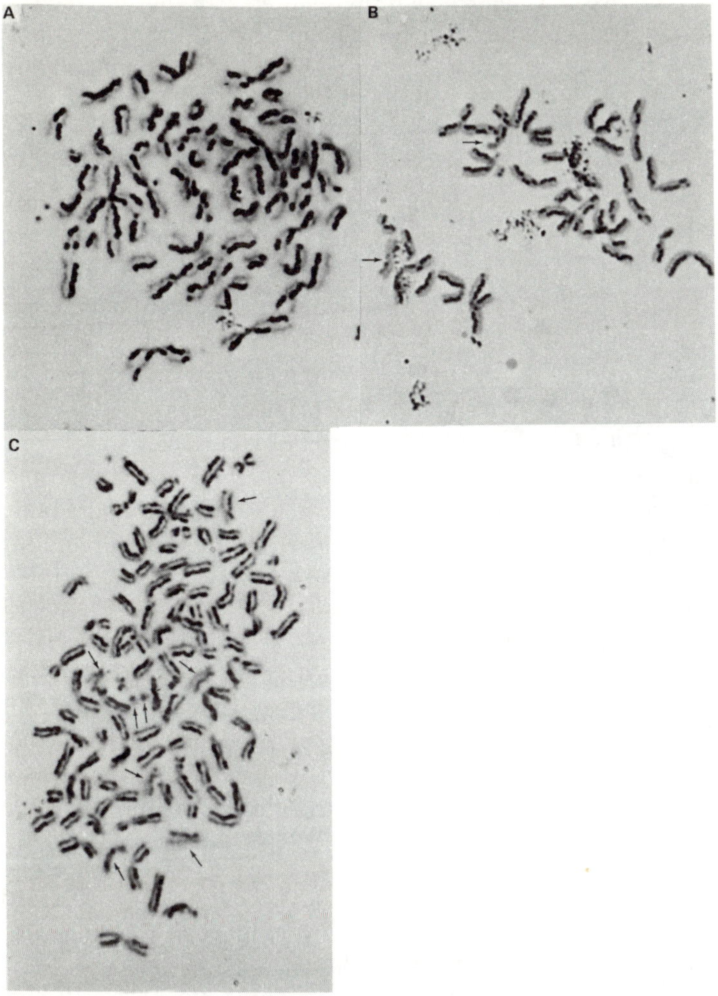

FIG. 4. Complementation for the high rate of SCE of the murine parental cells in somatic cell hybrids with normal human diploid cells:
A. Parental murine cell (L-A9-2) showing the high incidence of SCE typical for this cell line (0.79 ± 0.03).

B. Metaphase from cell hybrid line L-A9/HRBC2 (Balazs *et al.* 1978) showing the sole retention of the human X chromosome and a fragment of autosome No. 2. The rate of SCE (0.89 ± 0.04) is as high as in the parental L-A9 cells.

C. Metaphase from hybrid cell line L-A9/MAS-10-4 which has retained between 8 and 15 human chromosomes. The suppression of the rate of SCE among the murine chromosomes is obvious (0.28 ± 0.01). In the original preparation, the human chromosomes (with arrow) stain pale blue and the murine ones stain magenta with the combined FPG-Giemsa 11 method (Alhadeff *et al.* 1977).

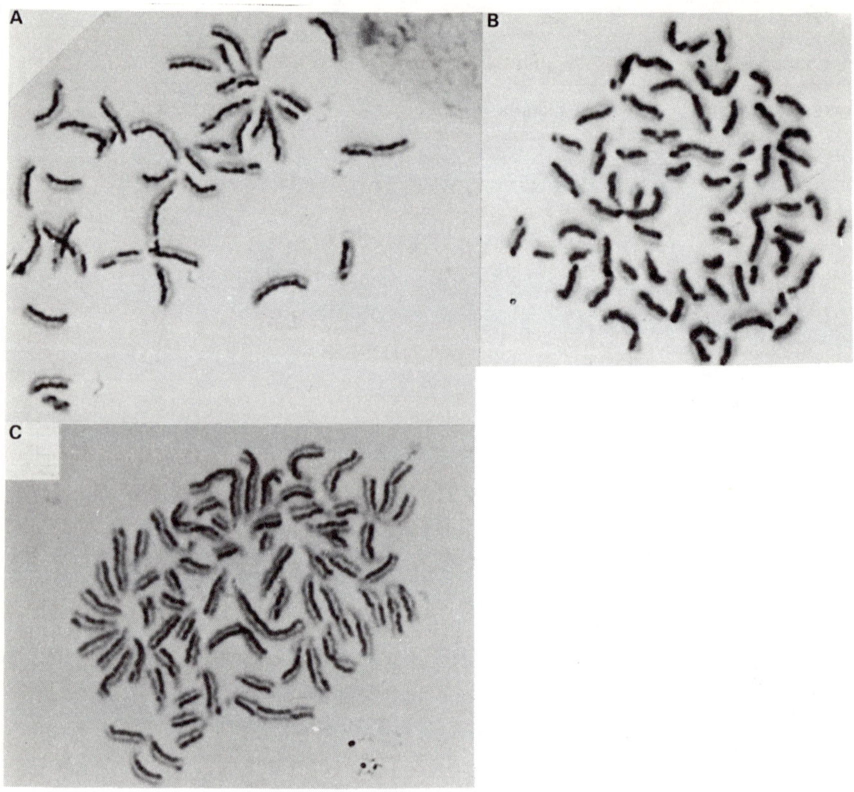

FIG. 5. Metaphases from C3H diploid fibroblasts (A), L cells (B) and Rag cells (C) showing normal rates of SCE. Further comments in the text.

incidence of SCE two- to fourfold higher than the one usually observed among cultured mammalian cells. This finding prompted us to change the directions and the experimental strategies of the original research plan.

As illustrated in Figs. 4 and 5 and Table 3, the L-A9 murine cells exhibit an unquestionably higher rate of spontaneous SCE per chromosome (0.78 ± 0.04) than do L cells (0.41 ± 0.02) and diploid fibroblasts (0.04 ± 0.02) of C3H mice, the live strain from which the L cells and consequently the A9 cells originated. Another murine cell line (Rag), also deficient in HGPRT but derived from BALB/c mice, was found to exhibit a normal rate of SCE (Fig. 5*c*), thus ruling out the possibility that the high SCE might have been correlated with deficiency in the X-linked enzyme. Somatic hybrid cells originating from the fusion of L-A9 cells and SCE-normal human cells were found to fall into three distinct groups: *(a)* those with a homogeneous population of cells with low incidence of SCE (Table 3, no. 5)

TABLE 3

Distribution of SCE/chromosome in different murine and hybrid cells

Denomination of cell line	*No. of human chromosomes retained*	*Total no. cells analysed*	*Average SCE/chromosome*	*Variances*	S.D.	S.E.
1. L-A9-1 (SKI)	–	50	0.78	0.07	0.26	0.04
2. L-A9-2 (AE)	–	50	0.79	0.05	0.22	0.03
3. L cells	–	50	0.41	0.02	0.14	0.02
4. C3H	–	50	0.44	0.03	0.17	0.02
5. L-A9/MAS-10-4	(8→15)	50	0.28	0.01	0.08	0.01
6. L-A9/HRBC2	1+ fragment	50	0.89	0.08	0.28	0.04
7. L-A9/MAS-10-5	(2→10)	25	0.36	0.12	0.35	0.07

The rates of SCE/chromosome reported in this table were calculated by averaging the counts of SCE/chromosome obtained in at least 50 metaphases of each cell line. It is clear that the two lines of L-A9 (cf. Fig. 4A), one from our cell library (SKI) and the other (AE) kindly provided by Dr S. Shin, exhibit essentially the same higher rate of SCE than L cells (cf. Fig. 5B) and C3H fibroblasts (cf. Fig. 5A). The rate of SCE is homogeneously suppressed in the hybrid clone L-A9/MAS-10-4 (with 8 to 15 human chromosomes per metaphase (cf. Fig. 4C), but reverts again to a high level of SCE after extreme loss of human chromosomes as in L-A9/HRBC2 (a hybrid cell line from a different fusion experiment which practically retains only the human X chromosome) (cf. Fig. 4B and Balazs *et al.* 1978).

It is of interest that another primary clone, L-A9/MAS-10-5, has retained a variable number of human chromosomes per cell and exhibits a variable degree of SCE, as shown by the higher variance of the estimate obtained from 25 metaphases.

indistinguishable from the rate of SCE observed in normal diploid metaphases; *(b)* those with a homogeneous population of cells with an incidence of SCE as high as that found in murine parental cells (Table 3, no. 6); *(c)* those with a bimodal distribution of cells with a high or low rate of SCE (Table 3, no. 7). Cytogenetic studies have shown that hybrid cells of the first group have on average a consistently higher number of human chromosomes (Fig. 4C) than those of the second group (Fig. 4B), while those of the third group have a mixed distribution of metaphases with a high or low number of human chromosomes. These findings suggest that the high rate of SCE in L-A9 cells is suppressed in the presence of the human genome and reverts to the level typical for the murine parental cells after extensive loss of human chromosomes. At this stage of the investigation, we cannot say whether one or more human chromosomes are involved in this complementation effect. However, we can already anticipate that a group C chromosome seems to be definitely involved, whereas the human X is certainly not, as suggested by the high rate of SCE found in the A9/HRBC2 hybrid cell line, which had regularly retained this chromosome in all of its metaphases.

One of the questions that we are trying to answer is whether

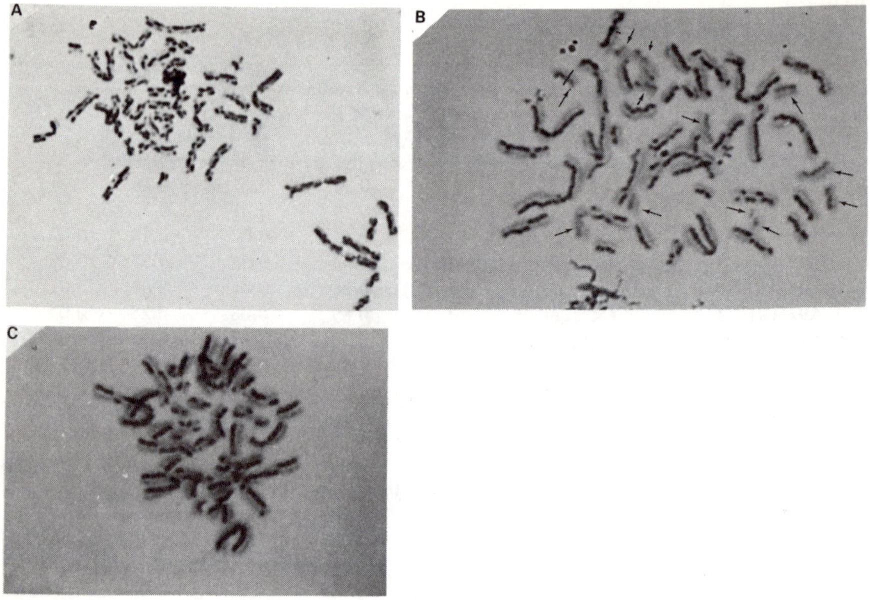

FIG. 6. Complementation of high SCE rate of Bloom's cells in hybrids with hamster cells CHO-YH21.

A. A typical metaphase in the diploid fibroblastic cells of a patient with Bloom syndrome (GM 1492).

B. A metaphase of the hybrid cell line CHO-YH21 × Bloom's fibroblasts showing suppression of SCE in Bloom's chromosomes (with arrow). The SCE of the parental rodent cells has remained the same. The chromosome intercepted by a bar is the result of a human-rodent translocation: the portion with the arrow is the human one.

C. A typical metaphase from the parental rodent cell CHO-YH21.

complementation for murine SCEs will occur in somatic cell hybrids between the L-9 cells and the high SCE Bloom's cells. Unfortunately, we have not yet obtained this type of hybrid, but we have isolated several hybrid clones between Bloom's diploid fibroblastic cells (strain GM 1492 supplied by the Genetic Repository of Camden, New Jersey) and the Chinese hamster cells (CHO-YH21) which, as stated previously, are deficient in HGPRT and G6PD and show a moderately high rate of SCE (0.4 ± 0.02). Interestingly enough, in these types of hybrid cells the murine chromosomes continue to exhibit the parental rate of SCE (if not a higher one), but the Bloom's chromosomes definitely do not (Fig. 6). It might be argued that the latter phenomenon might be due to the loss of the critical part of the genome of Bloom's cells which induces excessive SCE. In this case the increased SCE should be

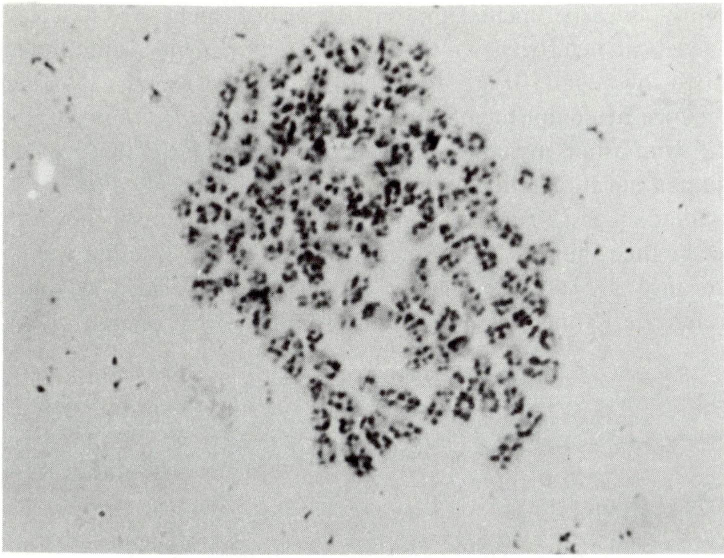

FIG. 7. A metaphase from a human malignant melanoma (SKI-Mel-28) stained with the FPG–Giemsa 11 method of Alhadeff *et al.* (1977). The chromosomal pattern suggests a dramatically increased rate of SCE. See note at end of text for further comments.

regarded as a dominant feature and it is rather awkward to think of a recessive autosomal mutation behaving as a dominant marker in hybrid cells, when it is known that somatic cells from obligatory heterozygotes for this mutant have a perfectly normal rate of SCE (Chaganti *et al.* 1974). Moreover, Dr Alhadeff now has about 200 early hybrid metaphases of this type (derived from at least four separate fusion events and still possessing a high number of human chromosomes) and has never found one with high SCE human chromosomes or with low SCE hamster chromosomes. From all this, we can tentatively conclude that: *(i)* the Chinese hamster genome is evidently able to complement the Bloom's chromosomes for SCE even though its own chromosomes continue to exhibit the moderately high rate of SCE typical of this animal cell line; *(ii)* unlike the situation in the hybrid *L-A9* × *normal human fibroblasts,* the rate of SCE of the parental murine line is never lowered in the hybrid *CHO-YH21* × *Bloom's fibroblasts,* thus suggesting that the Bloom's genome may be unable to reciprocate the complementing effect of the CHO-YH21 genome even when large numbers of its chromosomes are still present in the hybrid metaphases.

These findings indicate that, as expected, more than one type of lesion can upset the normal mechanism of chromatid replication of mammalian cells and that, consequently, the experimental approach described can be used to screen for possible genetical heterogeneity among different patients with Bloom syndrome and to investigate, in general, the biology and genetics of sister chromatid exchange by complementation analysis. *(Note added in proof:* At the time of the symposium my colleagues and I were convinced that we had identified a human melanoma line with a dramatic increase in SCE (Fig. 7). Subsequent studies suggest that the peculiar chromosomal pattern shown in Fig. 7 may not be the result of SCE. Since the nature of this finding and its biological relevance are not yet clearly established, I decided to omit any further reference to the melanoma cells from this final edition of my paper.)

References

ALHADEFF, B., VELIVASAKIS, M. & SINISCALCO, M. (1977) Simultaneous identification of chromatid replication and of human chromosomes in metaphases of man-mouse somatic cell hybrids. *Cytogenet. Cell Genet. 19,* 236-239

BALAZS, I., SZABO, P. & SINISCALCO, M. (1978) Hybridization properties of human X-chromosomal RNA transcripts from murine-human hybrids. *Somatic Cell Genet. 4,* 617-631

BOBROW, M. & CROSS, J. (1974) Differential staining of human and mouse chromosomes in interspecific cell hybrids. *Nature (Lond.) 251,* 77-79

BLOOM, D. (1966) The syndrome of congenital telangectatic erythema and stunted growth. *J. Pediatr. 68,* 103-113

BODMER, W.F. (1979) Gene clusters and the HLA system, in this symposium, pp. 205-223

BUCK, D.W. & BODMER, W.F. (1975) The human species specific antigen on chromosome 11, in *Human Gene Mapping 2* (Bergsma, D., ed.) *(S. Karger Ser. vol. 11),* pp. 87-89, The National Foundation, March of Dimes, New York

CHAGANTI, R.S.K., SCHONBERG, S. & GERMAN, J. (1974) A manyfold increase in sister chromatid exchanges in Bloom's syndrome lymphocytes. *Proc. Natl. Acad. Sci. U.S.A. 71,* 4508-4512

CHAN, T.S., CREAGAN, R.P. & REARDON, M.P. (1978) Adenosine kinase as a new selective marker in somatic cell genetics. *Somatic Cell Genet. 4,* 1-12

CHASIN,L. (1975) Isolation of mammalian cell mutants deficient in glucose-6-phosphate-dehydrogenase activity: linkage to hypoxanthine phosphoribosyl transferase. *Proc. Natl. Acad. Sci. U.S.A. 72,* 493-497

FIALKOW, P.J., LISKER, R., GIBLETT, E.R., ZAVALA, C., COBO, A. & DETTER, J.C. (1972) Genetic markers in chronic myelocytic leukaemia: evidence opposing autosomal inactivation and favouring 6-PGD-Rh linkage. *Ann. Hum. Genet. 35,* 321-326

GOSS, S.J. & HARRIS, H. (1975) New method for mapping genes in human chromosomes. *Nature (Lond.) 255,* 680-684

GOSS, S.J. & HARRIS, H. (1977*a*) Gene transfer by means of cell fusion. I. Statistical mapping of the human X-chromosome by analysis of radiation-induced gene segregation. *J. Cell Sci. 25,* 17-37

GOSS, S.J. & HARRIS, H. (1977*b*) Gene transfer by means of cell fusion. II. The mapping of 8 loci on human chromosome 1 by statistical analysis of gene assortment in somatic cell hybrids. *J. Cell Sci. 25,* 39-57

JONES, C., WUTHER, P., KAO, F.T. & PUCK, T.T. (1972) Genetics of somatic mammalian cells. XV. Evidence for linkage between human genes for lactic dehydrogenase B and serine hydroxymethylase. *J. Cell. Physiol. 80,* 291-298

KAO, F.T. & PUCK, T.T. (1972) Genetics of somatic mammalian cells: demonstrations of a human esterase activator gene linked to ade-B gene. *Proc. Natl. Acad. Sci. U.S.A. 69,* 3273-3277

KIT, S., DUBBS, D.R., PIEKARSKI, L.J. & HSU, T.C. (1963) Deletion of thymidine kinase activity from L cells resistant to bromodeoxyuridine. *Exp. Cell Res. 31,* 297-312

KÖHLER, G. & MILSTEIN, C. (1975) Continuous cultures of fused cells secreting antibody of predefined specificity. *Nature (Lond.) 256,* 495-497

KUSANO, T., LONG, C. & GREEN, H. (1971) A new reduced human-mouse somatic cell hybrid containing the human gene for adenine phosphoribosyltransferase. *Proc. Natl. Acad. Sci. U.S.A. 68,* 82-86

LATT, S.A. (1973) Microfluorometric detection of deoxyribonucleic acid replication in human metaphase chromosomes. *Proc. Natl. Acad. Sci. U.S.A. 70,* 3395-3399

LEVAN, G. & MITELMAN, F. (1977) Chromosomes and etiology of cancer, in *Chromosome Today,* vol. 6 (de la Chapelle, A. & Sorsa, M., eds.), pp. 363-371, North-Holland, Amsterdam

MCKUSICK, V.A. (1977) in *The Human Gene Map. Newsletter,* 21 November, 1977

MCKUSICK, V.A. & RUDDLE, F.H. (1977) The status of the gene map of the human chromosomes. *Science (Wash. D.C.) 196,* 390-405

MILLER, O.J., COOK, P.R., MEERA KHAN, P., SHIN, S. & SINISCALCO, M. (1971) Mitotic separation of two human X-linked genes in man-mouse somatic cell hybrids. *Proc. Natl. Acad. Sci. U.S.A. 68,* 116-120

MILLER, O.J., SINISCALCO, M. & SANGER, R. (1978) Report of the committee on the genetic constitution of the X and Y chromosomes, in *Human Gene Mapping 4* (Bergsma, D., ed.) *(S. Karger Ser. vol. 14),* pp. 124-128, The National Foundation, March of Dimes, New York

MILSTEIN, C. et al. (1979) Monoclonal antibodies and cell surface antigens, in this symposium, pp. 251-266

OLD, L.J., BOYSE, E.A. & CAMPBELL, H.A. (1968) L-asparagine and leukemia. *Sci. Am. 219*(2), 34-40

PATERSON, M.C., SMITH, B.P., LOHMAN, P.H.M., ANDERSON, A.K. & FISHMAN, L. (1976) Defective excision repair of x-ray-damaged DNA in human (ataxia telangectasia) fibroblasts. *Nature (Lond.) 260,* 444-447

PENROSE, L.C. (1959) Genetics analysis in man, in *Biochemistry of Human Genetics (Ciba Found. Symp.),* pp. 9-22, Churchill, London

PERRY, P. & EVANS, H.J. (1975) Cytological detection of mutagen-carcinogen exposure by sister chromatid exchange. *Nature (Lond.) 258,* 121-125

PERRY, P. & WOLFF, S. (1974) New Giemsa method for the differential staining of sister chromatids. *Nature (Lond.) 251,* 156-158

PONTECORVO, G. (1959) in *Biochemistry of Human Genetics (Ciba Found. Symp.),* p. 285, Churchill, London

PUCK, T.T. (1973) *The Mammalian Cell as a Microorganism,* pp. 56-79, Holden-Day, San Francisco

SIMINOVITCH, L. (1976) On the nature of hereditable variation in cultured somatic cells. *Cell 7,* 1-11

SINISCALCO, M. (1970) Somatic cell hybrids as tools for genetic studies in man. *Symp. Int. Soc. Cell Biol. 9,* 205-231

STERN, C. (1960) *Principles of Human Genetics,* 2nd edn., Freeman and Company, San Francisco

SZYBALSKI, W. & SZYBALSKA, E. (1962) Drug sensitivity as a genetic marker for human cell lines. *Univ. Mich. Med. Bull. 28,* 277-293

WEISS, M.C. & GREEN, H. (1967) Human-mouse hybrid cell lines containing partial complements of human chromosomes and functioning human genes. *Proc. Natl. Acad. Sci. U.S.A. 58,* 1104-1111

WOLFF, S. (1977) Sister chromatid exchanges. *Annu. Rev. Genet. 11,* 183-201

Discussion

Henry Harris: In August 1977 I received a letter from Jon Jonasson in which he told me of his finding that the malignant cells he had looked at had a higher rate of sister chromatid exchange than normal. I sent him some matched hybrid pairs in which malignancy was suppressed and in which it later reappeared. Although many tumours did show a high rate of sister chromatid exchange, as you have found, tumorigenicity and the correlation between high rate of exchange did not survive the tests with matched hybrid pairs.

Peto: Is sister chromatid exchange a marker showing that mitotic recombination is also going on to a considerable extent? If it were, it would indicate that cells became malignant when something recessive was expressed on both chromosomes by some kind of mitotic recombination accident. If the forces that lead to sister chromatid exchange also lead to mitotic recombination it could be that malignancy and sister chromatid exchange have a common cause.

Pontecorvo: The question is whether sister chromatid exchange has anything to do with segregation. German has claimed on purely cytological grounds that some of the configurations ('quadriradial') are somatic crossing-over events. But individuals with Bloom syndrome do not have patches of pigmented skin, do they? They should be patchy if there was a lot of somatic crossing over.

Polani: They are more patchy than normal people because they show areas of pigmentation of different sizes. Also, they may have—and one of our patients had—small patches of pigmented and non-pigmented skin quite close to each other, rather like the *Drosophila* twin spots. The patches are about one-2000th to one-3000th or even less of the body surface. I have also seen a beautiful example of the same thing in a patient with Fanconi anaemia. Furthermore German *et al.* (1977) described families with Bloom syndrome where 2–50%, I think, of the cells of the affected subjects did not show the characteristic excess of sister chromatid exchanges and did not have chromosome aberrations. This again could be taken to be the result of mutation at the somatic level.

Peto: If this abnormality is turning up in all melanomas, it would suggest that that is the normal mechanism whereby melanomas arise. What about other things such as carcinomas?

Siniscalco: As far as I know, there is no published report on sister chromatid exchange (SCE) in solid tumours. A recent abstract by Otter *et al.* (1978) claims that the rate of SCE in patients with acute lymphoblastic leukaemia is twice as high as that in normals. However, they do not say whether the patients had received chemotherapy. It is known that some chemotherapeutic agents do increase SCE (Perry & Evans 1975).

Polani: To return to the question of mutation and the origin of melanoma raised by Mr Peto: there is an interesting report which related the incidence of melanoma in man to the occurrence of sun spots (Houghton *et al.* 1978).

Siniscalco: I don't believe, for the time being, that the high rate of SCE is specific for melanomas. It may still be a matter of a chance association due to the very aberrant chromosomal structure of these malignant cells. Perhaps some genes involved in the normal mechanism of chromatid replication have been lost. It is too early to attribute a specific significance to these findings. However, human lines with high SCE can be very precious tools with which to investigate the biology and genetics of SCE by complementation analysis.

Peto: It could be more important than that. There are various reasons for wanting there to be several different stages in the production of cancer. There are also good reasons for not wanting too many of these to involve mutagens. It seems that the mutagen-involved stages have to come first. This is all based on a rather unreliable mixture of animal and human cancer data. We want to try to think of processes which are involved in the production of a particular phenotype but which can only occur in a certain order. If you are making mutations on one strand and then making some change which allows that mutation to be expressed, this would fit a lot of data which are otherwise completely inexplicable. That is why I keep on about mitotic recombination.

Siniscalco: The fact that a homozygous recessive condition such as Bloom syndrome shows high SCE in 100% of individuals, even if not necessarily in 100% of the cells, means that at least one major gene (which is defective in patients with Bloom syndrome) must be involved in the normal mechanism of chromatid replication. It is of interest to speculate about the possible causal relationship between high SCE and malignancies in general. However, I suppose that the zero hypothesis for the time being must be that the correlation between high SCE rate and tumours is due to chance; we shall see where we end up when more data on melanoma and on solid tumours in general become available. (And see note on p. 306.)

References

GERMAN, J., SCHONBERG, S., LOUIE, E. & CHAGANTI, R.S.K. (1977) Bloom's syndrome IV. Sister-chromatid exchanges in lymphocytes. *Am. J. Hum. Genet. 29,* 248-255

HOUGHTON, A., MUNSTER, E.W. & VIOLA, M.V. (1978) Increased incidence of malignant melanoma after peaks of sunspot activity. *Lancet 1,* 759-760

OTTER, M., PALMER, C. & BAEHNER, R. (1978) *The Cancer Letter 4,* (18), 7

PERRY, P. & EVANS, H.J. (1975) Cytological detection of mutagen-carcinogen exposure by sister chromatid exchange. *Nature (Lond.) 258,* 121-125

Some recent progress in the analysis of malignancy by cell fusion

HENRY HARRIS

Sir William Dunn School of Pathology, University of Oxford

Abstract Malignancy, as measured by the ability of cells to grow progressively *in vivo,* is intimately linked to the presence of a structural abnormality in the polysaccharide moiety of one particular membrane glycoprotein. This abnormality is present in a wide range of different tumours; it co-segregates with malignancy in all crosses between malignant and non-malignant cells that have so far been tested; and it remains linked to malignancy in a stringent new test in which non-malignant variants are selected from tumour cell populations by the use of a lectin. Current work indicates that the abnormal membrane glycoprotein is, in all probability, one of the glucose transport proteins.

For some years my colleagues and I have been applying cell fusion techniques to the analysis of the genetic basis of malignancy, which, in the present context, I define as the ability of cells to grow progressively and kill their host. We have found, for a wide range of mouse tumours, that when the malignant cells are fused with normal diploid cells, the hybrids initially formed are, in general, unable to grow progressively in immunosuppressed genetically compatible hosts (Wiener *et al.* 1971, 1974*a*; Harris 1971; Jonasson *et al.* 1977). This appears also to be true for crosses between malignant and normal human cells (Stanbridge 1976). The literature does contain one or two apparent exceptions to this rule, but I expect that, when these apparent exceptions are examined in more detail, they will not provide support for the notion that the genetic lesions determining malignancy act in a dominant fashion. In any case, I do not propose in the present communication to discuss the general problem of the dominance or recessivity of these genetic lesions, but to describe how hybrids between malignant and normal cells have been used to identify a phenotypic marker that is intimately linked to progressive growth *in vivo.*

As I say, the great majority of crosses between malignant and normal diploid cells are unable to grow progressively *in vivo*; but, on continued cultivation of the hybrid cells *in vitro,* segregation of genetic elements occurs, usually in the form of chromosome losses, and malignant cells reappear in the population (Jonasson *et al.* 1977). This disappearance of malignancy and its reappearance as a result of genetic segregation provides us with a method for testing whether any biochemical, or other, marker is closely linked to malignancy or not (Harris 1975). Any such marker must be present in the malignant parent cells, must disappear when malignancy is suppressed by fusion of the malignant cell with a non-malignant one, and must reappear in all independently derived segregant tumours that the hybrid cell population subsequently generates. In our laboratory this genetic test has been used to examine the relevance of a number of properties associated with 'transformation' of cells *in vitro*: growth rate, growth in semi-solid medium, 'contact inhibition', production of plasminogen-activating enzyme (Straus *et al.* 1977), and the organization of cytoplasmic microfilaments and microtubules (Watt *et al.* 1978). All of these markers in mouse cells could easily be separated by genetic segregation from the ability of the cells to grow progressively *in vivo*; and similar results have been reported for human cells tested for their tumorigenicity in athymic mice (Stanbridge & Wilkinson 1978). About three years ago, M.E. Bramwell and I began a systematic examination of the chemical composition of the cell membranes of hybrids in which malignancy was suppressed and of malignant segregants derived from these hybrids. Our joint work has led to the observation that malignancy, as measured by the ability of cells to grow progressively *in vivo*, is intimately linked to the presence of a structural abnormality in the polysaccharide moiety of one particular membrane glycoprotein. This abnormality is present in a wide range of different tumours; it co-segregates with malignancy in all crosses between malignant and non-malignant cells that we have so far tested; and it remains linked to malignancy in a stringent new test that we have devised in which non-malignant variants are selected from tumour cell populations by the use of a lectin (Bramwell & Harris 1978).

We have looked at the following mouse tumours: PG19, a hypoxanthine guanine phosphoribosyl transferase-deficient derivative of a spontaneous melanoma of the C57 Black mouse; TA3(Hauschka), an immuno-resistant mammary carcinoma of the strain A mouse; SEWA, a polyoma virus-induced sarcoma of the strain ASW mouse; MSWBS, a methylcholanthrene-induced sarcoma of the strain ASW mouse; YACIR, an immuno-resistant Moloney-virus-induced lymphoma of the strain A/Sn mouse; A9HT, a fibrosarcoma derived by passage of the A9 cell line through syngeneic C3H mice. Details of these tumours are given elsewhere (Klein *et al.* 1971; Bregula *et al.* 1971; Harris 1971; Wiener *et al.* 1973, 1974*a, b*; Jonasson *et al.* 1977; Jonasson & Harris 1977); they all give 100% take incidences in syngeneic animals from small inocula. We have cursorily examined the human carcinoma cell line, HeLa, and a malignant human lymphoma of the Burkitt type, Daudi. The normal diploid cells with which the malignant cells were fused were mouse embryo fibroblasts, mouse lymphocytes and human lymphocytes. We have so far examined the following hybrid cell populations:

PG19 × *human lymphocyte hybrid clone* 19 6TG. A cross between PG19 and a human lymphocyte, in which only the human X chromosome was retained. This was then eliminated by back selection in 6-thioguanine, so that the hybrid finally contained no identifiable human chromosomes. It was none the less very much less tumorigenic in athymic mice than the parental PG19 cell (Jonasson & Harris 1977).

PG19 × CBAT6T6 *fibroblast hybrid clone* 100A. A cross between PG19 and a diploid fibroblast taken from a mouse bearing the T6 translocation in the homozygous form (Jonasson *et al.* 1977).

PG19 × T13HT13H *fibroblast hybrid clones* 7, 8 and 10 are independent crosses between PG19 and diploid fibroblasts from mice bearing the T13H translocation in the homozygous form (Jonasson *et al.* 1977).

YACIR × CBAT6T6 *fibroblast hybrid clone* 1G. A cross between the YACIR lymphoma and the diploid mouse fibroblast (Jonasson *et al.* 1977; Straus *et al.* 1977).

A9HT × C57 Black *lymphocyte hybrid clone* A. A cross between the A9HT sarcoma and a diploid lymphocyte taken from a C57 Black mouse (Wiener *et al.* 1974*a*).

TA3(Hauschka) × ACA *fibroblast hybrid clone* 1. A cross between TA3(Hauschka) and a diploid fibroblast taken from an ACA mouse (Harris 1971).

In all cases comparisons were made between hybrids in which malignancy was suppresssed and malignant segregants derived from these hybrids.

The membrane proteins were prepared by standard procedures from purified cell membranes and from Triton X-100 extracts of whole cells and tissues (Atkinson 1973; Butters & Hughes 1974). The proteins were separated in one dimension by electrophoresis in sodium dodecyl sulphate–polyacrylamide gels; and in two dimensions by isoelectric focusing in the first dimension and electrophoresis in sodium dodecyl sulphate gels in the second (O'Farrell 1975). About 160 membrane proteins could be resolved by these techniques. Glycoproteins were located by affinity labelling on the gels with the radioactive lectins concanavalin A and wheat-germ agglutinin (WGA)

(Tanner & Anstee 1976; Burridge 1976). The amounts of labelled concanavalin A and WGA bound to the glycoproteins were measured. Details of all the chemical methods are given in Bramwell & Harris (1978).

Although Coomassie Blue-stained preparations of proteins separated in the two-dimensional gels revealed numerous differences between tumour cells and normal diploid cells and occasional differences between hybrids in which malignancy was suppressed and malignant segregants were derived from these hybrids, no membrane change detectable by staining with Coomassie Blue consistently co-segregated with the ability of the cell to grow progressively *in vivo*. We began our detailed study of the membrane glycoproteins by examining the PG19 × human lymphocyte hybrid clone 19 6TG and a tumour derived from it. It was thought possible that this pair would be especially informative because all the identifiable chromosomes from the human diploid lymphocyte had been eliminated, and the karyotype of the hybrid cell in which malignancy remained suppressed was indistinguishable from that of the parental PG19 tumour cell. There was therefore reason to hope that in this case irrelevant differences between the non-malignant hybrid cells and the malignant derivatives would be largely screened out, and that any residual differences would have some significance. As might be expected from two such closely similar cell lines, there were very few differences in the pattern of Coomassie Blue staining or lectin binding of the separated membrane proteins; but one striking difference was observed that proved on further examination in other crosses to be of especial interest. As shown in Fig. 1, a component of an estimated molecular mass of about 100 000 binds much more concanavalin A in the preparation from the hybrid tumour cell than in the preparation from the hybrid in which malignancy was suppressed. The

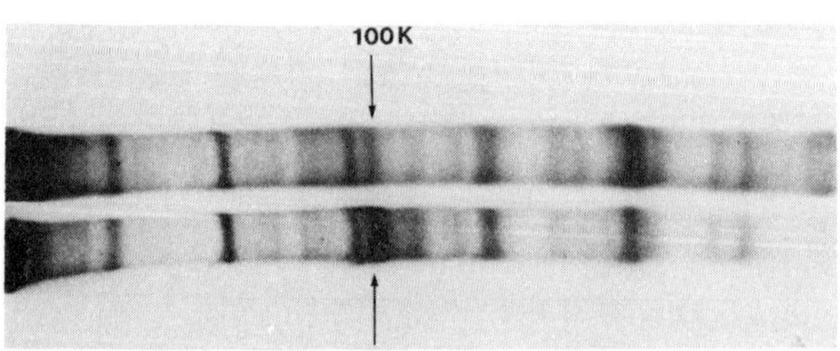

FIG. 1. Autoradiographs of gels labelled by affinity binding with [^{125}I]concanavalin A. The upper gel shows a preparation of membrane glycoproteins from PG19 × human lymphocyte hybrid clone 19 6TG, in which malignancy is suppressed; the lower gel shows a similar preparation from a tumour derived from this hybrid clone. Note that in the 100 000 molecular mass region, marked by arrows, there is a band that is much more heavily labelled in the tumour cell population than in the hybrids in which malignancy is suppressed.

parental PG19 tumour cells also show a high level of concanavalin A binding at this site; phytohaemagglutinin-stimulated diploid human lymphocytes show a low level (Fig. 2). When the binding pattern of labelled WGA was examined, the reciprocal situation was found: much more WGA was bound to the 100 000 molecular mass component in the non-malignant hybrids and the diploid cells than in the malignant segregants and the tumour cells (Fig. 3). Resolution of the membrane proteins in the two-dimensional system confirmed that both the concanavalin A and the WGA were binding to the

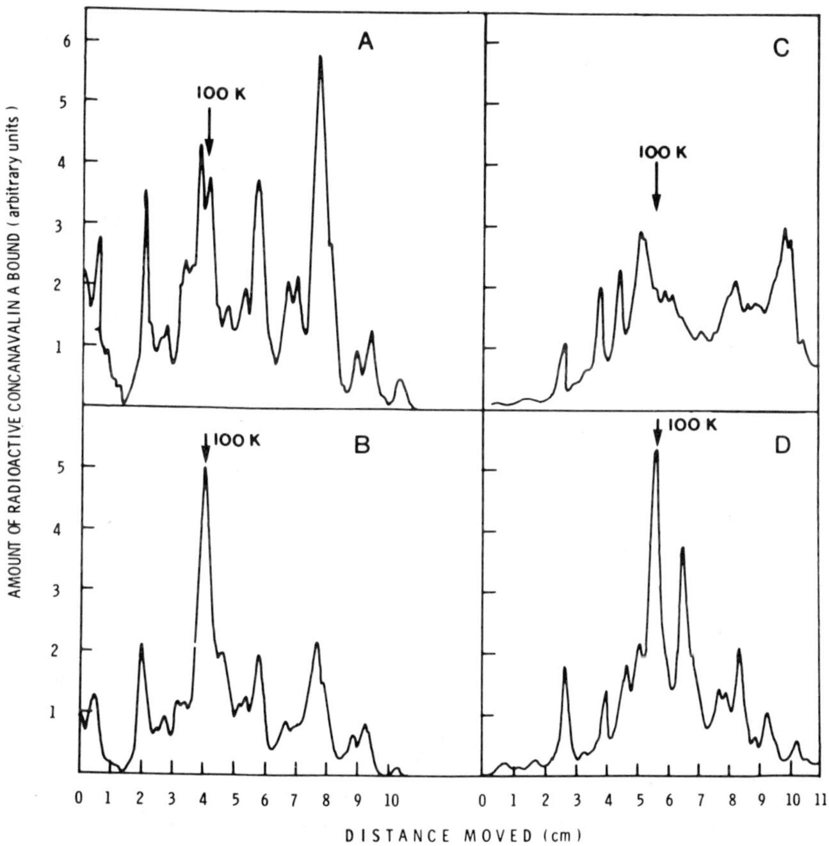

FIG. 2. Microdensitometer tracings of autoradiographs of gels labelled with [^{125}I]concanavalin A. (A) Membrane protein preparations from PG19 × human lymphocyte hybrid clone 19 6TG. (B) Membrane protein preparation from a tumour derived from hybrid clone 19 6TG. (C) Triton extract of phytohaemagglutinin-stimulated human lymphocytes. (D) Membrane protein preparation from PG19 tumour cells. Note that the 100 000 molecular mass component binds much more labelled concanavalin A in the two tumour cell populations (B and D) than in the hybrids in which malignancy is suppressed or in the normal diploid cells (A and C).

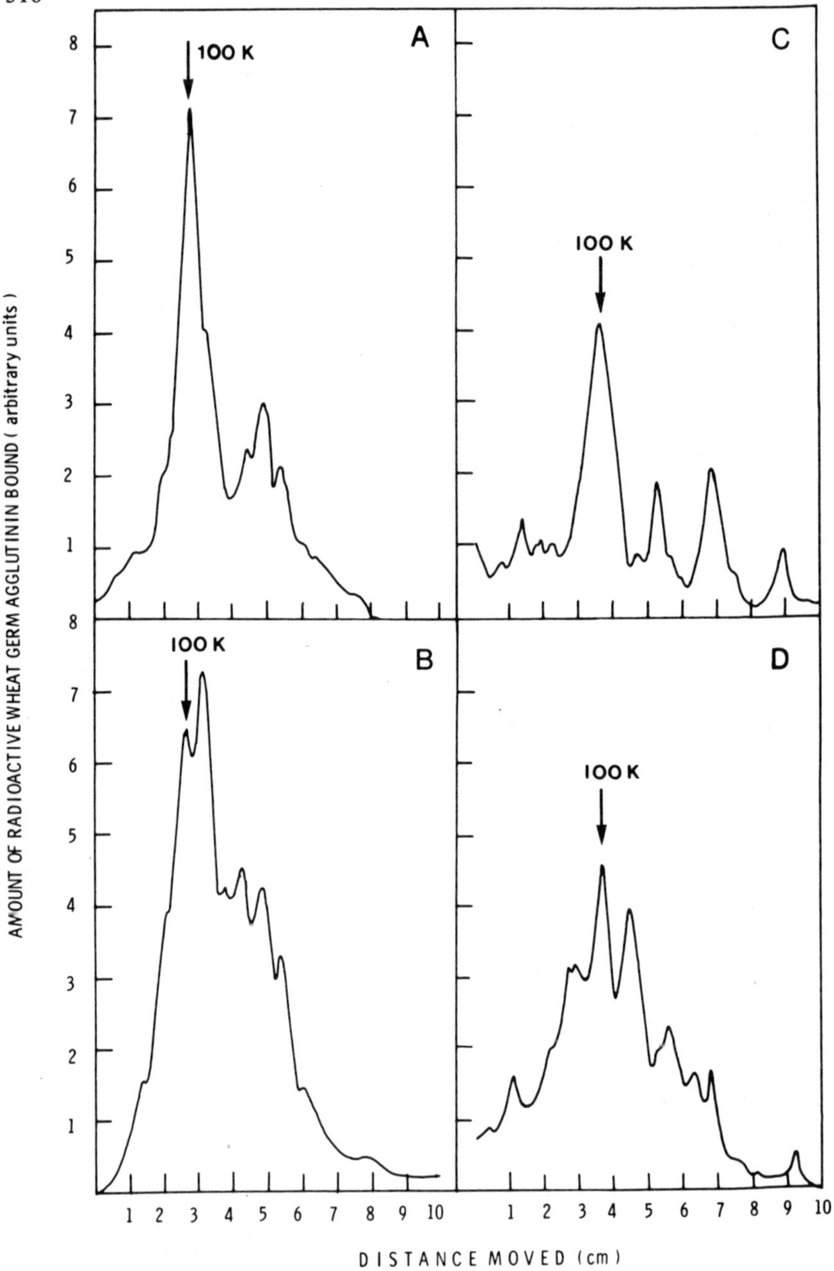

FIG. 3. Microdensitometer tracings of autoradiographs of gels labelled with [^{125}I]wheat-germ agglutinin. (A) Membrane protein preparation from PG19 × human lymphocyte hybrid clone 19 6TG. (B) Membrane protein preparation from a tumour derived from hybrid clone 19 6TG. (C) Membrane protein preparation from T6T6 diploid fibroblasts. (D) Membrane protein preparation from PG19 tumour cells. Note that the 100 000 molecular mass component binds much less labelled WGA in the two tumour cell populations (B and D) than in the hybrids in which malignancy is suppressed or in the normal diploid cells (A and C).

same component, which was found to have an apparent isoelectric point of about 4.0. We can therefore conclude that the polysaccharide moiety of this 100 000 molecular mass glycoprotein in the malignant cells differs in some way from the polysaccharide moiety of the same glycoprotein in the non-malignant cells. Since the major affinity of concanavalin A is for mannose and glucose residues and the major affinity of WGA is for *N*-acetylglucosamine residues, it seems probable that the nature of the change in the polysaccharide is that in the malignant cells there are more accessible mannose or glucose residues than normal and fewer accessible *N*-acetylglucosamine residues. The lectin-binding pattern of the glycoprotein was not greatly affected by cultural conditions *in vitro*: growing and confluent cultures gave essentially the same results.

We found these same changes in lectin-binding pattern in *all* matched pairs of PG19 crosses that we examined. Fig. 4 shows the relative amounts of concanavalin A bound to the 100 000 molecular mass glycoprotein in a series of non-malignant PG19 hybrids and in the malignant segregants derived from them. It will be seen that in all the hybrids in which malignancy is suppressed the amount of concanavalin A bound of the 100 000 molecular mass

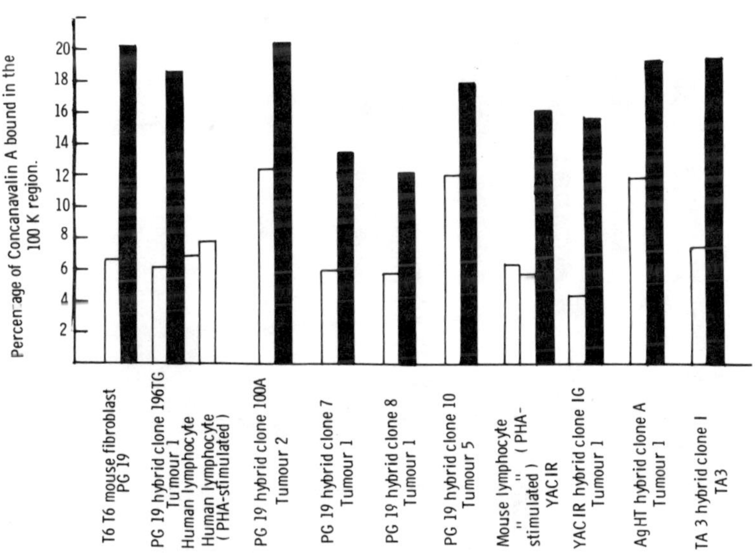

FIG. 4. Histogram showing the relative amounts of [^{125}I]concanavalin A bound to the 100 000 molecular mass membrane glycoprotein in a range of tumour × diploid cell hybrids in which malignancy is suppressed and in segregant tumours derived from these hybrids. □, Non-malignant cells; ■, malignant cells. Note that the malignant cells bind much more labelled concanavalin A than the non-malignant ones.

glycoprotein is much lower than in the malignant PG19 parent cells and is comparable in most cases with the levels seen in diploid cells. In the malignant segregants the amount of concanavalin A bound to this protein is substantially increased. The amount of WGA bound varies inversely with the amount of concanavalin A bound (Fig. 5). Since each of these hybrid tumours, being derived from a separate clone of cells, represents an independent segregation event, the association between the aberrant lectin-binding pattern and malignancy cannot be fortuitous.

Fig. 4 also shows the results of crosses between diploid cells and the Moloney-virus-induced lymphoma YACIR, the mammary carcinoma

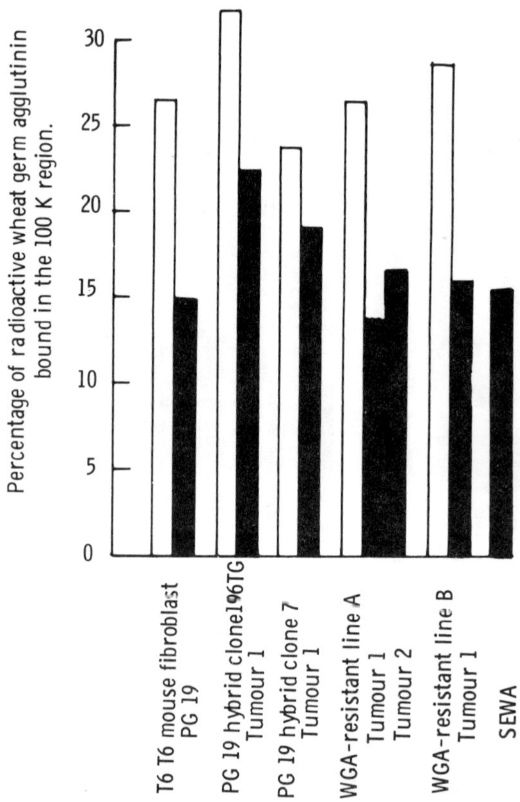

FIG. 5. Histogram showing the relative amounts of [^{125}I]wheat-germ agglutinin bound to the 100 000 molecular mass membrane glycoprotein. PG19 × diploid cell hybrids in which malignancy is suppressed are compared with segregant tumours derived from these hybrids. Non-malignant PG19 derivatives resistant to wheat-germ agglutinin are compared with malignant variants. □, Non-malignant cells; ■, malignant cells. Note that the malignant cells bind much less labelled WGA than the non-malignant ones.

TA3(Hauschka) and the spontaneous sarcoma A9HT. All three tumours show the high concanavalin A binding pattern in the 100 000 molecular mass region; all three crosses in which malignancy is suppressed show a greatly reduced level of concanavalin A binding in this region; and the segregant tumours derived from the YACIR × diploid fibroblast cross and the A9HT × diploid lymphocyte cross show the reappearance of the high concanavalin A binding pattern. (The TA3(Hauschka) × ACA fibroblast cross has not yet produced any tumours.) The polyoma-virus-induced sarcoma SEWA and the methylcholanthrene-induced sarcoma MSWBS also showed the high concanavalin A binding pattern in the 100 000 molecular mass zone (Fig. 6).

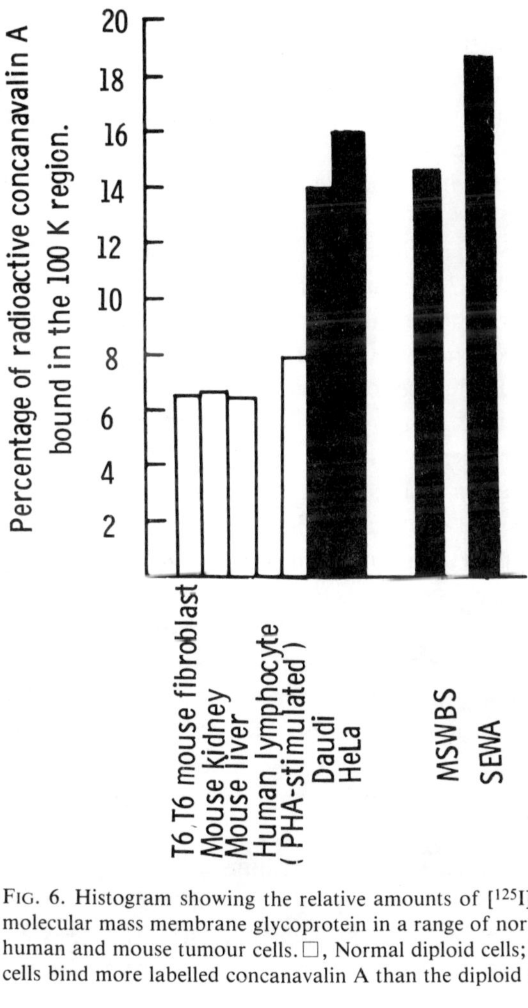

FIG. 6. Histogram showing the relative amounts of [^{125}I]concanavalin A bound to the 100 000 molecular mass membrane glycoprotein in a range of normal diploid cells or tissues and in some human and mouse tumour cells. □, Normal diploid cells; ■, tumour cells. Note that the tumour cells bind more labelled concanavalin A than the diploid cells.

The membrane proteins of the crosses between these two tumours and diploid cells were not, however, analysed because these hybrids generate malignant segregants at a very high frequency *in vitro* and do not therefore constitute favourable material for analysis.

Some of the mouse tumour cells had broader or partly resolved multiple concanavalin-binding peaks in the 100 000 molecular mass region. It seemed likely that these were functionally homologous variants of the same glycoprotein, for they all showed the same behaviour in crosses between the tumour cells and diploid cells. The human HeLa and Daudi tumour cells (Fig. 6) had a high concanavalin A-binding component that migrated a little more rapidly in electrophoresis than the 100 000 molecular mass component seen in the mouse tumour cells. Diploid human cells bound much less concanavalin A in this region. I shall present further evidence later that these concanavalin A-binding components with slightly different electrophoretic mobilities in the 100 000 molecular mass region are indeed functionally homologous. It is, in any case, apparent that the high concanavalin A, low WGA-binding pattern is intimately linked with malignancy in a wide variety of different kinds of tumour cell.

The very large differences between malignant and non-malignant cells in the patterns of concanavalin A and WGA binding to the 100 000 molecular mass component suggested that it might be worth while to examine whether there were any large differences between the two sorts of cells in their susceptibility to killing by lectins. We screened a number of different lectins for their ability to discriminate between matched pairs of non-malignant hybrids and segregant tumours derived from them. Only WGA showed a pronounced differential effect: at a concentration of 10 μg/ml, WGA was, in many cases, much more toxic to the malignant segregants than to the non-malignant hybrids from which they were derived; and most of the parental tumour cells were more susceptible to the toxic effects of this lectin than diploid fibroblasts or epithelial cells. We therefore thought it might be useful to select WGA-resistant variants from the tumour cell populations in order to see whether the acquisition of WGA-resistance might not, in some cases, be associated with a loss of, or reduction in, tumorigenicity. For, if this occurred, we would have an additional, and independent, test of the link between abnormalities in the 100 000 molecular mass glycoprotein and the ability of cells to grow progressively *in vivo*.

Bulk cultures of PG19 cells, which are killed within 72 h by 5 μg/ml of WGA, were exposed to increasing concentrations of the lectin ranging from 2–10 μg/ml over a period of six weeks. Four independent WGA-resistant subpopulations (A-D) were isolated. These survived 5 μg/ml of WGA and were capable of multiplication at a slow rate in this concentration of the lectin. All the WGA-resistant lines were grown in medium without lectin for several days before being inoculated into test animals. (Their growth rate in lectin-free medium was comparable to that of PG19 cells.) Preparations of membrane proteins were made at the time of inoculation. The take incidences given by these WGA-resistant lines are shown in Table 1. Lines C and D produced no tumours. A and B produced some tumours, but they did not give the 100% take incidence characteristic of PG19 with inocula of this size; and the latent periods preceding the appearance of the tumours produced by the WGA-resistant lines were substantially longer than that of PG19. Our previous experience has shown that a reduction in take incidence coupled with a prolongation of the latent period indicates that the tumours are not produced by progressive growth of all the cells inoculated, but by selective overgrowth of minority subpopulations. This was confirmed in the present case by cytological examination of the tumours. In all cases, the tumour cell populations had modal chromosome numbers that differed significantly from those of the WGA-resistant cells injected; and the distributions of metacentric markers in the two cell populations were also different. Figs. 5 and 7 show the lectin-binding patterns of the WGA-resistant lines and of the tumours arising from them. It will be seen that all four WGA-resistant lines had lower levels of concanavalin A binding to the 100 000 molecular mass glycoprotein than the PG19 cells from which they were derived. Lines C and D, which produced no tumours, showed levels comparable to diploid cells; lines A and B, which were mixed cell populations, showed slightly higher levels. However, the tumours arising from the inoculation of the WGA-resistant cell populations

TABLE 1

Growth of wheat-germ agglutinin-resistant melanoma derivatives *in vivo*.

Cell type	*Genotype of host**	*Inoculum size*	*No. of takes*	*Mean latent period, days*	*Latent period range, days*
PG19 melanoma	C57BL/6J	5×10^4	23/23	35	27-39
WGA-resistant derivatives					
A	,,	,,	3/12	48	40-54
B	,,	,,	16/18	50	39-56
C	,,	,,	0/17	–	–
D	,,	,,	0/12	–	–

*All animals were syngeneic newborns given 4 J kg^{-1} of total body irradiation.

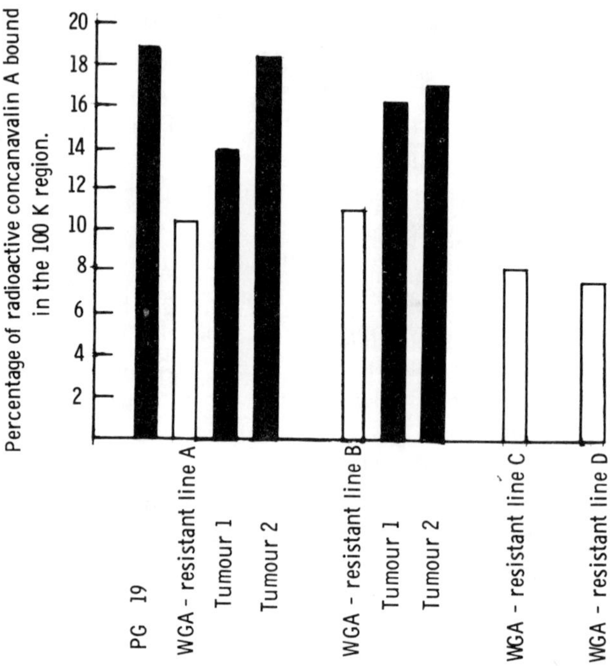

FIG. 7. Histogram showing the relative amounts of [^{125}I]concanavalin A bound to the 100 000 molecular mass membrane glycoprotein. Non-malignant PG19 derivatives resistant to wheat-germ agglutinin are compared with malignant variants. □, Non-malignant cells; ■, malignant cells. Note that the malignant cells bind more labelled concanavalin A.

all showed the high levels of concanavalin A binding characteristic of malignant cells. It was thus apparent that the WGA-resistant cells that showed the lectin-binding characteristics of non-malignant cells were indeed incapable of progressive growth *in vivo*; the tumours that arose from the inoculation of these cells were produced by selective overgrowth of cells having the lectin-binding characteristics of malignant cells. The experiments with the WGA-resistant cells thus confirmed in a completely independent way the intimate association between the structural abnormality in the polysaccharide moiety of the 100 000 molecular mass glycoprotein and the ability of the cells to grow progressively *in vivo*.

We have now to consider what the function of the 100 000 molecular mass glycoprotein might be and how the change in polysaccharide structure might be brought about. The physical properties of the glycoprotein are somewhat unusual: it has a molecular mass of about 100 000, an apparent isoelectric point of about 4.0 and it binds both concanavalin A and WGA. In scanning the published work we found that there were very few membrane glycoproteins that met these specifications even approximately; but we came across two plausible candidates which might indeed be related to each other. The 'insulin receptor' recently purified by Jacobs *et al.* (1977) from rat adipocyte membranes was found to have an isoelectric point of about 4.0, to bind both concanavalin A and WGA, and to have an estimated molecular mass of about 135 000. Photoaffinity labelling of the insulin receptor in rat adipocytes indicates that its molecular mass might be about 125 000, as estimated from electrophoretic migration in gels (Yip *et al.* 1978). The difference in molecular mass between these insulin receptors and our glycoprotein is not telling, because the molecular mass of glycoproteins cannot be accurately estimated from electrophoresis gels, especially when gels of different kinds are compared. However, the low isoelectric point and the lectin-binding characteristics do suggest that the molecule isolated by Jacobs *et al.* (1977) may correspond to the glycoprotein that we have been studying.

The second possibility that we uncovered in the published work was that our glycoprotein might be one of those that has been implicated in the transport of glucose into the cell. There are at present two membrane glycoproteins that have been proposed as glucose transport proteins, one having a molecular mass of about 100 000, and the other about 55–70 000. The 100 000 molecular mass protein has been studied in some detail and has been shown to have the unusual property of undergoing dimerization when intrinsic SH groups are catalytically oxidized (Kahlenberg 1976). It is likely that this protein exists as a dimer in the cell membrane and is reduced to the monomeric form in the presence of the dithiothreitol used in the electrophoretic systems. Our glycoprotein also behaves in this unusual way. When the membrane proteins are prepared and separated by electrophoresis in the absence of dithiothreitol, or if the proteins in the cell membrane are subjected to mild oxidation, the concanavalin A-binding component having an estimated molecular mass of 100 000 disappears and is replaced by a new concanavalin A-binding component with an estimated molecular mass of about 200 000. This property of the 100 000 molecular mass glycoprotein can be used to achieve its complete separation from all other membrane glycoproteins. If the membrane proteins are resolved in a two-dimensional electrophoretic system in which the buffer systems are identical in the first and second dimensions except that dithiothreitol is added to the latter, then virtually all the membrane proteins are aligned along a simple diagonal; but, in the second dimension, the dimerized 100 000 molecular mass glycoprotein is dissociated into its monomers by the dithiothreitol and consequently moves to a new position far removed from the diagonal (Fig. 8). This complete

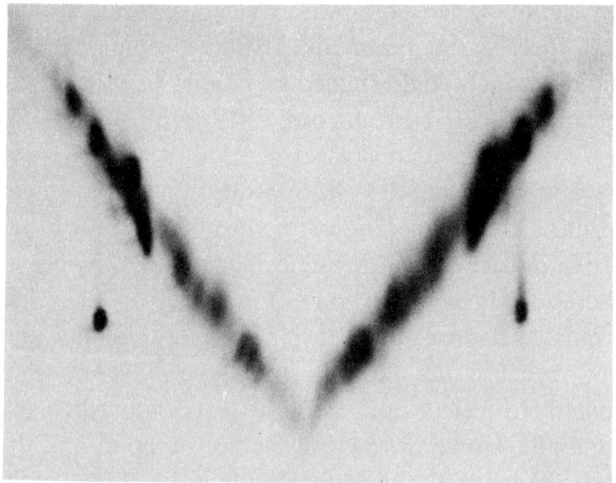

FIG. 8. Autoradiograph of gels labelled by affinity binding with [^{125}I]concanavalin A. The two gels are duplicates of a preparation of membrane glycoproteins from HeLa cells. The buffer system was the same in both dimensions except that dithiothreitol was added in the second dimension. The dimerized form of the 100 000 molecular mass glycoprotein is dissociated into its constituent monomers in the second dimension and is thus separated from all the other concanavalin A-binding membrane glycoproteins.

separation of the 100 000 molecular mass glycoprotein from all the others enabled us to show decisively that the concanavalin A and the WGA were indeed bound by the same component; and it also enabled us to test the behaviour of the apparently homologous components in human malignant cells that bound both lectins but had slightly different electrophoretic mobility. These components in the human cells were also found to be present in the cell membrane as dimers which were dissociated under reducing conditions (Fig. 8). The broader or partially resolved multiple concanavalin A-binding peaks found in the 100 000 molecular mass region in some of the mouse tumours also behaved in this way. It thus appears very probable that, despite minor differences in electrophoretic mobility, the concanavalin A- and WGA-binding components in the 100 000 molecular mass region are functionally homologous.

We do have some evidence that this dimerized membrane glycoprotein is indeed involved in some way in glucose transport. It has been shown that fluorodinitrobenzene inhibits glucose transport in erythrocytes (Krupka 1971; Jung 1974), and that the binding of this compound to the glucose receptor in the cell membrane is facilitated by glucose (Jung & Carlson 1975). (The precise mechanism responsible for this facilitation is not fully understood, but apparently the binding of glucose to the receptor alters its conformation and enhances its interaction with the fluorodinitrobenzene: Krupka 1971.) Fluorodinitrobenzene also binds to the dimerized form of our glycoprotein, and the binding is substantially increased by glucose. We therefore believe that we are dealing with one of the glucose transport proteins, and we are at present doing experiments to see whether this protein might not also have specific binding sites for insulin.

Our present position is then that in a wide variety of tumours (including some human tumours) there is something about one of the glucose transport proteins that results in its having an abnormal polysaccharide structure; and, in a range of exacting genetic tests, we have not been able to segregate this abnormality from the ability of the cell to grow progressively *in vivo*. That many malignant cells do show abnormalities in glucose transport and in their growth responses to insulin is, of course, well known.

We now come to the question of how the abnormality in the polysaccharide structure of this glycoprotein might be brought about. It might, in principle, be brought about either by a change in one or more of the relevant glycosyl transferases or by a change in the protein component that results in an abnormal glycosylation pattern even though the glycosyl transferases might themselves not be abnormal. Changes in the activities of glycosyl transferases have been described in lectin-resistant cells (Gottlieb *et al.* 1975; Stanley *et al.* 1975; Narasimhan *et al.* 1977); but if a defective glycosyl transferase were responsible for the aberrant lectin-binding pattern that we see in the 100 000 molecular mass glycoprotein, it would be reasonable to expect similar aberrations in other membrane glycoproteins. While we cannot exclude the possibility that minor changes might have occurred elsewhere, the dramatic differences that we report here were found only in this one component. It is possible that this particular glycoprotein might have its own specific set of glycosyl transferases; but one might perhaps more easily envisage a structural change in the protein moiety that results in some modification of the receptor site for glycosyl transferases, such that an aberrant pattern of glycosylation results.

We are at present engaged in the large-scale isolation and purification of the membrane glycoprotein with a view to obtaining a clearer idea of the structure of the normal and aberrant homologues. We naturally propose to see whether the aberrant forms behave differently from the normal ones in their ability to bind glucose or insulin or other related growth-controlling polypeptides. We are also making monoclonal antibodies to the glycoprotein in order to help with purification and to form the basis of an accurate method for measuring the amount of the glycoprotein present in different cell types. And finally, we are seeking to immunize mice with the partially purified protein that we have already isolated from malignant cells, in order to see whether such mice will resist the growth of tumours. Since many tumours apparently carry a structurally aberrant form of the same membrane glycoprotein, the possibility exists that purified preparations of the aberrant glycoprotein, or modifications of it, might provide the basis for achieving immunity to a wide range of different tumours. One must, after all, do something to pass the time of day.

References

Atkinson, P.H. (1973) *Methods in Cell Biology,* vol. 7 (Prescott, D.M., ed.), pp. 157-188, Academic Press, New York

Bramwell, M.E. & Harris, H. (1978) An abnormal membrane glycoprotein associated with malignancy in a wide range of different tumours. *Proc. R. Soc. Lond. B Biol. Sci. 201,* 87-106

Bregula, U., Klein, G. & Harris, H. (1971) The analysis of malignancy by cell fusion. II. Hybrids between Ehrlich cells and normal diploid cells. *J. Cell Sci. 8,* 673-680

Burridge, K. (1976) Changes in cellular glycoproteins after transformation: identification of specific glycoproteins and antigens in sodium dodecyl sulphate gels. *Proc. Natl. Acad. Sci. U.S.A. 73,* 4457-4461

Butters, T.D. & Hughes, R.C. (1974) Solubilization and fractionation of glycoproteins and glycolipids of KB cell membranes. *Biochem. J. 140,* 469-478

Gottlieb, C., Baenziger, J. & Kornfield, S. (1975) Deficient uridine diphosphate-*N*-acetylglucosamine: glycoprotein *N*-acetylglucosaminyltransferase activity in a clone of Chinese hamster ovary cells with altered surface glycoproteins. *J. Biol. Chem. 250,* 3303-3309

Harris, H. (1971) Cell fusion and the analysis of malignancy. The Croonian Lecture. *Proc. R. Soc. Lond. B Biol. Sci. 179,* 1-20

Harris, H. (1975) Cell fusion and the analysis of malignancy. *Cancer Research Campaign 53rd Annual Report,* pp. 40-45, London

Jacobs, S., Shechter, Y., Bissell, K. & Cuatrecasas, P. (1977) Purification and properties of insulin receptors from rat liver membranes. *Biochem. Biophys. Res. Commun. 77,* 981-988

Jonasson, J. & Harris, H. (1977) The analysis of malignancy by cell fusion. VIII. Evidence for the intervention of an extra-chromosomal element. *J. Cell Sci. 24,* 255-263

Jonasson J., Povey, S. & Harris, H. (1977) The analysis of malignancy by cell fusion. VII. Cytogenetic analysis of hybrids between malignant and diploid cells and of tumours derived from them. *J. Cell Sci. 24,* 217-254

Jung, C.Y. (1974) Inactivation of glucose carriers in human erythrocyte membranes by 1-fluoro-2, 4-dinitrobenzene. *J. Biol. Chem. 249,* 3568-3573

Jung, C.Y. & Carlson, L.M. (1975) Glucose transport carrier in human erythrocyte membranes. *J. Biol. Chem. 250,* 3217-3220

Kahlenberg, A. (1976) Partial purification of a membrane protein from human erythrocytes involved in glucose transport. *J. Biol. Chem. 251,* 1582-1590

Klein, G., Bregula, U., Wiener, F. & Harris, H. (1971) The analysis of malignancy by cell fusion. I. Hybrids between tumour cells and L cell derivatives. *J. Cell Sci. 8,* 659-672

Krupka, R.M. (1971) Evidence for a carrier conformational change associated with sugar transport in erythrocytes. *Biochemistry 10,* 1143-1148

Narasimhan, S., Stanley, P. & Schachter, H. (1977) Control of glycoprotein synthesis. *J. Biol. Chem. 252,* 3926-3933

O'FARRELL, P.H. (1975) High resolution two-dimensional electrophoresis of proteins. *J. Biol. Chem. 250,* 4007-4021

STANBRIDGE, E.J. (1976) Suppression of malignancy in human cells. *Nature (Lond.) 260,* 17-20

STANBRIDGE, E.J. & WILKINSON, J. (1978) Analysis of malignancy in human cells: malignant and transformed phenotypes are under separate genetic control. *Proc. Natl. Acad. Sci. U.S.A. 75,* 1466-1469

STANLEY, P., NARASIMHAN, S., SIMINOVITCH, L. & SCHACTER, H. (1975) Chinese hamster ovary cells selected for resistance to the cytotoxicity of phytohemagglutinin are deficient in a UDP-*N*-acetylglucosamine-glycoprotein *N*-acetylglucosaminyltransferase activity. *Proc. Natl. Acad. Sci. U.S.A. 72,* 3323-3327

STRAUS, D.S., JONASSON, J. & HARRIS, H. (1977) Growth *in vitro* of tumour cell × fibroblast hybrids in which malignancy is suppressed. *J. Cell Sci. 25,* 73-86

TANNER, R.M.J. & ANSTEE, D.J. (1976) A method for the direct demonstration of the lectin-binding components of the human erythrocyte membrane. *Biochem. J. 153,* 265-270

WATT, F.M., HARRIS, H., OSBORN, M. & WEBER, K. (1978) The distribution of actin cables and microtubules in hybrids between malignant and non-malignant cells and in tumours derived from them. *J. Cell Sci. 32,* 419-432

WIENER, F., KLEIN, G. & HARRIS, H. (1971) The analysis of malignancy by cell fusion. III. Hybrids between diploid fibroblasts and other tumour cells. *J. Cell Sci. 8,* 681-692

WIENER, F., KLEIN, G. & HARRIS, H. (1973) The analysis of malignancy by cell fusion. IV. Hybrids between tumour cells and a malignant L cell derivative. *J. Cell Sci. 12,* 253-261

WIENER, F., KLEIN, G. & HARRIS, H. (1974*a*) The analysis of malignancy by cell fusion. V. Further evidence of the ability of normal diploid cells to suppress malignancy. *J. Cell Sci. 15,* 177-183

WIENER, F., KLEIN, G. & HARRIS, H. (1974*b*) The analysis of malignancy by cell fusion. VI. Hybrids between different tumour cells. *J. Cell Sci. 16,* 189-198

YIP, C.C., YEUNG, C.W.T. & MOULE, M.L. (1978) Photoaffinity labelling of insulin receptor of rat adipocyte plasma membrane. *J. Biol. Chem. 253,* 1743-1745

Discussion

Brenner: Is it only this protein which shows a difference in pattern of sugars?

Henry Harris: As I have said, this is the only protein that shows such a dramatic difference. We cannot exclude much smaller changes in other proteins.

Brenner: But if you label the protein rather than the sugars, is that protein present in all cells, whether they are malignant or not?

Henry Harris: Yes, all cells have the protein. It is not a new, adventitious protein, but a modified form of a protein normally present. We haven't yet got good figures for the amounts of the protein present in different cells, because it is present in rather small amounts. There is more of the protein in HeLa cells than in diploid human fibroblasts, because we can stain the protein easily with Coomassie Blue in HeLa cells. But HeLa has multiple copies of several autosomes, so this observation may be trivial.

Scriver: Have you been able to look at the presumed function of the protein and to study glucose transport in your cells?

Henry Harris: Glucose transport seems to be kaput in all the malignant cells that have been looked at.

Harry Harris: What do you mean by kaput?

Henry Harris: Most tumours are insensitive to growth stimulation by insulin, but they nonetheless transport much more glucose into the cell than they should. It looks as if a gate that is normally regulated by insulin or by other growth factors is jammed open in malignant cells. This may provide a plausible explanation for Otto Warburg's findings of a generation ago. In the tumours he worked with there was a high level of aerobic glycolysis. He therefore proposed that there was some impairment of normal aerobic respiration in tumour cells. But later, other people showed that there could be a high level of aerobic glycolysis in cells which were not malignant. Leucocytes, for example, convert most of their glucose to lactate even in the presence of oxygen. Malignant cells might convert the excess glucose that they take in to lactate rather than send it through the normal respiratory pathway. There is evidence that increasing the glucose concentration induces cells in culture to produce more lactate.

Scriver: Is glucose transport in this type of cell sodium-dependent?

Henry Harris: I don't know the answer to that.

Ruddle: What you are postulating is that this particular protein has some quantitative or qualitative differences in malignant and non-malignant cells?

Henry Harris: I can't say much about quantitation but the polysaccharide component of the glycoprotein in malignant cells is in some way different from normal.

Ruddle: But you haven't been able to demonstrate what the difference is?

Henry Harris: Not yet, but we now have methods for obtaining microgram quantities of this material in relatively pure form. We are going to do the sugar analyses and eventually the amino acid sequences. At the moment the phenomenon rests on a profoundly different lectin-binding pattern, which must represent some difference in the polysaccharides.

Ruddle: Can you be sure that the spot in the two-dimensional gel that pulls out from the diagonal is the same material that accounts for the difference in the binding system from the one-dimensional gel?

Henry Harris: Yes, we are pretty sure of that. There are only five or six glycoproteins in the cell membrane that bind the two lectins in considerable amounts. Only two of these have isoelectric points in the region of pH 4, and only one of these two has a molecular mass anywhere near 100 000. So we can identify our glycoprotein pretty accurately, even in the one-dimensional system. Of course, it is conceivable that there are two proteins, one binding concanavalin A and one binding wheat-germ agglutinin, but then the two proteins must have the same isoelectric point, the same molecular mass and the same curious pattern of dimerization. This seems very unlikely.

Jacob: How far did you look in teratocarcinomas? The spectrum of surface glycoproteins in teratocarcinoma is widely different from those in adult cells.

Henry Harris: We simply looked at one malignant teratocarcinoma line that we obtained from Christopher Graham's laboratory, and we found our malignant marker on it. People throw us odd tumours from time to time and we look at them when we can. Friend's erythroleukaemia was interesting. When we grew it *in vitro* we obtained rather equivocal results, but when we put the cells into an animal, the tumour produced showed the marker dramatically. We often see the marker come up much more clearly in the tumour than in the cell population growing *in vitro*, which I suspect is due to the selection of tumorigenic cells from mixed populations.

Peto: How much was it a quantitative difference and how much was it qualitative? The numerical results from your one-dimensional plots don't show whether it was a two-fold change. If it is that, could it just be that the cells that are dividing rapidly enough to behave malignantly have certain characteristics, such as taking up a lot of glucose? If you could fix all the malignant and non-malignant cells that you want to study so that none of them are growing, would you then find the same surface differences? Is it a marker of increased cellular activity rather generally?

Henry Harris: All of these matched pairs, that is, the suppressed hybrids, and the malignant segregants, grow very well *in vitro*. They usually have roughly comparable growth rates. Sometimes the non-malignant hybrids grow a little more quickly than the malignant segregants and sometimes the reverse is the case. Our marker has nothing to do with growing or non-growing cells or with growth rate.

Peto: The chief thing is whether it is qualitative or quantitative.

Henry Harris: I am not sure what that means.

Peto: There is a difference between a two or threefold ratio and a 20- or 30-fold ratio.

Henry Harris: There is about a threefold or fourfold difference in the amount of concanavalin A bound to the glycoprotein, but I don't think you can read too much into these numbers. One or two *N*-acetylglucosamine residues near the end of the polysaccharide chain might make a very big difference to the amount of lectin bound, or it might not. It is not simply the composition of the polysaccharide that determines how much lectin is bound; the conformation is also relevant. The difference in lectin-binding is not a small effect; but it could conceivably be produced by a small change in polysaccharide structure.

Do you have any information about polysaccharide composition in your teratomas, François?

Jacob: In glycoproteins of adult cells, there are two main types, the 'mannose-rich' type and the 'complex' type with a molecular weight between 2000 and 3500. The mannose-rich seems to be the precursor of the complex one. In the polysaccharide moieties of glycoproteins in the embryonal carcinoma cells of teratomas, there is no trace of the so-called complex structure and only a small amount of the mannose-rich one. The bulk of the glycopeptide obtained from these cells is of a much bigger size. It has a molecular weight around 15–20 000. It is not found in the adult, but it is present in the embryo during early stages.

Henry Harris: There is certainly a 'mannose-rich' glycoprotein of a molecular mass of about 100 000 on the teratocarcinoma we looked at.

Jacob: Yes, there is the mannose-rich structure.

Gardner: Is this particular glycoprotein one that is shared by both malignant cells and normal early embryonic cells? What is your explanation for the behaviour of the hybrids in which you selected against retention of the human X-chromosome after other human chromosomes appeared to have been lost? As I understand, you found in some cases that you still did not get tumours and the protein change found in the malignant parent did not show up.

Henry Harris: What we call normal fibroblasts are actually embryonic fibroblasts. We have not done a systematic comparison between adult and embryonic tissues. Most of the normal organs that I have shown—liver, kidney and so on—are adult material. Whether at some stage in embryonic life one might find something different, I don't know. It is worth looking at sometime.

The results Jonasson and I obtained with highly segregated human–mouse hybrids were strange and I felt that they should be discussed (Jonasson & Harris 1977). These hybrids had apparently thrown out all, or almost all, of their diploid human chromosomes, leaving an intact mouse chromosome set derived from a malignant parent cell; but they failed to grow in *vivo* nonetheless. I suggested that there might be an extrachromosomal element suppressing the malignancy. However, with respect to our glycoprotein marker, these hybrids and the tumours derived from them behaved in the same way as our usual intraspecific mouse crosses. I found it difficult to suppose that our membrane glycoprotein was extrachromosomally determined, so I began to seek some other explanation. I therefore made a series of crosses between the PG19 melanoma and very heavily irradiated human diploid cells—the sort of crosses with which Stephen Goss and I did the mapping experiments (Goss & Harris 1977*a, b*). In this way we generated a whole range of hybrids which were essentially PG19 containing odd chips of human chromosomes or sometimes none that we could see. About half of these hybrids didn't grow in nude mice. Those that didn't grow had the non-malignant type of lectin-binding patterns in the 100 000 molecular mass glycoprotein, and those that did grow had the malignant lectin-binding pattern. So the results originally described by Jonasson and myself were not a chance aberration.

If you don't think that the suppression of malignancy is due to an extrachromosomal element, what sort of model can you have? What might be happening is an analogue of a phenomenon described by Yvonne Boyd and myself for the thymidine kinase gene (Boyd & Harris 1973). If you fuse a mouse cell with a chick erythrocyte, you get extreme segregation of the chick chromosomes. In fact you don't see them at all in most cases. When we fused TK^- mouse cells with chick erythrocytes and selected for TK^+ cells, we expected to find retention of the chick TK gene. Instead we found an increased level of mouse TK. This has also been described for HPRT in human–mouse crosses (Watson *et al.* 1972). The HPRT activity that comes up again is from the species that was originally $HPRT^-$. Some of the TK^- and $HPRT^-$ mutants may not be structural mutants, but may be regulatory mutants, for want of a better word. You might be able by cell fusion to throw a switch that restores normal function; and that is what might be happening when malignancy is suppressed in these highly segregated hybrids where it is very unlikely that one would retain in all cases one particular unselected human gene.

Bodmer: That has been done with mutants that in other respects we know are quite stable. The notion that in one system they may be regulatory and in another not seems a little odd. Or maybe I didn't understand what you meant by 'throwing a switch'. It seemed to me that you implied that the initial mutation that makes a cell $HPRT^-$ is a control mutation.

Henry Harris: There might be all sorts of mutations making a cell $HPRT^-$.

Bodmer: The Edinburgh results from John Evans' group were done with the IR line which we have used a lot and which is like A9. In other hands it just doesn't behave in the way you described.

Henry Harris: With the TK^- mouse cells fused with chick erythrocytes, we certainly got back mouse TK as judged by temperature sensitivity and electrophoresis. There are other examples of this phenomenon (Siciliano *et al.* 1978). They occur with a very much higher frequency than one would expect for revertants. In some of the cell lines, no revertants have been reported.

Brenner: A likely explanation is that there are multiple isozymes of glucose transport which happen to have different glycosylation patterns and that one particular one has been switched on in tumour cells. I cannot see any other way of looking at the results as long as you assure us that the change in glycosylation is not global. It could depend in some way on the amino acid sequence of the protein itself.

Henry Harris: I would think it does.

Milstein: Not necessarily. It could be the ratio of the different glycosylating enzymes.

Brenner: But then I think you would find that many proteins would be altered. Perhaps this is a special isozyme with insulin resistance or other physiological properties. Perhaps it is needed by the cell for malignant growth in animals.

Henry Harris: On some of the tumours one certainly gets what people call microheterogeneity in that region. You get multiple peaks, and there is evidence that they may be due to differences in the polysaccharide components of the proteins.

Milstein: In some cases they are.

Henry Harris: Yes, but in this case we don't know. We do, however, have evidence that these multiple peaks are functionally homologous.

Harry Harris: Have you ever looked at two-dimensional gels after treating the preparations with neuraminidase? You are talking about a low isoelectric point and if there are a lot of sialic acids, these could largely determine the isoelectric point. If you remove them, you might see differences.

Henry Harris: Mike Bramwell has tried neuraminidase, but not a really pure neuraminidase. When cells are treated with commercial neuraminidase you lose a great deal of the material that binds wheat-germ agglutinin on the cell surface. But you do not convert the high wheat-germ agglutinin binding in the 100 000 molecular mass protein to a high concanavalin-binding configuration. So the difference between malignant and non-malignant cells is not merely due to terminal sialic acid.

References

BOYD, Y. & HARRIS, H. (1973) Correction of genetic defects in mammalian cells by the input of small amounts of foreign genetic material. *J. Cell Sci. 13,* 841-861

GOSS, S.J. & HARRIS, H. (1977*a*) Gene transfer by means of cell fusion. I. Statistical mapping of the human X-chromosome by analysis of radiation-induced gene segregation. *J. Cell Sci. 25,* 17-37

GOSS, S.J. & HARRIS, H. (1977*b*) Gene transfer by means of cell fusion. II. The mapping of 8 loci on human chromosome 1 by statistical analysis of gene assortment in somatic cell hybrids. *J. Cell Sci. 25,* 39-53

JONASSON, J. & HARRIS, H. (1977) The analysis of malignancy by cell fusion. VIII. Evidence for the intervention of an extra-chromosomal element. *J. Cell Sci. 24,* 255-263

SICILIANO, M.J., BORDELON, M.R. & KOHLER, P.O. (1978) Expression of human adenosine deaminase after fusion of adenosine deaminase-deficient cells with mouse fibroblasts. *Proc. Natl. Acad. Sci. U.S.A. 75,* 936-940

WATSON, B., GORMLEY, I.P., GARDINER,S.E., EVANS, H.J. & HARRIS, H. (1972) Reappearance of murine hypoxanthine guanine phosphoribosyl transferase activity in mouse A9 cells after attempted hybridisation with human cell lines. *Exp. Cell Res. 75,* 401-409

The role of viral transformation and cytogenetic changes in viral oncogenesis

GEORGE KLEIN

Department of Tumour Biology, Karolinska Institutet, Stockholm, Sweden

Abstract A wide variety of DNA viruses and a more restricted family of RNA viruses can transform normal cells *in vitro*. Transformation means either immortalization and/or the appearance of certain phenotypic changes. Although it has been often inferred that *in vitro* transformation can be essentially equated with malignant transformation, increasing evidence indicates that the latter, reflected by tumorigenicity *in vivo*, requires additional cytogenetic changes. The evidence will be reviewed for EB virus-associated human malignancy (Burkitt's lymphoma) and the role of the 14q+ translocation marker in human B-cell neoplasia. These findings point to an initiating role of viral transformation, reflected by *in vitro* immortalization, followed by a cytogenetic evolution where chromosome 14-associated changes are essential for the liberation of B lymphocytes from superimposed controls.

The contrast of tissue-associated, specific chromosomal changes that bring about malignant transformation after the initiating impact of different agents will be illustrated experimentally for murine T-cell lymphoma. Here, X-ray, DMBA and different virus (RadLV, Gross virus)-induced T lymphomas show the same chromosomal change: trisomy 15.

It may be questioned whether viral transformation can ever lead to neoplasia in the absence of subsequent cytogenetic changes.

The topic that has been assigned to me, expressed in the title of this talk, is much too large to cover in general terms. Rather than attempting it, I shall focus on a case where the interactions of a virus with environmental factors and the role of cytogenetic changes in causing neoplasia are becoming more and more apparent. I am referring to the interaction of Epstein-Barr (EB) virus with the human B lymphocyte, particularly as it relates to Burkitt's lymphoma. I shall also briefly deal with the role of viral and other inducing agents, as contrasted to cytogenetic changes, in the causation of murine T-cell lymphoma.

Time–space clustering of a highly malignant childhood lymphoma that appeared with a remarkably high frequency in the hot and humid regions of tropical Africa and had characteristic histological, cytological and clinical features led Burkitt to propose that the disease was due to an insect-transmitted virus (see Burkitt 1967). Subsequently, a tumour-associated virus was identified in a suspension culture derived from Burkitt lymphoma tissue, first by electron microscopy (Epstein *et al.* 1964) and later by serological tests (Henle & Henle 1966). Nucleic acid hybridization (zur Hausen *et al.* 1970) and the detection of a virally determined nuclear antigen, EBNA (Reedman & Klein 1973), conclusively showed that 97% of the African Burkitt lymphomas from the highly endemic regions carry multiple copies of the viral genome in a latent form (Klein 1975). EBNA is reminiscent of the T antigen of the small oncogenic DNA viruses. It is the only viral product regularly expressed in all virus–DNA-positive cells *in vitro* and *in vivo.* Virus production has never been seen in the tumour tissue *in vivo.* After *in vitro* explantation, the viral cycle is switched on in a minority of cells in some lines whereas others remain non-producers (Nadkarni *et al.* 1970). In some non-producer lines, it is possible to induce the viral cycle by the halogenated pyrimidines, bromodeoxyuridine and iododeoxyuridine (Gerber 1972; Klein & Dombos 1973), and by exposure to certain tumour promoters, such as phorbol esters (zur Hausen *et al.* 1978).

The isolation of EB virus from Burkitt lymphoma lines and its serological identification as a previously unknown herpes virus that failed to cross-react with any other member of the herpes group (Henle & Henle 1967) has led to the rapid development of EB virus research. It has revealed some important facts about the biology of the virus and its interaction with its host, the B lymphocyte. Those facts will be summarized briefly, since they are essential for a discussion of the aetiological role of a virus in Burkitt's lymphoma.

BIOLOGY OF EB VIRUS

Epstein-Barr virus can infect human and some non-human primate B lymphocytes *in vitro* but no other tissues so far tested. *In vivo,* the virus has only been found in one tissue that was not derived from B lymphocytes, namely the epithelial cells of nasopharyngeal carcinoma (Andersson-Anvret *et al.* 1978).

EB virus receptors have only been found on B lymphocytes where they are intimately associated with the complement receptors (Yefenof *et al.* 1976;

Yefenof & Klein 1977). All laboratory and natural isolates of EB virus, with the exception of one laboratory strain (Menezes *et al.* 1975), 'transform' normal lymphocytes into permanently growing, EB virus–DNA and EBNA-positive lines (Miller 1971).

Viral adsorption to B cells is followed by penetration and induction of the nuclear antigen, EBNA, as the earliest detectable event (Leibold *et al.* 1975; Einhorn & Ernberg 1978), occurring after 24–48 h. This is followed by polyclonal activation of immunoglobulin synthesis (Rosén *et al.* 1977), and cellular DNA synthesis that appears to represent a true S phase (Einhorn & Ernberg 1978; Robinson & Miller 1975). Subsequently, large cell clumps are formed and, in one or two weeks, a 100% EBNA-positive cell line grows out. Established lines carry multiple viral genomes, as a rule, similar to the Burkitt tumour *in vivo* (Nonoyama & Pagano 1973; zur Hausen & Schulte-Holthausen 1970). In both types of cells, a few copies of the viral genome are covalently integrated with the host cell DNA, whereas the majority are present in the form of free plasmids with covalently closed circular DNA (Adams 1979). The interrelationships of the integrated and the plasmid forms are not known.

In vitro EB-virus-transformed lymphoblastoid cell lines (LCL) are initially purely diploid (Zech *et al.* 1976). As long as they remain diploid, they fail to grow in nude mice and either do not plate in agarose or plate only at a very low efficiency (Nilsson *et al.* 1977). After serial *in vitro* passage for several months or years, they tend to become aneuploid and acquire the ability to grow in agarose and cause tumours in nude mice. In contrast, Burkitt lymphoma biopsies and derived lines have not been found to be purely diploid in any case so far studied (Zech *et al.* 1976). Most of them carry the same characteristic chromosome aberration (see below). They are regularly tumorigenic in nude mice and grow in agarose. This implies that *in vitro* immortalization cannot be equated with *in vivo* tumorigenicity, although it may be a prerequisite for it. For malignant growth *in vivo,* additional changes appear to be required, at the cytogenetic level.

Phenotypic changes involved in EB virus transformation have been most extensively studied by comparisons between EB-virus-negative but EB-virus-convertible B-lymphoma lines and their *in vitro* converted EBNA-positive sublines. Changes detected include increased cellular resistance to saturation conditions (Steinitz & Klein 1975), decreased serum requirement (Steinitz & Klein 1976), independence of a dialysable serum factor (Steinitz & Klein 1977), reduced lateral mobility (capping) of membrane constituents (Yefenof & Klein 1976), increased lectin agglutinability (Yefenof *et al.* 1977) and an increased ability to activate the alternative complement pathway (McConnel *et al.* 1978). These changes are reminiscent of the events that are known to accompany the transformation of monolayer cultures by oncogenic DNA (polyoma, SV40) and RNA (Rous) viruses. It is therefore likely that the *in vitro* immortalization that can be induced by different viruses and in very diverse target cells is accompanied by a characteristic set of phenotypic changes.

EB virus is oncogenic in certain New World monkeys, notably cotton top marmosets (Shope *et al.* 1973) and owl monkeys (Epstein *et al.* 1973*a, b;* 1975). It is important to note that no Old World monkeys so far tested were susceptible to the oncogenic effect of the virus. The reasons for this may be sought in the fact that all large apes and Old World monkeys tested carry EB-related viruses that induce the formation of EB-virus-neutralizing cross-reactive antibodies. New World monkeys do not carry EB-related viruses, although they have lymphotropic herpes viruses of their own (for review see Deinhardt *et al.* 1974). Three EB-related Old World monkey viruses have been studied: an EB-related chimpanzee virus (Gerber *et al.* 1977), a baboon virus, herpes virus papio (Falk *et al.* 1977), and an orang-outang herpes virus (Rabin *et al.* 1979). Virus-carrying B-cell lines have been isolated from seropositive normal animals in all three systems. All three viruses show partial DNA homology and partial antigenic cross-reactivity with human EB virus but virus–cell relationships are remarkably similar with only some differences of detail (Rabin *et al.* 1978; Ohno *et al.* 1977; Klein *et al.* 1978).

While EB virus fails to induce tumours in the Old World monkeys, the susceptible New World monkey species respond with the development of EB virus–DNA and EBNA-positive lymphomas (Frank *et al.* 1976). The success rate is not 100% and self-limiting lymphoproliferative lesions have been noted. While the detailed mechanisms of the oncogenic process remain to be studied further, the experiments already at hand clearly show the oncogenic potential of the virus.

EB-VIRUS-RELATED DISEASES

Three diseases are clearly associated with EB virus: infectious mononucleosis, Burkitt's lymphoma (BL) and nasopharyngeal carcinoma.

Infectious mononucleosis

It has been proved beyond doubt that the heterophile positive form of infectious mononucleosis (and a proportion of the heterophile negatives) is caused by EB virus. The evidence is based on retrospective (Henle *et al.* 1968;

Evans *et al.* 1968) and prospective (Niederman *et al.* 1970) seroepidemiological data, together with the ready isolatability of infectious (transforming) EB virus from the throat washings of patients with infectious mononucleosis (Gerber *et al.* 1976; Golden *et al.* 1973; Niederman *et al.* 1976), the easy establishment of permanent EB-virus-carrying lines from the peripheral blood of patients with infectious mononucleosis (Diehl *et al.* 1968), and the demonstration of EBNA-positive cells as a minority fraction of the circulating B cells (Klein *et al.* 1976; Hinuma & Katsuki 1978). A fatal case of infectious mononucleosis showed extensive infiltration of the lymphoid tissues with EBNA-positive B cells (Britton *et al.* 1978). The apparently 'irrelevant' antibody formation in patients with infectious mononucleosis against a variety of unrelated antigens (including the heterophile reaction) may be due to the polyclonal B-cell activation that occurs during a primary EB virus infection.

During acute infectious mononucleosis, EB-virus-specific killer T cells appear in the peripheral circulation (Svedmyr & Jondal 1975; Bakács *et al.* 1978). They disappear in convalescence (E. Svedmyr *et al.* 1979). They kill EB-virus-carrying lines and EB-virus-positive Burkitt biopsies without any apparent HLA restriction but fail to kill EB-virus-negative cell lines. It may be surmised that the killer T cells play an important role in the rejection of the EB-virus-transformed B blasts. More or less malignant forms of mononucleosis, including chronic infectious mononucleosis, have been described and may be related to a deficiency of the normal rejection mechanism against EB-virus-transformed cells (Purtilo *et al.* 1974, 1978).

Burkitt's lymphoma

As already mentioned, 97% of the African Burkitt lymphomas examined were found to be EB virus–DNA and EBNA-positive (Klein 1975). In contrast, only 22% of the non-African (sporadic) BL cases were found to carry the viral genome (Ziegler *et al.* 1976). This raises the important question of whether the virus-carrying and the virus-negative BL represent the same disease entity. Since the histopathological diagnosis of BL is not based on sharply defined criteria, it is conceivable that the virus-positive and negative forms represent two different diseases (zur Hausen 1975). The possibility that they may represent the same disease is discussed in connection with the cytogenetic evidence below. Whichever theory one favours, the fact that Burkitt's lymphoma was defined as a clinical and histopathological entity before the discovery of EB virus implies that the question needs reexamination, preferably by unbiased histopathological and prospective clinical studies.

Burkitt's lymphoma is a uniclonal tumour, as conclusively shown by the glucose-6-phosphate dehydrogenase method (Fialkow *et al.* 1970), studies of immunoglobulin synthesis (Fialkow *et al.* 1973), and the cytogenetic evidence discussed below. Fialkow *et al.* 1970) have also approached the question of whether recurring tumours represent regrowth of the same original clone or, at least occasionally, a new induction event. The original clone recurred in all studied cases except one. In one case, a tertiary tumour that arose in a patient exposed to heavy chemotherapy was derived from a different clone, probably due to a second induction event.

BL lines differ from EB-virus-transformed lymphoblastoid cell lines of non-neoplastic origin with regard to multiple phenotypic properties (Nilsson & Pontén 1975). While different BL lines tend to be quite different among themselves, each line is usually very uniform within itself. In contrast, lymphoblastoid cell lines are rather heterogeneous within each line, but different lymphoblastoid cell lines are very similar. These differences may reflect the uniclonal nature of BL compared with the polyclonal lymphoblastoid cell lines.

Cytogenetic features

Approximately 90% of the *in vivo* BL tumours studied were found to contain the same 14q+ marker (Zech *et al.* 1976; Manolov & Manolova 1972; Jarvis *et al.* 1974). The marker arises by the addition of an extra band at the end of the long arm of one No. 14 chromosome. Zech *et al.* (1976) showed that the extra material is consistently derived from the distal part of the long arm of one No. 8 chromosome. The remaining 10% BL that had no M14 had other chromosomal anomalies.

14q+ markers have been also found in EB-virus-negative Burkitt lymphomas (Zech *et al.* 1976). Also, 14q+ markers of similar or more or less similar appearance were found in non-Burkitt lymphomas of both lymphocytic and histiocytic types (Zech *et al.* 1976; Reeves 1973; Prigogina & Fleischman 1975; Mark 1975; Fukuhara *et al.* 1976; Fleischman & Prigogina 1977; Mark 1977; Yamada *et al.* 1977). These studies raised interesting questions concerning the origin of the extra material on No. 14 in these cases, and the cytogenetic relationship between different histopathological types of lymphomas.

Very recently, Mark *et al.* (1978) studied the banding patterns in six histiocytic lymphomas. The authors reviewed their results together with data from 32 further non-Hodgkin and non-Burkitt lymphomas in the literature. They have drawn the following conclusions: *(i)* No. 14 was most commonly affected, usually by structural deviations involving its long arm, preferentially band q32; this resulted in the formation of 14q+ markers, as a rule, found in 17 of the 45 cases: the extra material on No. 14 showed inconsistent derivation, in contrast to Burkitt lymphomas. *(ii)* Nos. 3, 7, 8 and 11 were next in frequency of involvement; structural deviations predominated for Nos. 3 and 11, and some recurrent marker types were seen; Nos. 7 and 8 were mostly affected by numerical deviations, usually gains for No. 7 and losses for No. 8. *(iii)* Structural deviations often affected Nos. 1, 6, 9 and 13 and recurrent marker types related to Nos. 6 and 13 were seen. *(iv)* The centromeric and the light-staining regions were preferentially affected by the breakpoints.

While the significance of the chromosome 14 anomaly for Burkitt's and other lymphomas is not known, it is tempting to speculate that it may be related to the liberation of the normal human lymphocyte from superimposed growth control. In this respect, it is of interest that McCaw *et al.* (1975) found a remarkable instability of chromosome 14 in ataxia telangiectasia, a condition that predisposes for leukaemias, lymphomas and other neoplasias. Recently, Cohen and co-workers identified a 14q+ marker that closely resembled the BL-associated marker, in mitogen-stimulated peripheral lymphocytes of patients with ataxia telangiectasia (M. M. Cohen, personal communication). The interpretation of the chromosome 14 anomaly in human lymphomas may be facilitated by considering some recent findings on murine lymphomas. The majority of T-cell-derived mouse lymphomas of spontaneous (Dofoku *et al.* 1975), X-ray-induced (Chang *et al.* 1977), viral (Wiener *et al.* 1978*a*) and chemical (DMBA) origin (Wiener *et al.* 1978*b*) had the same anomaly, chromosome 15-trisomy. T-cell lymphomas with rearranged (translocated) karyotypes had a cryptic 15-trisomy (Wiener *et al.* 1978*b*). In contrast, murine non-T cell lymphomas had no 15-trisomy.

AETIOLOGY OF BURKITT'S LYMPHOMA

Several questions can now be asked that may illuminate the aetiology of BL. Most of them are answerable by available evidence and/or by further studies.

(a) Is EB virus a passenger in BL that is picked up after the tumour has been induced by other agents or is it present from the inception of the tumour?

EB-virus-negative but *in vitro* EB-virus-convertible B-cell lymphomas have been found in EB-virus-seropositive (i.e. infected) patients (Klein *et al.* 1975; Menezes *et al.* 1975). This shows that an EB virus infection does not readily 'jump' onto lymphomas that arise for other reasons. This is understandable since persons chronically infected with EB virus have high neutralizing antibody levels that can stop *in vitro* infection with great efficiency (de Schryver *et al.* 1974; de Thé & Geser 1976). It appears more likely that the EB-virus-carrying monoclonal lymphomas arise originally from an EB-virus positive cell. It is clear that the virus can remain latent indefinitely in B lymphocytes. *In vitro,* such lymphocytes are immortal and they probably have a long lifespan *in vivo* as well. Under certain conditions, e.g. prolonged growth stimulation, they may undergo secondary changes and occasionally give rise to an autonomous clone.

(b) What is the EB-virus-infection status of an African child before BL development, in comparison with healthy controls from the same area?

This question is now partially answerable by the recent prospective seroepidemiological study of de Thé & Geser (1976). The anti-EB virus (viral capsid antigen; VCA) antibody titres of children who developed BL 7–21 months after the bleeding were significantly higher than in controls from the same region. Invariably, they showed signs of chronic EB virus infection with IgG anti-VCA and anti-EBNA antibodies but no IgM anti-VCA. This is clearly different from the serological picture seen in acute primary infection (in the course of infectious mononucleosis) where IgM anti-VCA is the predominating antibody and anti-EBNA antibodies are absent, as a rule. It is therefore clear that Burkitt's lymphoma clone arises in children chronically infected with EB virus, in contast to infectious mononucleosis. It is likely that the pre-BL child carries a higher virus load than controls. This would fit with the role postulated for EB-virus-carrying preneoplastic cells in the genesis of BL, as postulated under *(a)* above.

(c) What is the explanation for the peculiar geographical pathology of high endemic BL?

Burkitt (1969) has suggested that an insect-transmitted co-factor, probably chronic holoendemic malaria, precipitates the development of BL. His reasoning is suggestive but not conclusive. It is quite conceivable that chronic malaria or other insect-vectored diseases act as 'promoters', if the word is used in the same sense as in experimental carcinogenesis. Promoting agents would be expected to stimulate chronic growth of the 'dormant' or preneoplastic cells. In addition, they may delay or inhibit differentiation and also act as immunosuppressants. In the actual context, the heavy, chronic proliferation of holoendemic malaria would urge the long-lived EB-virus-carrying, preneoplastic B lymphocytes to proliferate. This would increase the chances for the generation of chromosomal aberrations. When the 'right' cytogenetic change arises by chance, postulated as the ultimate cause of the malignant growth, autonomous neoplasia would arise.

In conclusion, the genesis of Burkitt's lymphoma is visualized as proceeding in three steps:

I. Primary EB virus infection affects the young child, probably at a relatively high multiplicity. It immortalizes a certain number of B lymphocytes *in vivo.*

II. This is followed by the impact of an environmental promoting agent, perhaps chronic holoendemic malaria, providing a chronic stimulus to the proliferation of the EB-virus-carrying preneoplastic cells.

III. Chromosomally abnormal variants appear in the stimulated tissue by chance. After certain types of changes, particularly the 8 to 14 translocation that leads to the 14q+ marker, the affected B lymphocyte would no longer obey the negative feedback controls that would otherwise restrict its proliferation *in vivo.*

Since III is regarded as the essential, ultimate change, it follows that I and II are only facultative. This may explain the existence of EB-virus-negative Burkitt lymphomas where initiation may have been due to other viral or non-viral agents. It also explains the fact that the high-endemic, time–space-clustered cases are not the only form of the disease. Sporadic cases occur all around the world, presumably through the impact of other promoters. Finally, the chromosomal aberrations (III) are also subject to certain variations, as shown by the 10% BL that carry no 14q+ marker but have other anomalies. It is well known, however, that the same phenotypic change can be brought about by multiple genetic changes.

THE ROLE OF CHROMOSOMAL CHANGES IN THE CAUSATION OF MURINE T-CELL LYMPHOMA

The role of chromosomal changes will be discussed only briefly, since available evidence is less extensive than on the EB virus–Burkitt system.

Dofoku *et al.* (1975) have shown that most 'spontaneous' AKR lymphomas are characterized by the same chromosomal anomaly, trisomy 15. The relatively regular occurrence of this anomaly is remarkable. It is known that the AKR strain, selected for high leukaemia incidence by continued brother–sister breeding and preferential mating of the progeny from leukaemic animals, has fixed at least five different genetic systems, all of which favour high leukaemia incidence, namely two integrated copies of the Gross virus genome, leukaemogenic in itself, the Fv-1^n amplification system, a relative immunological unresponsiveness to Gross-virus-induced leukaemia cells, reflected by the Rgv-1^s allele, and a relatively high tendency of the target T-cell system itself to undergo neoplastic transformation. This, taken together with the fact that AKR mice are viraemic, as a rule, leads one to expect that the leukaemia will appear very promptly and presumably by a relatively direct induction mechanism. The fact that the preleukaemic development takes several months, together with the regular occurrence of the cytogenetic change, suggests, however, that viral transformation is probably the initiating, but not the final, definite cause of leukaemia development. Recently, recombination between ecotropic and xenotropic viruses has been put forward as the explanation for the leukaemia outgrowth. Alternatively, one might envisage the cytogenetic change, leading to trisomy 15+, as bearing the main responsibility.

The role of the trisomy 15+ was further emphasized when Wiener *et al.* (1978*a*) found that leukaemias of C57BL mice induced by the radiation-dependent or the radiation-independent autonomous variants of the radiation leukaemia virus (RadLV) were also characterized by trisomy 15 in the vast majority of cases studied. Even more surprisingly, however, T-cell lymphomas induced by dimethylbenzanthracene, a chemical carcinogen where the causation of leukaemia is not believed to involve the activation of leukaemia virus, is even more regularly characterized by trisomy 15 (Wiener *et al.* 1978*b*). Even in cases where the karyotype of the primary leukaemia was rearranged, trisomy 15 occurred, although it was cryptic and localized within translocations.

The T6T6 marker is due to a translocation from chromosome 15 to 14. Wiener *et al.* (1978*c*) have examined dimethylbenzanthracene-induced leukaemias in T6T6 mice. They found that the large, translocated 14;15 chromosome was present in the trisomic state, whereas the small, centromeric T6 marker was present in two copies. This localizes the gene or genes responsible for the leukaemia-associated trisomy to the distal, longer part of chromosome 15. By using additional translocations, it ought to be possible to localize the relevant area more critically.

F. Wiener *et al.* (unpublished work) have also studied non-T lymphomas induced by Rauscher, Friend, and Duplan viruses, as well as a number of reticulum cell sarcomas. It is not yet clear whether these neoplasias have chromosomal anomalies of their own, or not, but it is already clear that they do not have a chromosome 15-trisomy.

Thus, in the murine T-cell system we face another example of what might be termed 'convergent evolution'. Tumours are initiated by different agents, viral and non-viral, in a certain target-cell type. In the ultimate tumour, the same chromosomal anomaly is found, no matter what the initiating agent was. The parallelisms with the findings in the human lymphoma system are obvious, even though the murine system focuses on T-cell lymphomas and has trisomy as its main feature, as contrasted to a translocation.

References

ADAMS, A. (1979) in *The Epstein-Barr Virus* (Epstein, A. M. & Achong, R., eds.), in press

ANDERSSON-ANVRET, M., FORSBY, N. & KLEIN, G. (1978) Nasopharyngeal carcinoma, in *Progress in Experimental Tumor Research,* pp. 100-106, Karger, Basel

BAKÁCS, T., SVEDMYR, E., KLEIN, E., ROMBO, L. & WEILAND, O. (1978) EBV-related cytotoxicity of FC receptor negative T lymphocytes separated from the blood of infectious mononucleosis patients. *Cancer Lett. 4,* 185-189

BRITTON, S., ANDERSSON-ANVRET, M., GERGELY, P., HENLE, W., JONDAL, M., KLEIN, G., SANDSTEDT, B. & SVEDMYR, E. (1978) Epstein-Barr virus immunity and tissue distribution in a fatal case of infectious mononucleosis. *N. Engl. J. Med. 298,* 89-92

BURKITT, D. (1967) Chemotherapy of jaw tumors, in *Treatment of Burkitt's Tumour* (Burchenal, J. H. & Burkitt, D, P., eds.) *(U.I.C.C. Monogr. 8),* pp. 94-101, Springer, Berlin

BURKITT, D. (1969) Etiology of Burkitt's lymphoma—an alternative hypothesis to a vectored virus. *J. Natl. Cancer Inst. 42,* 19-28

CHANG, T.D., BIEDLER, J.L., STOCKERT, E. & OLD, L.J. (1977) Trisomy of chromosome 15 X-ray induced mouse leukemia. *Proc. Am. Assoc. Cancer Res. 18,* 225

DEINHARDT, F. W., FALK, L. A. & WOLFE, L. G. (1974) Simian herpesviruses and neoplasia. *Adv. Cancer Res. 19,* 167-205

DE SCHRYVER, A., KLEIN, G., HEWETSON, J., FOÇCHI, G., HENLE, W., HENLE, G., MOSS, D. J. & POPE, J. H. (1974) Comparison of EBV neutralization tests based on abortive infection or transformation of lymphoid cells and their relation to membrane reactive antibodies (anti-MA). *Int. J. Cancer 13,* 353-362

DE THÉ, T. & GESER, A. (1976) Études épidemiologiques prospectives et nature de l'association entre le virus d'Epstein-Barr (E.B.V.) et le lymphôme de Burkitt (B.L.). *C.R. Acad. Sci. (D) (Paris) 282,* 1387-1390

DIEHL, V., HENLE, G., HENLE, W. & KOHN, G. (1968) Demonstration of a herpes group virus in cultures of peripheral leukocytes from patients with infectious mononucleosis. *J. Virol. 2,* 663-669

DOFOKU, R., BIEDLER, J. L., SPENGLER, B. A. & OLD, L. J. (1975) Trisomy of chromosome 15 in spontaneous leukemia of AKR mice (mouse thymomas/karyotype analysis/trypsin-Giemsa banding). *Proc. Natl. Acad. Sci. U.S.A. 72,* 1515-1517

EINHORN, L. & ERNBERG, I. (1978) Induction of EBNA precedes the first cellular S-phase after EBV-infection of human lymphocytes. *Int. J. Cancer 21,* 157-160

EPSTEIN, M. A., ACHONG, B. G. & BARRY, Y. M. (1964) Virus particles in cultured lymphoblasts from Burkitt's lymphoma. *Lancet 1,* 702-703

EPSTEIN, M. A., HUNT, R. D. & RABIN, H. (1973*a*) Pilot experiments with EB virus in owl monkeys *(Aotus trivirgatus).* I. Reticuloproliferative disease in an inoculated animal. *Int. J. Cancer 12,* 309-318

EPSTEIN, M.A., RABIN, H., BALL, G., RICKINSON, A. B., JARVIS, L. & MELENDEZ, L. V. (1973*b*) Pilot experiments with EB virus in owl monkeys *(Aotus trivirgatus)* II. EB virus in a cell line from an animal with reticuloproliferative disease. *Int. J. Cancer 12,* 319-332

EPSTEIN, M. A., ZUR HAUSEN, H., BALL, G. & RABIN, R. (1975) Pilot experiments with EB virus in owl monkeys *(Aotus trivirgatus).* III. Serological and biochemical findings in an animal with reticuloproliferative disease. *Int. J. Cancer 15,* 17-22

EVANS , A. S., NIEDERMAN, J. C. & MCCOLLUM, R. W. (1968) Seroepidemiologic studies of infectious mononucleosis with EB virus. *N. Engl. J. Med. 279,* 1121-1127

FALK, A. L., HENLE, G., HENLE, W., DEINHARDT, F. & SCHUDEL, A. (1977) Transformation of lymphocytes by herpesvirus papio. *Int. J. Cancer 20,* 219-226

FIALKOW, P. J., KLEIN, G., GARTLER, S. M. & CLIFFORD, P. (1970) Clonal origin for individual Burkitt tumours. *Lancet 1,* 384-386

FIALKOW, P. J., KLEIN, E., KLEIN, G., CLIFFORD, P. & SINGH, S. (1973) Immunoglobulin and glucose-6-phosphate dehydrogenase as markers of cellular origin in Burkitt lymphoma. *J. Exp. Med. 138,* 89-102

FLEISCHMAN, E. W. & PRIGOGINA, E. L. (1977) Karyotype peculiarities of malignant lymphomas. *Humangenetik 35,* 269-279

FRANK, A., ANDIMAN, W. A. & MILLER, G. (1976) Epstein-Barr virus and nonhuman primates: natural and experimental infection. *Adv. Cancer Res. 23,* 171-201

FUKUHARA, S., SHIRAKAWA, S. & UCHINO, H. (1976) Specific marker chromosome 14 in malignant lymphomas. *Nature (Lond.) 259,* 210-211

GERBER, P. (1972) Activation of Epstein-Barr virus by 5-bromodeoxyuridine in 'virus-free' human cells (complement-fixing antigen/immunofluorescence/leukocytes). *Proc. Natl. Acad. Sci. U.S.A. 69,* 83-85

GERBER, P., NKRUMAH, F. K., PRITCHETT, R. & KIEFF, E. (1976) Comparative studies of Epstein-Barr virus strains from Ghana and the United States. *Int. J. Cancer 17,* 17-81

GERBER, P., KALTER, S. S., SCHIDLOVSKY, G., PETERSON, W. D. Jr. & DANIEL, M. D. (1977) Biologic and antigenic characteristics of Epstein-Barr virus-related herpesviruses of chimpanzees and baboons. *Int. J. Cancer 20,* 448-459

GOLDEN, H. D., CHANG, R. S., PRESCOTT, W., SIMPSON, E. & COOPER, T. Y. (1973) Leukocyte-transforming agent: prolonged excretion by patient with mononucleosis and excretion by normal individual. *J. Infect. Dis. 127,* 471-473

HENLE, G. & HENLE, W. (1966) Immunofluorescence in cells derived from Burkitt's lymphoma. *J. Bacteriol. 91,* 1248-1256

HENLE, G. & HENLE, W. (1967) Immunofluorescence, interference, and complement fixation technics in the detection of the herpes-type virus in Burkitt tumor cell lines. *Cancer Res. 27,* 2442-2446

HENLE, G., HENLE, W. & DIEHL, W. (1968) Relation of Burkitt's tumor-associated herpes-type virus to infectious mononucleosis. *Proc. Natl. Acad. Sci. U.S.A. 59,* 94-101

HINUMA, Y. & KATSUKI, T. (1978) Colonies of EBNA-positive cells in soft agar from peripheral leukocytes of infectious mononucleosis patients. *Int. J. Cancer 21,* 426-431

JARVIS, J. E., BALL, G., RICKINSON, A. B. & EPSTEIN, M. A. (1974) Cytogenetic studies on human lymphoblastoid cell lines from Burkitt's lymphomas and other sources. *Int. J. Cancer 14,* 716-721

KLEIN, G. (1975) Studies on the Epstein-Barr virus genome and the EBV-determined nuclear antigen in human malignant disease. *Cold Spring Harbor Symp. Quant. Biol. 39,* 783-790

KLEIN, G. & DOMBOS, L. (1973) Relationship between the sensitivity of EBV-carrying lymphoblastoid lines to superinfection and the inducibility of the resident viral genome. *Int. J. Cancer 11,* 327-337

KLEIN, G., GIOVANELLA, B., WESTMAN, A., STEHLIN, J. G. & MUMFORD, D. (1975) An EBV-genome-negative cell line established from an American Burkitt lymphoma; receptor characteristics. EBV infectability and permanent conversion into EBV-positive sublines by in vitro infection. *Intervirology 6,* 319-334

KLEIN, G., SVEDMYR, E., JONDAL, M. & PERSSON, P. O. (1976) EBV-determined nuclear antigen (EBNA)-positive cells in the peripheral blood of infectious mononucleosis patients. *Int. J. Cancer 17,* 21-26

KLEIN, G., FALK, L. & FALK, K. (1978) Antigen-inducing ability of herpesvirus papio in human and baboon lymphoma lines, compared to Epstein-Barr virus. *Intervirology 10,* 153-164

LEIBOLD, W., FLANAGAN, T. D., MENEZES, J. & KLEIN, G. (1975) Brief communication: induction of Epstein-Barr virus-associated nuclear antigen during in vitro transformation of human lymphoid cells. *J. Natl. Cancer Inst. 54,* 65-68

MANOLOV, G. & MANOLOVA, Y. (1972) Marker band in one chromosome 14 from Burkitt lymphomas. *Nature (Lond.) 237,* 33-34

MARK, J. (1975) Histocytic lymphomas with the marker chromosome 14q+. *Hereditas 81,* 289-292

MARK, J. (1977) Chromosomal abnormalities and their specificity in human neoplasms: an assessment of recent observations by banding techniques. *Adv. Cancer Res. 24,* 165-222

MARK, J., EDEDAHL, C. & DAHLENFORS, R. (1978) Characteristics of the banding patterns in non-Hodgkin and non-Burkitt lymphomas. *Hereditas 88,* 229-242

MCCAW, B. K., HECHT, F., HARNDEN, D. & TEPLITZ, R. L. (1975) Somatic rearrangement of chromosome 14 in human lymphocytes (leukemia/lymphoproliferation/ataxia-teleangiectasia). *Proc. Natl. Acad. Sci. U.S.A. 72,* 2071-2075

MCCONNELL, I., KLEIN, G., LINT, T. F. & LACHMANN, P. J. (1978) Activation of the alternative complement pathway by human B cell lines is associated with Epstein-Barr virus (EBV) transformation of the cells. *Eur. J. Immunol. 8,* 453-458

MENEZES, J., LEIBOLD, W. & KLEIN, G. (1975) Biological differences between Epstein-Barr virus (EBV) strains with regard to lymphocyte transforming ability, superinfection and antigen induction. *Exp. Cell Res. 92,* 478-484

MENEZES, J., LEIBOLD, W., KLEIN, G. & CLEMENTS, G. (1975) Establishment and characterization of an Epstein-Barr virus (EBV)-negative lymphoblastoid B cell line (BJAB) from an exceptional EBV genome-negative African Burkitt's lymphoma. *Biomedicine (Paris) 22,* 276-284

MILLER, G. (1971) Human lymphoblastoid cell lines and Epstein-Barr virus: a review of their interrelationships and their relevance to the etiology of leuko-proliferative states in man. *Yale J. Biol. Med. 43,* 358-383

NADKARNI, J. S., NADKARNI, J. J., KLEIN, G., HENLE, W., HENLE, G. & CLIFFORD, P. (1970) EB viral antigens in Burkitt tumor biopsies and early cultures. *Int. J. Cancer 6,* 10-17

NIEDERMAN, J. C., EVANS, A. S., SUBRAHAMANYAN, M. S. & MCCOLLUM, R. W. (1970) Prevalence and persistence of EB virus antibody in young adults. *N. Engl. J. Med. 282,* 361-365

NIEDERMAN, J. C., MILLER, G., PEARSON, H. A., PAGANO, J. S. & DOWALIBY, J. M. (1976) Infectious mononucleosis Epstein-Barr-virus shedding in saliva and the oropharynx. *N. Engl. J. Med. 294,* 1355-1359

NILSSON, K. & PONTÉN, J. (1975) Classification and biological nature of established human hematopoietic cell lines. *Int. J. Cancer 15,* 321-341

NILSSON, K., GIOVANELLA, B. C., STEHLIN, J. S. & KLEIN, G. (1977) Tumorigenicity of human hematopoietic cell lines in athymic nude mice. *Int. J. Cancer 19,* 337-344

NONOYAMA, M. & PAGANO, J. S. (1973) Detection of Epstein-Barr viral genome in nonproductive cells. *Nature New Biol. 233,* 103-106

OHNO, S., LUKA, J., FALK, L. & KLEIN, G. (1977) Detection of a nuclear, EBNA-type antigen in apparently EBNA-negative herpesvirus papio (HVP)-transformed lymphoid lines by the acid-fixed nuclear binding technique. *Int. J. Cancer 20,* 941-946

PRIGOGINA, E. L. & FLEISCHMAN, E. W. (1975) Marker chromosome 14q+ in two non-Burkitt lymphomas. *Humangenetik 30,* 109-113

PURTILO, D. T., CASSEL, C. & YANG, J. P. S. (1974) Fatal infectious mononucleosis in familial lymphohistocytosis. *N. Eng. J. Med. 291,* 736

PURTILO, D. T., HUTT, L., BHAWAN, J., YANG, J. P. S., CASSEL, C., ALLEGRA, S. & ROSEN, F. S. (1978) Immunodeficiency and the Epstein-Barr virus in the X-linked recessive lymphoproliferative syndrome. *Clin. Immunol. Immunopathol. 9,* 147-156

RABIN, H., NEUBAUER, R. H., HOPKINS, R. F. & NONOYAMA, M. (1979) Further characterization of a herpesvirus positive orangutan cell line and comparative aspects of an in vitro transformation with lymphotropic Old World primate herpesviruses. *Int. J. Cancer,* in press

REEDMAN, B. M. & KLEIN, G. (1973) Cellular localization of an Epstein-Barr virus (EBV)-associated complement-fixing antigen in producer and non-producer lymphoblastoid cell lines. *Int. J. Cancer 11,* 499-520

REEVES, B. R. (1973) Cytogenetics of malignant lymphomas. Studies utilising a Giemsa-banding technique. *Humangenetik 20,* 231-250

ROBINSON, J. & MILLER, G. (1975) Assay for the Epstein-Barr virus based on a stimulation of DNA synthesis in mixed leukocytes from human umbilical cord blood. *J. Virol. 15,* 1065-1072

ROSÉN, A., GERGELY, P., JONDAL, M., KLEIN, G. & BRITTON, S. (1977) Polyclonal Ig production after Epstein-Barr virus infection of human lymphocytes in vitro. *Nature (Lond.) 267,* 52-54

SHOPE, T., DECHAIRO, D. & MILLER, G. (1973) Malignant lymphoma in cottontop marmosets after inoculation with Epstein-Barr virus. *Proc. Natl. Acad. Sci. U.S.A. 70,* 2487-2491

STEINITZ, M. & KLEIN, G. (1975) Comparison between growth characteristics of an Epstein-Barr virus (EBV)-genome-negative lymphoma line and its EBV-converted subline in vitro. *Proc. Natl. Acad. Sci. U.S.A. 72,* 3518-3520

STEINITZ, M. & KLEIN, G. (1976) Epstein-Barr (EBV)-induced change in the saturation sensitivity and serum dependence of established, EBV-negative lymphoma lines in vitro. *Virology 70,* 570-573

STEINITZ, M. & KLEIN, G. (1977) Further studies on the differences in serum dependence in EBV negative lymphoma lines and their in vitro EBV converted, virus-genome carrying sublines. *Eur. J. Cancer 13,* 1269-1275

SVEDMYR, E. & JONDAL, M. (1975) Cytotoxic effector cells specific for B cell lines transformed by Epstein-Barr virus are present in patients with infectious mononucleosis. *Proc. Natl. Acad. Sci. U.S.A. 72,* 1622-1626

SVEDMYR, E., JONDAL, M., HENLE, W., WEILAND, O., ROMBO, L. & KLEIN, G. (1979) EBV-specific T cells and serologic responses after onset of infectious mononucleosis. *Clin. Lab. Immunol.,* in press

WIENER, F., OHNO, S., SPIRA, J., HARAN-GHERA, N. & KLEIN, G. (1978*a*) Chromosomal changes (trisomy 15 and 17) associated with tumor progression in leukemias induced by radiation leukemia virus. *J. Natl. Cancer Inst. 61,* 227-238

WIENER, F., SPIRA, J., OHNO, S., HARAN-GHERA, N. & KLEIN, G. (1978*b*) Chromosome changes (trisomy 15) in murine T-cell leukemia induced by 7,12-dimethylbenz(a)anthracene (DMBA) *Int. J. Cancer 22,* 447-453

WIENER, F., OHNO, S., SPIRA, J., HARAN-GHERA, N. & KLEIN, G. (1978*c*) Cytogenetic mapping of the trisomic segment of chromosome 15 in murine T-cell leukaemia. *Nature (Lond.) 275,* 658-660

YAMADA, K., YOSHIOKA, M. & OAMI, H. (1977) A 14q+ marker and a late replicating chromosome # 22 in brain tumor. *J. Natl. Cancer Inst. 59,* 1193-1194

YEFENOF, E. & KLEIN, G. (1976) Difference in antibody induced redistribution of membrane IgM in EBV-genome free and EBV positive human lymphoid cells. *Exp. Cell Res. 99,* 175-178

YEFENOF, E. & KLEIN, G. (1977) Membrane receptor stripping confirms the association between EBV receptors and complement receptors on the surface of human B lymphoma lines. *Int. J. Cancer 20,* 347-352

YEFENOF, E., KLEIN, G., JONDAL, M. & OLDSTONE, M. B. A. (1976) Surface markers on human B- and T-lymphocytes. IX. Two-color immunofluorescence studies on the association between EBV receptors and complement receptors on the surface of lymphoid cell lines. *Int. J. Cancer 17,* 693-700

YEFENOF, E., KLEIN, G., BEN-BASSAT, H. & LUNDIN, L. (1977) Differences in the Con A-induced redistribution and agglutination patterns of EBV genome-free and EBV-carrying human lymphoma lines. *Exp. Cell Res. 108,* 185-190

ZECH, L., HAGLUND, U., NILSSON, K. & KLEIN, G. (1976) Characteristic chromosomal abnormalities in biopsies and lymphoid-cell lines from patients with Burkitt and non-Burkitt lymphomas. *Int. J. Cancer 17,* 47-56

ZIEGLER, J. L., ANDERSSON, M., KLEIN, G. & HENLE, W. (1976) Detection of Epstein-Barr virus DNA in American Burkitt's lymphoma. *Int. J. Cancer 17,* 701-706

ZUR HAUSEN, H. (1975) Oncogenic herpes viruses. *Biochim. Biophys. Acta 417,* 25-53

ZUR HAUSEN, H. & SCHULTE-HOLTHAUSEN, H. (1970) Presence of EB virus nucleic acid homology in a 'virus-free' line of Burkitt tumor cells. *Nature (Lond.) 227,* 245-248

ZUR HAUSEN, H., SCHULTE-HOLTHAUSEN, H., KLEIN, G., HENLE, W., HENLE, G., CLIFFORD, P. & SANTESSON, L. (1970) EBV DNA in biopsies of Burkitt tumors and anaplastic carcinomas of the nasopharynx. *Nature (Lond.) 28,* 1056-1058

ZUR HAUSEN, H., O'NEILL, F. J., FEESEN, U. K. & HECKER, E. (1978) Induction of persisting genomes of oncogenic herpesviruses by the tumor promoter TPA. *Nature (Lond.) 272,* 373-375

Discussion

Koprowski: Are the 'common' receptors for EB virus and for complement present on all the cell lines you have studied?

My second question concerns the uniclonal origin of Burkitt lymphomas. Are the recurrences of Burkitt lymphomas also of uniclonal origin and was the recurrent clone the same as the original clone?

In connection with your studies of similar chromosomal changes after exposure to various carcinogenic agents, I recall a study of Levan and his associates (cited in Mitelman & Levan 1976) on chromosomal changes in cells of rats bearing tumours induced either by Rous sarcoma virus (RSV) or a chemical carcinogen, dimethylbenzanthracene (DMBA). These authors described chromosomal aberrations *specific* for each type of tumour. When carcinogen treatment became leukogenic the leukaemic cells showed chromosomal changes identical to those observed in the cells of the solid tumour but still markedly different from those induced by RSV.

Finally, although malignant tumours developing from these papillomas do not show the presence of papilloma virus, carcinomas of the oesophagus or stomach developing from papillomas in cattle feeding on ferns in Scotland do show the presence of the virus.

Klein: Concerning the relationship between EB virus receptor and complement receptor: the history of the Jijoye-P3HR-1 line is of particular interest. The original line, Jijoye, releases a transforming virus and has EB virus and complement receptors. A clone derived from Jijoye, P3HR-1, releases cytopathic virus that is apparently a mutant or a defective variant (defective with regard to the ability to induce DNA synthesis). P3HR-1 cells have lost both the EB virus and complement receptor, apparently because reinfection with a cytopathic virus would be growth-inhibitory. It is strongly inhibitory for the original Jijoye line. Virus non-producer variants have been derived from the P3HR-1 line. After they have been non-producers for long periods, over several years, they regain EB virus and complement receptors, both simultaneously and in approximately the same proportion. Recent biological experiments show a close functional relationship between the two receptors. However, I would like to emphasize that it is clearly *not* the complement-fixing site of the receptor that binds the EB virus but some other site on the same structure.

Concerning the chromosomal anomalies, our studies on both human B and mouse T-cell neoplasia suggest that non-random chromosomal changes relate to the tissue (target cell) of origin, rather than the aetiological agent. The two hypotheses are not mutually exclusive, however, and particularly in the case of strong carcinogens, like DMBA, specific 'hot spots' are known to occur that may be responsible for the common chromosomal changes induced in different tissues, as observed by Levan's group.

Concerning virus integration and derived tumour cells, the usual situation is that the virus integrates and, once this has happened, it is very difficult for the host cell to lose it. Therefore, the Shope papilloma, although it is an exception, is an interesting one, because it shows that the virus can be lost, at least in this system, when the papilloma progresses to carcinoma or soon afterwards.

The recurrence of Burkitt lymphoma was studied extensively by Fialkow, in collaboration with us, by the G6PD method. In all cases that were tested, except one, the recurrent clone was the same as the original clone. The exception was a little girl who had extensive abdominal Burkitt lymphoma. She was treated by chemotherapy, was in remission for about one year and then developed a maxillary tumour which was still the same clone. She had further chemotherapy and developed another jaw tumour on the other side. That was a different clone.

Peto: I thought Fialkow had reported more examples than that and finished up with a sort of general rule that if there was a recurrence within six months it was from the same clone as the primary tumour, whereas later it was different.

Klein: No, you are probably thinking of Ziegler's work, extrapolations from the clinical sensitivity of early and late recurrences, not directly clonality tests. All the recurrences studied by Fialkow were found to be uniclonal (Fialkow *et al.* 1972, 1973).

Koprowski: I also thought Fialkow had claimed that the recurrences were not uniclonal.

Klein: Fialkow's study which shows the hard evidence is for one case. The general speculative extrapolation from therapeutic sensitivity is not direct evidence. Fialkow and I have collaborated on all this work.

Nossal: I want to try to create a link between your talk and Walter Bodmer's. The point of departure is the immunological surveillance which is so spectacularly successful for T lymphocytes recognizing the EB-virus-transformed immortalized B cells in humans, and so spectacularly unsuccessful in the New World monkey. T cell recognition of antigens is special in the sense that the recognition somehow involves both a foreign antigen such as a virus and some component of the major histocompatibility complex (MHC). So one can construct a general theory of immune responsiveness that would go something as follows: you will be genetically non-responsive to any antigen if you possess MHC determinants that fail to be able to make an association of some sort with that antigen. Is there something about the HLA equivalent antigens of these New World monkeys that causes some kind of failure of association with EB virus? Have any cytotoxicity studies been done in the monkeys to see whether their T cells can properly and appropriately polish off the transformed cells, or is there some other explanation?

Klein: In our experience there is no HLA restriction as far as EB virus specific killing with purified T cells of acute infectious mononucleosis patients is concerned. They kill autologous and allogeneic EB-virus-carrying lines (in the case of Burkitt, even autologous biopsies), apparently without any systematic distinction (although the lines differ in sensitivity). *In vitro* sensitization of normal T cells against *autologous* EB-virus-carrying lymphoblastoid cell lines can also generate powerful cytotoxic cells. They kill both EB-virus-positive and EB-virus-negative lines non-specifically but, again, there is no syngeneic restriction. All this refers to our own (human) Old World species. I am afraid there is no information yet about the reaction of EB-virus-transformed cell lines derived from New World monkeys. It would indeed be interesting to know whether they, too, will exert the same stimulatory effect and generate the same type of cytotoxic cells, when confronted with autologous T lymphocytes.

Bodmer: The lack of polymorphic restriction does not necessarily imply that the HLA–ABC products are not involved in T cell recognition. The phenomenon of restriction in H-2 is a product of the way the system is studied, which is unnatural. Allogeneic combinations do not usually occur within the individual. In Daudi, which lacks the HLA–ABC determinant, surely one can claim that the T cell killing involving EB virus is not effective? One must in the end analyse the role of these products not by the polymorphic difference but by their presence or absence in the cell.

Klein: I agree with your general points, but we do not agree with Tursz *et al.* (1977) about Daudi. While Daudi is a fairly resistant cell to both NK and T-cell killing, there is no doubt that it *can* be killed by the EB-virus-specific killer T cells isolated from mononucleosis blood.

Henry Harris: We have some results for SV40 that are completely concordant with George Klein's view that there is a shadow that falls between transformation *in vitro* and the ability of cells to make a tumour. I decided to have a look at SV40 because of some interesting results with interspecific hybrids from the Wistar Institute (Koprowski & Croce 1977). These results were interpreted as evidence that SV40 was operating as a dominant determinant of malignancy, an idea that would have disturbed me a little. So I decided to start at the beginning and see whether primarily transformed mouse cells, that is, cells studied as soon as possible after transformation *in vitro*, could produce tumours in sublethally irradiated syngeneic newborn mice or in nude mice. Strange as it might seem, I don't think this experiment has actually been done. As far as I can see, the only systematic tests done so far were done on adult mice that were not immunosuppressed. Jonathan Gee, in my laboratory, has isolated a number of primarily transformed clones from mouse fibroblast populations exposed to SV40 *in vitro*. Transformation was assessed by the usual *in vitro* criteria—growth in semi-solid medium, etc. The transformed cells did not produce any tumours until about the 20th passage *in vitro*. We explanted one of the tumours and thus derived a malignant line that produced tumours readily on injection. It carried the SV40 virus and expressed the SV40 T antigen. We fused these malignant cells with diploid fibroblasts in our standard procedure. The hybrids showed profound suppression of tumorigenicity, like most other such crosses that we have studied. In this system, therefore, SV40 behaves in exactly the same way as George Klein and I have previously described for polyoma virus (Klein & Harris 1972). I think this is the only experiment so far done to test whether primarily transformed cells can straight away make tumours in immunosuppressed animals. The *prima facie* evidence is that they can't.

Koprowski: Since it is almost impossible to distinguish between the chromosomes of the two parental cells in intraspecific cell hybrids, it would be very difficult to determine whether a reduction of the tumorigenicity of a hybrid cell was due to the presence of chromosomes responsible for suppression of malignancy or to the loss of chromosomes responsible for the expression of malignancy. Thus, the statement that re-expression of malignancy is due to the loss of chromosomes of the normal parental cell cannot be proven experimentally. In addition, the published work on the expression of malignancy in intraspecific mouse hybrids is contradictory. In fact, the same types of hybrid have been reported to be both normal (Wiener *et al.* 1971, 1973) and malignant (Jami & Ritz, 1975*a, b*; Aden & Knowles 1977: Ziegler 1978). Results of experiments concerning tumorigenicity in hybrids between malignant mouse cells and normal human cells are also contradictory, since Jonasson *et al.* (1977) and R. Kucherlapati & S. Shin (personal communication, 1978) have reported opposite results with the same type of hybrid. In order to prove conclusively that the expression of malignancy is dominant in DNA virus-transformed cells, we have hybridized freshly isolated mouse cells (normal cells derived directly from mice) and SV40-transformed human cells (Croce & Koprowski 1974, 1975). As a rule, this type of hybrid retains the entire complement of murine chromosomes and segregates chromosomes of the virus-transformed human cell (Croce & Koprowski 1974, 1975; Croce 1977). As we have shown (Croce *et al.* 1975; Koprowski & Croce 1977), such hybrids are capable of inducing tumours in nude mice even though they retain the entire chromosome complement of the normal mouse parental cell. These results conclusively prove that the normal mouse genome is incapable of suppressing the expression of malignancy in this hybrid. In addition, recent experiments using HT-1080 human fibrosarcoma cells (Croce 1976) for hybridization with either normal mouse or normal human cells have indicated that the expression of malignancy is dominant in such hybrids (Croce *et al.* 1978).

In conclusion, it is clear that malignancy can be expressed as a dominant trait. It is possible, however, that in certain forms of cancer malignancy might be recessive, but as yet no experimental approach has been able to show conclusively that this is the case.

Bodmer: Monocytes have the complement receptor. Would you say that the monocyte complement receptor is not the appropriate one for the B cells?

Klein: We have tested monocytes, granulocytes and mouse B cells. They all have complement receptors, but do not bind EB virus. EB virus is exclusively adapted to the complement receptor of the B-cell lymphocytes of human and non-human primates.

Bodmer: So it is something more specific than that. The notion that a genetic event, namely a specific translocation, is required is a very interesting one. But I wonder about the lag period and the length of time. In the end it should be possible to know the rate at which such specific events occur and see whether that rate is compatible with the lag period between the initial stages and the final development of the tumour. It seems surprising that that could be such a rare event and take such a long time so that it would account for the rareness of Burkitt's lymphoma itself in an area where malaria is endemic. The promoter is there all the time and the EB virus is virtually universal. The search for genetic factors involving HLA has, so far, proved negative. But it seems that something else may still be needed to explain the rareness of the tumour.

Klein: I would be unable to guess the risk figures for Burkitt lymphoma. This would be easier to do on a mouse system of an inbred population that has a high leukaemia risk, a known latency period and a controlled environment. In Burkitt lymphoma, the approximate incidence is 5 per 100 000 per year, in the high-risk zone and age, and I would not know whether to regard this as a high or low probability event in comparison with the occurrence of the specific 8–14 translocation in the EB-virus-infected cells—which incidentally, in an EB-virus-transformed lymphoblastoid cell line *in vitro*, was never seen to appear and 'take over' under ordinary conditions of propagation.

Bodmer: That surely depends on whether the sites have specific properties. If the break-points on all chromosomes were totally random you would be right. But if there is something specific about those sites, the translocation may not be that rare. That ought to be established *in vitro*. There are probably two effects. Translocation is only one of them.

Lachmann: The complement receptor on the lymphocytes claimed to be associated with EB virus uptake is the C3d receptor. Although it has been reported to be present on monocytes, most people think it is present only on lymphocytes. The C3d receptor has no known function and interaction of C3d with the receptor produces no obvious effects. It occurred to us that EB virus might go into cells through the C3d receptor because the virus has C3d fixed upon it as a result of its shedding from cells. However, experiments to try to prove this by passing the virus through anti-C3d columns were unsuccessful. Although it seemed to us to be a good explanation, it seems to be untrue! The C3d receptor may be an irrelevance. It may really be an EB virus receptor and its reaction with complement may have no biological significance.

Malaria reduces complement levels quite sharply and we have been impressed by the fact that EB-virus-transformed cells acquired the capacity to activate the alternative pathway in the absence of antibody (McConnell *et al.* 1978). This may be a mechanism of surveillance over these transformed cells at a very early stage, before there are specific T cells or antibodies. If malaria makes this mechanism much less efficient, the load of transformed cells may be much higher in children who have malaria. This might go together with the finding that the pre-Burkitt children have raised antibody titres.

Ruddle: Human chromosome 8 seems to be involved in rearrangement or aneuploidy in the progression of a number of human tumours.

In the work that Harold Klinger and I have been doing in human × mouse hybrids, lines which retain chromosome 9, and particularly chromosomes 9 plus 8, grow very poorly in nude mice as tumours. There seems to be a relationship to growth that involves human chromosome 8. Recently Tom Shows (Roswell Park Memorial Institute, Buffalo, New York) has done somatic genetic studies on the mapping of the Lets protein. He thinks he can map the gene for Lets protein to human chromosome 8. Lets protein has been involved in mitotic regulation in dividing cells. It will be interesting to see if the Lets gene maps anywhere near your translocation break-point on chromosome 8.

With regard to trisomy 15, what chemical carcinogen do you use and what type of tumour emerges?

Klein: We have looked at DMBA-induced T-cell leukaemias in C57BL. So far, all known leukaemias with a trisomy 15 as the exclusive or main chromosomal change were found to be T-cell leukaemias.

Ruddle: What is the role of trisomy 15 in the population that gives rise to the tumour? Does trisomy follow an initial event or could trisomy itself be the primary event in the T cell transformation?

Klein: Our bias is that the initiating agent, whether a viral or a chemical carcinogen, leads to the formation of long-lived preleukaemic cells. This first step would be the common denominator between different agents. During continued proliferation, possibly under the influence of other (promoting) agents, trisomy 15 would appear at random in a preleukaemic cell. This would then lead to the autonomous malignant clone. There may be of course more steps in this, and the picture at the moment is very preliminary. Our study of the T6 translocation system suggests that specific regions of may be involved, not the whole chromosome.

Polani: Returning to Ruddle's point and chromosome 8 of man, in the developmental mosaics, who are trisomic for chromosome 8 in one cell line, there is no evidence of any malignancy as yet, whereas of course in some myeloproliferative and leukaemic disorders trisomy 8 mosaicism is a characteristic feature (Philip 1975).

Siniscalco: There is no question that malignant growth is associated with chromosomal unbalance. It cannot be doubted, either, that the distribution of which specific chromosomes are mutated (numerically or structurally) in given types of tumours is not random. However, in spite of all this, it may still be argued that the associations found so far could be the result of somatic selection in favour of the cell phenotype most favourable under the circumstances of tumour growth. I feel that it is still too early to conclude in favour of the biases you describe.

Klein: All my preformed biases would have said that viral transformation and/or immunological breakdown can explain everything and I was also doubtful about specific changes at the genetic level. However, the evidence comes out in favour of specific genetic changes as playing the ultimate, perhaps the crucial role. The great disorder in tumour cytogenetics that you refer to is probably aneuploid noise, particularly pronounced in carcinomas and sarcomas. The occurrence of this noise does not exclude the possibility of some hidden music behind it. It is only since the era of banding that specific tumour-associated, non-random chromosomal changes have been recognized. The reason why they are largely found in lympho- and myeloproliferative diseases could lie in the fact that there is a much smaller tendency for broad aneuploidization to occur in the haematopoietic tumours than in carcinomas and sarcomas. In the latter, it will probably be more difficult to find specific changes, if they exist.

Polani: When Fleischman & Prigogina (1977) looked at the chromosomes in a series of lymphomas, they found translocations between numbers 11 and 14; again a kind of specificity.

Peto: If your model of delay in AKR mice was a true analogy to the delay in man, one would expect the age distribution in Burkitt lymphoma to keep on increasing, whereas it reaches a peak by about the age of five years, and then goes down again. It is not clear, if one is just waiting for second events to happen, why that should be the case. It looks rather as though very few cells are available as a target. A very small number are present at some critical stage of embryonic development or early infancy but cease to exist as a population later on.

Klein: This is an interesting point. There are great differences, however, in the immune response of the AKR mouse and of the Burkitt lymphoma (BL) patients towards their tumours. In the former, there is virtually no defence at all against the leukaemic cells, once they arise. Ordinary immunization-rejection tests simply do not work. This may be related to the presence of Rgv-1^{s}, an immune unresponsiveness gene, fixed in AKR by inbreeding and selection for high leukaemia incidence. In contrast, the BL patient has considerable immune defence against the EB-virus-carrying lymphoma cells; all the therapeutic experiments suggest that the patient only needs to get *some* help from therapy in about 20–25% of the cases and can then reject the residual tumour cells by immune mechanisms. The fact that BL is a typical childhood disease may be related to the maturation of the immune system, in analogy with a number of other diseases where a common childhood disease disappears after puberty. Recurring warts would be an example.

One interesting point that can be made about BL is the primary localization: the jaws around the age of dentition, frequent primary tumours in the gonads around puberty, long-bone tumours in adolescence, and a single case of bilateral breast tumour in a lactating woman. The last example, incidentally, has regressed spontaneously. All this suggests that local proliferation in a tissue, as part of a normal physiological process, may somehow lead to local breakdown of restrictive, presumably immunological host control.

References

ADEN, D. P. & KNOWLES, B. B. (1977) Tumorigenicity of intraspecific somatic cell hybrids in nude mice. *J. Natl. Cancer Inst. 58,* 743-746

CROCE, C. M. (1976) Loss of mouse chromosomes in somatic cell hybrids between HT−1080 human fibrosarcoma cells and mouse peritoneal macrophages. *Proc. Natl. Acad. Sci. U.S.A. 73,* 3248-3252

CROCE, C. M. (1977) Assignment of the integration site for simian virus 40 to chromosome 17 in GM54VA, a human cell line transformed by simian virus 40. *Proc. Natl. Acad. Sci. U.S.A. 74,* 315-318

CROCE, C. M. & KOPROWSKI, H. (1974) Concordant segregation of the expression of SV40 T antigen and human chromosome 7 in mouse-human hybrid subclones. *J. Exp. Med. 139,* 1350-1353

CROCE, C. M. & KOPROWSKI, H. (1975) Assignment of gene(s) for cell transformation to human chromosome 7 carrying the simian virus 40 genome. *Proc. Natl. Acad. Sci. U.S.A. 72,* 1658-1660

CROCE, C. M., BARRICK, J., LINNENBACH, A. & KOPROWSKI, H. (1979) Expression of malignancy in hybrids between normal and malignant cells. *J. Cell Physiol.,* submitted for publication

CROCE, C. M., ADEN, D. P. & KOPROWSKI, H. (1975) Tumorigenicity of mouse-human diploid hybrids in nude mice. *Science (Wash. D.C.) 190,* 1200-1202

FIALKOW, P. J., KLEIN, G. & CLIFFORD, P. (1972) Second clone underlying a Burkitt-tumor exacerbation. *Lancet 2,* 629-631

FIALKOW, P. J., KLEIN, E., KLEIN, G., CLIFFORD, P. & SINGH, S. (1973) Immunoglobulin and glucose-6-phosphate dehydrogenase as markers of cellular origin in Burkitt lymphoma. *J. Exp. Med. 138,* 89-102

FLEISCHMAN, E. W. & PRIGOGINA, E. L. (1977) Karyotype peculiarities of malignant lymphomas. *Hum. Genet. 35,* 269-279

JAMI, J. & RITZ, E. (1975*a*) Nonmalignancy of hybrids derived from two mouse malignant cells. II. Analysis of malignancy of LM (TK^-) Cl 1D parental cells. *J. Natl. Cancer Inst. 54,* 117-122

JAMI, J. & RITZ, E. (1975*b*) Tumor-associated transplantation antigens in immune rejection of mouse malignant cell hybrids *Proc. Natl. Acad. Sci. U.S.A. 72,* 2130-2134

JONASSON, J., POVEY, S. & HARRIS, H. (1977) The analysis of malignancy by cell fusion. VII. Cytogenetic analysis of hybrids between malignant and diploid cells and of tumours derived from them. *J. Cell Sci. 24,* 217-254

KLEIN, G. & HARRIS, H. (1972) Expression of polyoma-induced transplantation antigen in hybrid cell lines. *Nature New Biol. 237,* 163-164

KOPROWSKI, H. & CROCE, C. M. (1977) Tumorigenicity of simian virus 40-transformed human cells and mouse-human hybrids in *nude* mice. *Proc. Natl. Acad. Sci. U.S.A. 74,* 1142-1146

MCCONNELL, I., KLEIN, G., LINT, T. F. & LACHMANN, P. J. (1978) Activation of the alternative complement pathway by human B cell lymphoma lines is associated with Epstein Barr virus (EBV) transformation of the cells. *Eur. J. Immunol. 8,* 453-458

Mitelman, F. & Levan, G. (1976) Clustering of aberrations to specific chromosomes in human neoplasms. II. A survey of 287 neoplasms. *Hereditas 82,* 167-174

Philip, P. (1975) Trisomy 8 in acute myeloid leukaemia. *Scand. J. Haematol. 14,* 140-147

Tursz, T., Fridman, W. H., Senik, A., Tsapis, A. & Fellous, M. (1977) Human virus-infected target cells lacking HLA antigens resist specific T-lymphocyte cytolysis. *Nature (Lond.) 269,* 806-808

Wiener, F., Klein, G. & Harris, H. (1971) The analysis of malignancy by cell fusion. III. Hybrids between diploid fibroblasts and other tumour cells. *J. Cell Sci. 8,* 681-692

Wiener, F., Klein, G. & Harris, H. (1973) The analysis of malignancy by cell fusion. IV. Hybrids between tumour cells and a malignant L cell derivative. *J. Cell Sci. 12,* 253-261

Ziegler, M. L. (1978) Phenotypic expression of malignancy in hybrid and cybrid mouse cells. *Somatic Cell Genet. 4,* 477-489

The genetics of learning disabilities

BARTON CHILDS and JOAN M. FINUCCI

Department of Pediatrics, The Johns Hopkins University School of Medicine, Baltimore

Abstract Rapid progress in the study of learning disabilities requires a common context within which investigators and professionals with very diverse backgrounds can work towards common goals. Much evidence points to the heterogeneity of origin and clinical expression of learning disabilities as well as to many other ways in which they resemble such multifactorial conditions as mental retardation, gout and hypertension. In these conditions, genetic and epidemiological methods have been used to sort out the constitutional and environmental factors which precipitate disease in susceptible persons. Such an approach to learning disabilities can be expected to be rewarding.

It has become fashionable to say that the burden of the diseases of middle life could be reduced, if not eliminated, by environmental control of noxious agents and more disciplined living. That is, it is implied that social and medical policies can be determined by observations of the average behaviour of whole populations in their social settings, ignoring the variability, either latent or overt, of the individuals that compose them. Such typological thinking is prevalent in medicine, but it is to be found in studies of other aspects of human affairs too, including the behavioural sciences. For example, it is common in psychological research to treat variability as a nuisance, or to ignore it. Or, when some variation is under study, to compare one population, called abnormal, with a matched group, called normal, and to allow the inference to be drawn of homogeneity within each population. It may be that when studies of behavioural differences are confined to the normal range, the individual differences are so subtle as to defy definition; but for medicine, or for psychiatry, that is for abnormal behaviour, typological classifications are self-defeating, since it is the individual with his private version of illness with whom the physician must deal.

The genetic analysis of behaviour has as its goal the elucidation of the genetic contribution to behavioural differences. At first, human behaviour genetics was concerned with the quantitative partitioning of genetic and other sources of normal variation, but more recent efforts have been devoted to disease: for example, family, twin and adoption studies have verified a genetic contribution to schizophrenia (Kety *et al.* 1975) and to affective disorders (Gershon *et al.* 1977). But despite this progress there is still much to be learned about the genetics of abnormal behaviour and despite the dawning recognition that most behavioural syndromes must be heterogeneous in origin, little progress has been made in describing and validating subtypes (Goldman 1977). Some of the most salient reasons for this were suggested by Penrose: the lack of definitions of behavioural states with which all can agree, the difficulty of discriminating between affected and unaffected persons, the lack of measurable properties that can be presented in distributions, and the failure to use epidemiological methods (Penrose 1971). The cognitive attributes that are required for learning are behavioural properties, and they have been the object of genetic study (Broadhurst *et al.* 1974). But it is the heritability of variation in normal skills that has been emphasized, rather than individual differences. These studies examine the variance of the whole of distributions of measures of such skills, rather than the extremes, but it is the latter that offer more promise for the differentiation of individuals. One set of behavioural properties that can be studied by looking at the outliers in distributions is learning disabilities. Learning disability is defined in US Public Law 94-142 as 'a disorder in one of the basic psychological processes involved in understanding or in using spoken or written language. These may be manifested in disorders of listening, talking, thinking, reading, writing, spelling or arithmetic.' It should be added that for children with normal intelligence, 'learning disability' is a misnomer because such people can learn in the general sense as readily as others; it is only that they cannot do it by conventional means.

Attempts to discover the causes of learning disabilities suffer at least one serious constraint, namely the lack of a unifying context in which to study them—one that can accommodate the variety of disciplines of the workers in the field. Learning-disabled children are seen and managed by paediatricians, neurologists, psychiatrists and psychologists, ophthalmologists and optometrists, educators and social workers, and each approaches the problem from a different viewpoint, with different ideas and techniques for research strategy and management. For example, there are said to be over 90 names for learning disabilities, reflecting the variety of the disciplines of the workers engaged in dealing with affected children (Sulzbacher 1975). And because learning disabilities are so strongly associated in everyone's mind with school, little attention has been given to the later social and psychological lives of such children as adults, or to their occupational outcomes; that is, the natural history of learning disabilities is unknown.

Studies of the familial properties of learning disabilities have been inhibited for the same reasons, but aggregation of cases in families is generally acknowledged and the idea that something is inherited is gaining acceptance (Cantwell 1976). Furthermore the recognition of phenotypic heterogeneity has led to various classifications, but the genetic method has never been used to validate the classes.

This matter of classification is of supreme importance because it can lead to insights into causal mechanisms which in turn are useful in designing management. So it may be that genetics can provide the necessary unifying context through which the various disciplines can be brought efficiently to bear on this, so far, intractable problem. Learning disabilities are, in fact, similar to such other multifactorial disorders as mental retardation, hypertension and gout, and should be studied accordingly. An apposite example of this approach is Penrose's Colchester study in which 1280 mentally retarded patients were examined and information on nearly 7000 relatives was obtained (Penrose 1938). The result was the first comprehensive classification of the types of mental retardation. Something of the sort should be done for learning disabilities. A large sample of cases should be studied using standard methods. Measurements of many attributes should be made on all the members of the families, and the number of families should be great enough to embrace both frequent and infrequent subclasses.

One form of learning disability to be examined in this multifactorial context is specific reading disability (SRD).

Specific reading disability (SRD) is defined as inability to learn to read with ease despite a normal IQ, intact central nervous system, normal vision and hearing, normal emotional state and ample educational and cultural opportunity (Eisenberg 1966). This is a definition of exclusion; this is what is left after cases due to known causes (of which lack of opportunity is the chief) have been removed. In this, as in many other ways, SRD is similar to other multifactorial conditions and it may be useful to observe how and where it shares the qualities that characterize those conditions. In what follows we shall refer from time to time to our own work. The sample consists of 47 families, selected from a larger group simply because all the members of all families have been examined and tested. Details of ascertainment of the index cases and families and of the battery of reading, spelling and other tests have been published elsewhere (Finucci *et al.* 1976).

PROBLEMS OF DEFINITION

Distributions of scores of reading skills look like those of blood pressure readings and serum uric acid levels; they are continuous and markedly skewed towards one end, in this case the low end (Yule *et al.* 1974). That is, there are more cases occupying positions below the mean than can be accounted for in a perfectly Gaussian distribution. Also like these others, SRD presents with one or both of two phenotypes: low scores in tests of reading and spelling, and overt difficulty in school. Low scores are highly correlated with school problems, and for educators that is the phenotype that counts, but like blood pressure rises that do not lead to strokes, or high levels of serum uric acid that do not lead to gout, there are borderline scores in reading and spelling that need not lead to much difficulty in school. So it is difficult to decipher unequivocally a point within a narrow area in the distribution above which no reader would have school troubles and below which all would do so. Reading scores are conditioned by IQ, sex, age, school attainment, and no doubt other qualities of temperament, so misclassification can be reduced by recourse to quotients or regression procedures that reveal the degree to which the scores reflect these influences, but it is only minimized thereby, not eliminated. In addition, something depends upon the premium accorded to reading by parents, schools and society; in a non-reading society, reading disability does not exist. Classification for adults is complicated by the ambiguities of historical information about school problems, which may be preferentially forgotten or in less strenuous times did not exist. Recognizing these hazards we have calculated reading quotients (Myklebust 1968) (we are using quotients pending assembly of scores of enough randomly selected children to develop regression equations) for children which take into account IQ, age and school level; for adults we have calculated regression equations based on IQ, sex and school achievement. For purposes of analysis we have chosen a quotient of 0.9 as a point to break the distribution for children since we have observed no one with a higher score to have trouble in school, and for adults we have chosen two standard errors of prediction below the predicted level of performance. As it happens, the latter correlates rather well with a history of significant difficulty in learning to read when young.

SECULAR CHANGES

Multifactorial conditions represent individual responses to experiences, so we may expect to see secular changes in frequency, and perhaps in intensity, due to variations in those environmental circumstances that raise susceptibility to overt expression. Reading disability has become an educational and medical issue only recently, although SRD, called congenital word blindness, was described in a family nearly 80 years ago (Finucci 1978; Bryan 1978). Word blindness was thought then, and until recently, to be rare, but we now know it to be as frequent as several per cent in some places (Rutter & Yule 1975; Tarnopol & Tarnopol 1976). The reasons for this new focus have not been given formal attention, but it is plausible to suspect that it has to do with changes in the social value of reading. In the United States, reading has always been taught with cultural development in mind (Smith 1965). The earliest emphasis was moralistic and religious. After the revolution reading texts took on a patriotic and nationalistic tone, then turned to preparation for citizenship, and during the latter part of the 19th and early 20th centuries became preoccupied with reading as a cultural asset. Now, in a technological society in the midst of an exponential proliferation of knowledge, reading skill has become a part of the armamentarium needed for the pursuit of life's prizes. Without it school attainment is made difficult, and without school, many occupations are foreclosed. The educational system is the bottleneck through which all must pass before seeking fortune and status, and all must learn to read so as to learn by reading. If reading standards have been raised to meet the more rigorous requirements of the business and technological world, then people who managed to get along under less stringent conditions will fail to measure up to the new standard and so have trouble in school. Since the end of World War II, and even more so since the first Russian Sputnik, reading requirements in the United States have risen and it is evident that reading disability is now a more salient problem than it was a generation ago (Bryan 1978).

GEOGRAPHICAL AND ETHNIC DIFFERENCES

It is to be expected that conditions dependent upon both genes and experiences would vary in prevalence in different societies, and where the question has been raised with regard to most multifactorial conditions such differences have been found. It is clear that in countries where literacy ranks below survival as a matter for public concern, reading disability is no problem, but in the developed nations prevalence figures are given of from 2 to 25% of children of normal intelligence who are said to read inadequately (Tarnopol & Tarnopol 1976). It is probable that not all of these children fulfil the definition of SRD given above. Some of the differences may be social. For example, Rutter and his colleagues found three times as many 10-year-old children with reading retardation unaccounted for by low IQ in an inner

London borough as in the Isle of Wight (Berger *et al.* 1975). The London population was also characterized by much higher rates of marital discord and psychiatric disorder among the parents, by overcrowding and disorderly home conditions, and by schools with high teacher and pupil turnover and populated mainly by children from economically depressed backgrounds (Rutter *et al.* 1975). Such a relationship between depressed reading skill and inner city conditions exists also in the United States (Eisenberg 1966).

On the other hand, some of the differences may be related to the language itself. There are reports of especially low levels of SRD in Italy and Spain, and even lower in Japan (Faglioni *et al.* 1969; Makita 1968). This is attributed to the strong phonetic relationship between sounds of words and their symbols. In this regard English is among the most difficult languages for beginners to master, and if the figure of 3.9% of specific reading retardation among 10-years-olds in the Isle of Wight (Berger *et al.* 1975) is any criterion, then levels of SRD among English speakers are higher than those of Italy, Spain and Japan. Thus there is nothing to suggest that genetic differences account for variable freqencies of SRD, but some light might be shed on the possibility by a study of its prevalence among native-born Japanese-Americans living in California.

FAMILIAL TRANSMISSION

Familial aggregation of cases of reading disability measured by psychometric testing, significant school problems, or both, has been reported many times (Finucci 1978). The data, however, are often incomplete, consisting of historical or hearsay evidence, variable standards for diagnosis and only partially tested families. There are, however, some reliable studies which establish the fact of familial aggregation unequivocally (Hallgren 1950; Walker & Cole 1965; Owen *et al.* 1971; Finucci *et al.* 1976; Smith *et al.* 1977; DeFries *et al.* 1978).

Most studies have not been designed to test modes of hereditary transmission, but those that give pedigree information show no consistent pattern (Finucci 1978). Twin studies add some support to the genetic hypothesis by showing the expected differences in concordance between monozygotic and dizygotic twin pairs (Zerbin-Rüdin 1967). In all these respects SRD resembles other multifactorial disorders.

PHENOTYPIC HETEROGENEITY

The idea that a multifactorial diagnosis is actually the repository of several or many conditions sharing a common phenotype is gradually taking hold. Such syndromes may be fragmented into several or many clinically and genetically distinct conditions, usually infrequent, leaving a residuum of cases presumed to be the result of the effects of several or many genes responding to various environmental influences. Mental retardation is an example; some cases are due to trauma or infections, others to single gene diseases and chromosome abnormalities; and many more, especially the mild ones, are aggregated in families in such a way as to suggest multigenic inheritance. Table 1 shows that phenotypic heterogeneity is an idea with an honourable past in the reading field too. It will be seen that most of these schemes have an auditory type in which the disabled reader has trouble relating sound and visual symbols, a visual type in which perception or memory for visual symbols appears to be impaired, and a mixture of these, and there are also a few other types involving linguistic disorders, object and colour naming and the like (Denkla 1972). So it is probable that subtypes exist, but the validity of each of these is difficult to ascertain. For example, each subtype has been established with a different set of tests; none of the investigators have tried

TABLE 1

Subtypes of specific reading disability

Subtypes	*Investigator*	*Reference*
I Visuo-spatial Audiophonic	Johnson & Myklebust	1967
II Dysphonetic (audio) Dyseidetic (visual) Mixed	Boder	1971
III Visuo-spatial Audio phonic Mixed	Ingram *et al.*	1970
IV Language disorder Articulatory and graphomotor Visuo-spatial	Mattis *et al.*	1975
V Auditory memory Visual memory Mixed	Bateman	1968
VI Visual perceptual Intersensory integration Naming Temporal ordering	Doehring & Hoshko	1977

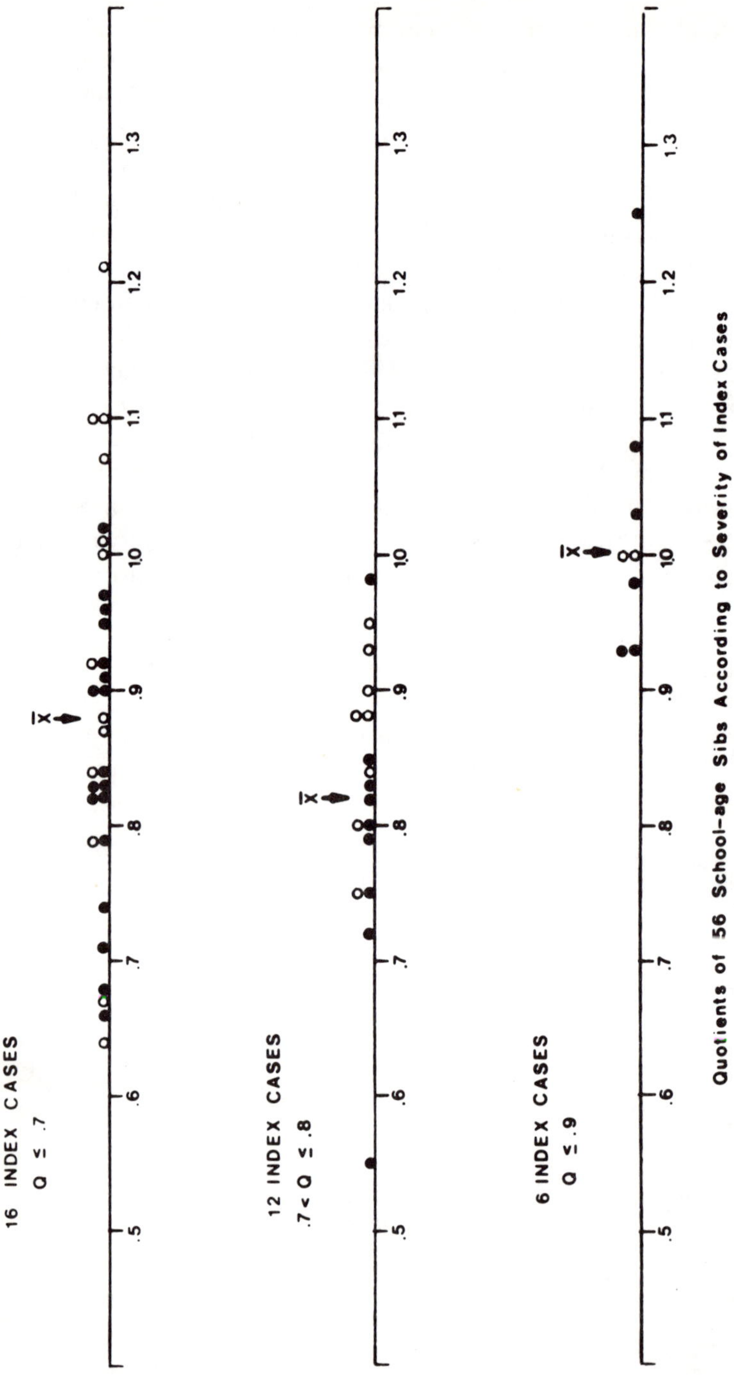

FIG. 1. Variations in reading quotients of sibs according to the reading skill of the index cases.

out their battery of tests on the others' subtypes. In addition there are differences in diagnostic criteria for inclusion of cases; different standards are used for the definition of reading disability, and quantitative variability is often ignored. Finally, even if we accept the entirely reasonable proposition that there are visual and auditory types of disabled reader, we cannot be sure that they are distinctive syndromes of separate origin, rather than the consequences of the same cause imposed on different individual developmental substrates. The genetic method could provide such a validation.

We have been examining our data with this in mind, using, to start with, the severity of the reading disability and the type of misspelling as criteria. Fig. 1 shows distinct differences in the reading quotients of the sibs of index cases according to the severity of reading disability of the index case. The numbers are small, and apparent relationships may disappear when more families are tested, but for the moment there is an apparent difference between the distributions; when the index cases are more severely affected, the sibs' reading scores vary most widely, and when the index case is mildly affected, the scores of the sibs fall into the mild or normal classes and the variation appears to be less. This is compatible with the idea of the presence of some single gene types among the severely affected cases, and multigenic influences among both severe and mildly affected classes. If all were multigenic the means of the three distributions ought to drop progressively as the severity increases. These distributions are reminiscent of the position in mental retardation, in which the rare monogenic types appear among the cases with the lowest IQs while the multigenic, and more frequent cases are more heavily represented among those with higher IQs.

A second approach takes advantage of differences in spelling mistakes. Most normal readers misspell phonetically, that is they use phonetic rules in trying to spell an unfamiliar word. Some severely disabled readers, lacking a visual image of an unfamiliar word, also misspell phonetically. Others are able to spell familiar words correctly but, lacking the power to translate what they hear into visual symbols, they are at a loss as to how to proceed. Tables 2 and 3 show examples of those two kinds of mistake. Boder (1971) divides spelling errors into two classes: (1) dysphonetic errors which, although they might resemble the correct word are not phonetically similar; and (2) dyseidetic errors which may be phonetically similar but do not look like the test word. Since not all dysphonetic errors in fact look like the test word, and not all dyseidetic errors are phonetic, we designed a simple set of rules for classifying errors in any word as phonetic or dysphonetic. Table 4 shows that while normal readers rarely spell dysphonetically, disabled readers often do

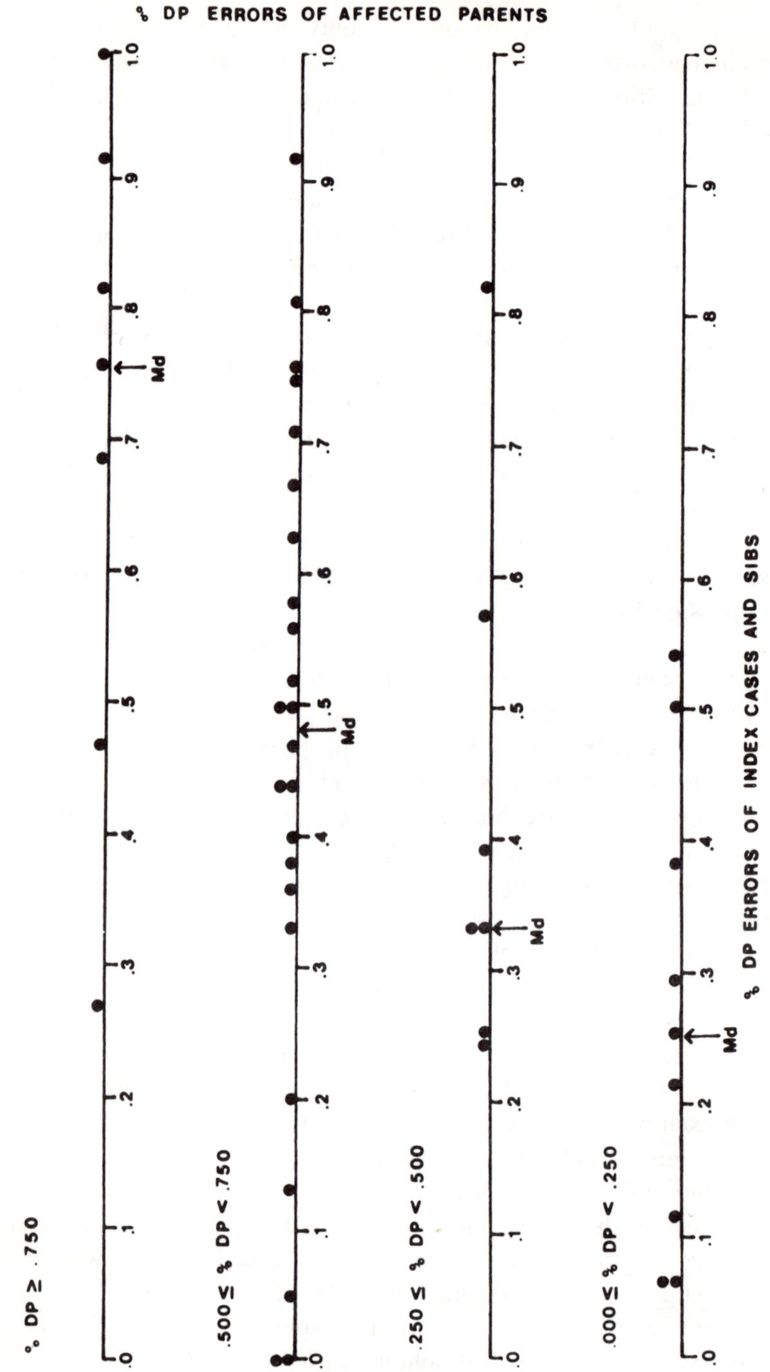

FIG. 2. Proportion of dysphonetic spelling errors according to spelling type of parents.

TABLE 2

An example of dysphonetic spelling: white male age 18 years, misspelled 44 of 46 words (WRAT)

Correct spelling	*Misspelling*
Courteous	Crulelus
Medieval	Meadivlold
Literature	Litlr
Majority	Mugord
Decision	Dsicun
Privilege	Pvlag
Irresistible	Arsittbal

TABLE 3

An example of phonetic spelling: white male age 15 years, misspelled 37 of 46 words (WRAT)

Correct spelling	*Misspelling*
Circle	Curcul
Precious	Preshus
Executive	Egzecutive
Exaggerate	Agzadurate
Ruin	Rouen
Majority	Magoraty
Physician	Phisision

so. Spelling and reading skills are highly correlated; no poor reader spells well, although some adequate readers may spell badly. We were interested in testing intrafamily likeness with regard to spelling type, and Fig. 2 shows the results of such a test: there is a strong relationship between the spelling type of disabled parents and their disabled children. This is not due to an association between severity of the reading disability and dysphonetic spelling, because when the proportion of dysphonetic spelling errors of index cases and sibs is considered in the light of the severity of reading disability in the parents, there is no correlation.

SEX DIFFERENCES

Multifactorial conditions often show sex differences in prevalence or age of onset; pyloric stenosis, hip dislocation and gout are examples. SRD shares this characteristic: males outnumber females (Critchley 1970). The sex ratio among index cases in our families was 40:7, and Table 5, which shows the sexes of sibs of these index cases, reveals that there are somewhat fewer females than males in the affected zone, and that the former tend to be more mildly affected. This is expected since females score higher than males on the verbal subtests of the Wechsler IQ tests, but the ratio of the sexes among the sibs is nothing like 40:7. The reason for this difference has not been settled, but may have to do with the greater severity of the handicap in males, their lower tolerance for frustration and higher parental expectation for sons than daughters. These may be among the reasons why boys come more frequently to the attention of teachers and other authorities than girls.

When multigenic diseases show sex differences, it has been observed that when the index case is of the less vulnerable sex, affected relatives tend to be more numerous and more severely affected (Carter 1976). In our families, this tendency is not fulfilled; the brothers of female index cases are neither more frequently, nor more severely, affected than in other families. Nor are there such differences between the offspring of disabled mothers and fathers.

TABLE 4

Proportion of dysphonetic (DP) spelling errors according to level of reading competence.

	Proportion DP errors		
	<0.50	≥0.50	Total
Good readers	43 (86%)	7 (14%)	50
Average readers	20 (87%)	3 (13%)	23
Disabled readers	13 (39%)	20 (61%)	33
	76	30	106

TABLE 5

Sex distributions of index cases and sibs with SRD.

Index cases	*Male* *40 (85%)*	*Female* *7 (15%)*	*Total* *47*
Total sibs	40	32	72
Borderline affected sibs	10 (25%)	8 (25%)	18 (25%)
Severely affected sibs	13 (33%)	6 (19%)	19 (26%)
All affected sibs	23 (58%)	14 (44%)	37 (51%)

THE NEUROLOGICAL BASIS OF READING DISABILITY

Benton (1975) has reviewed the literature on neurological correlates of reading disability and concludes that although existing empirical data do not establish any specific neurological basis for this handicap, such a hypothesis is more tenable than opposing theories which hold, with very little empirical evidence to support them, that bad habits or emotional disorder are at the root of it.

Theories of a neurological basis for specific reading disability are centred either on focal maldevelopment in the parietal lobes, or on defects in lateralization of cerebral functions. Although the literature is extensive in support of both these ideas, no conclusions can be drawn. Tests of visual evoked response have shown some promise in illuminating these theories (Preston 1978), but the results show average group differences rather than reliable individual differences which could be of help in the description of subtypes and in decisions about competing theories. It is possible that such decisions cannot be made, in this as in other aspects of reading and learning disabilities, until some consensus is reached about which subjects to study with which methods, until the techniques of study show reliable individual differences rather than group differences, until it is agreed that reading and learning disability are words that embrace, but do not define, a mixture of many conditions of diverse origin, and until some sort of context in which to do the studies can be agreed upon, so that paediatricians, neurologists, psychiatrists, psychologists, educators, disabled readers and their relatives and the general public can share their knowledge, thoughts and experiences. Until these conditions are met, the literature will continue to be a Babel of voices.

SOCIAL IMPLICATIONS

It is difficult for normal readers to appreciate the frustration, anxiety and often despair of children who are unable to do what comes so easily to others. For some, alienation and antisocial behaviour are the outcome (Critchley 1968; Poremba 1975). Others occupy socioeconomic positions beneath those predicted by their endowment, or compensate in various ways for their reading handicap and succeed in business or professionally (Rawson 1968). How many fill each of these levels, what factors account for individual fates and what might have been done to change the course of their lives is uncertain and should be studied. Questions like these are investigated as a matter of course in schizophrenia, affective disorder, alcoholism and other multifactorial conditions: why not in learning disabilities too? The answers are of overriding importance because these conditions represent the inability of some people to meet arbitrarily defined social conditions. Their handicap is not inevitable as Tay-Sachs disease is inevitable. If the educational establishment and society could accept diversity in the capacity to learn as they have accepted diversity in things to be learned, learning disabilities would largely disappear, and their victims would be freed of unnecessary social and psychological constraints.

ACKNOWLEDGEMENT

This work was supported by Grant HD 00486, National Institute of Child Health and Human Development.

References

BATEMAN, B. (1968) *Interpretation of The Illinois Test of Psycholinguistic Abilities,* Special Child Publication, Seattle

BENTON, A.L. (1975) Developmental dyslexia: neurological aspects. *Adv. Neurol. 7,* 1-47

BERGER, M., YULE, W. & RUTTER, M. (1975) Attainment and adjustment in two geographical areas. II. The prevalence of specific reading retardation. *Br.J. Psychiatry 126,* 510-519

BODER, E. (1971) Developmental dyslexia: prevailing diagnostic concepts and a new diagnostic approach. *Prog. Learn. Disabil. 2,* 293-321

BROADHURST, P.L., FULKER, D.W. & WILCOCK, J. (1974) Behavioral genetics. *Annu. Rev. Psychol. 25,* 389-414

BRYAN, T.S. (1978) Learning disabilities. *Pediatr. Ann. 7,* 306-313

CANTWELL, D.P. (1976) Genetic factors in the hyperkinetic syndrome. *J. Am. Acad. Child Psychiatry 15,* 214-233

CARTER, C.O. (1976) Genetics of common single malformations. *Br. Med. Bull. 32,* 21-26

CRITCHLEY, E.M.R. (1968) Reading retardation, dyslexia, and delinquency. *Br. J. Psychiatry 115,* 1537-1547

CRITCHLEY, M. (1970) *The Dyslexic Child,* Thomas, Springfield

DEFRIES, J.C., SINGER, S.M., FOCH, T.T. & LEWITTER, F.I. (1978) Familial nature of reading disability. *Br. J. Psychiatry 132,* 361-367

DENKLA, M.B. (1972) Performance on color task in kindergarten children. *Cortex 8,* 177-190

DOEHRING, D.G. & HOSHKO, I.M. (1977) Classification of reading problems by the Q-technique of factor analysis. *Cortex 13,* 281-294

EISENBERG, L. (1966) Reading retardation. I. Psychiatric and sociologic aspects. *Pediatrics 37,* 352-365

FAGLIONI, P., GATTI, B., PAGANONI, A.M. & ROBUTTI, A. (1969) A psychometric evaluation of developmental dyslexia in Italian children. *Cortex 5,* 15-26

FINUCCI, J.M. (1978) Genetic considerations in dyslexia. *Prog. Learn. Disabil. 4,* 41-63

FINUCCI, J.M., GUTHRIE, J.T., CHILDS, A.L., ABBEY, H. & CHILDS, B. (1976) The genetics of specific reading disability. *Ann. Hum. Genet. 40,* 1-23

GERSHON, E.S., TARGUM, S.D., KESSLER, L.R., MAZURE, C.M. & BUNNEY, W.E. JR. (1977) Genetic studies and biologic strategies in the affective disorders. *Prog. Med. Genet. 2,* 101-164

GOLDMAN, H. (1977) Editorial. *Schizophrenia Bull. 3,* 3

HALLGREN, B. (1950) Specific dyslexia: a clinical and genetic study. *Acta Psychiatr. Neurol. Scand. Suppl. 65,* 1-287

INGRAM, T., MASON, A. & BLACKBURN, I. (1970) A retrospective study of 82 children with reading disability. *Dev. Med. Child Neurol. 12,* 271-281

JOHNSON, D. & MYKLEBUST, H. (1967) *Learning Disabilities: Educational Principles and Practices,* Grune and Stratton, New York

KETY, S., ROSENTHAL, D., WENDER, P.H. SCHULSINGER, F. & JACOBSEN, B. (1975) Mental illness in the biological and adoptive families of adopted individuals who have become schizophrenic, *in Genetic Research in Psychiatry* (Fieve, R.R. *et al.,* eds.), pp. 147-165, Johns Hopkins Press, Baltimore

MAKITA, K. (1968) The rarity of reading disability in Japanese children. *Am. J. Orthopsychiatry 38,* 599-614

MATTIS, S., FRENCH, J.H. & RAPIN, I. (1975) Dyslexia in children and young adults: three independent neuropsychological syndromes. *Dev. Med. Child Neurol. 17,* 281-300

MYKLEBUST, H. (1968) Learning disabilities; definition and overview, in *Progress in Learning Disabilities,* vol. 1 (Myklebust, H., ed.), pp. 1-15, Grune & Stratton, New York

OWEN, F., ADAMS, P., FORREST, T., STOLZ, L. & FISHER, S. (1971) Learning disorders in children: sibling studies. *Monogr. Soc. Res. Child Dev. 36,* 1-77

PENROSE, L.S. (1938) A clinical and genetic study of 1280 cases of mental defect. *Special Report Series No. 229,* Medical Research Council, London

PENROSE, L.S. (1971) Psychiatric genetics. *Psychol. Med. 1,* 265-266

POREMBA, C.D. (1975) Learning disabilities, youth and delinquency: programs for intervention. *Prog. Learn. Disabil. 3,* 123-150

PRESTON, M.S. (1978) The use of evoked potential procedures in studies of reading disability, in *Evoked Brain Potentials and Behavior* (Begleiter, H., ed.), vol. 2, pp. 247-268, Plenum Press, New York

RAWSON, M.B. (1968) *Development Language Disability: Adult Accomplishments of Dyslexic Boys,* Johns Hopkins Press, Baltimore

RUTTER, M. & YULE, W. (1975) The concept of specific reading retardation. *J. Child Psychol. Psychiatry Allied Discipl. 16,* 181-197

RUTTER, M., YULE, B., QUINTON, D., ROWLANDS, O., YULE, W. & BERGER, M. (1975) Attainment and adjustment in two geographical areas. III. *Br. J. Psychiatry 126,* 520-533

SMITH, N.B. (1965) *American Reading Instruction,* International Reading Association

SMITH, S., KIMBERLING, W. & LUBS, H. (1977) Family studies in specific dyslexia. *Am. J. Hum. Genet. 29,* 101A

SULZBACHER, S.I. (1975) The learning disabled or hyperactive child. *J. Am. Med. Assoc. 234,* 938-941

TARNOPOL, L. & TARNOPOL, M. (1976) *Reading Disabilities: An International Perspective,* University Park Press, Baltimore

WALKER, L. & COLE, E. (1965) Familial patterns of expression of specific reading disability in a population sample. *Bull. Orton Soc. 15,* 12-24

YULE, W., RUTTER, M., BERGER, M. & THOMPSON, J. (1974) Over and under achievement in reading: distribution in the general population. *Br. J. Educ. Psychol. 44,* 1-12

ZERBIN-RÜDIN, E. (1967) Congenital word blindness. *Bull. Orton Soc. 17,* 47-54

Discussion

Wolstenholme: The low incidence of specific reading disability in Japan suggests that the whole process of reading Chinese and Japanese characters is quite different from reading in the western world. If children with a reading disability learn shorthand, do they read the shorthand symbols easily or do they continue to have difficulty?

Childs: I don't know. Is there a pictographic word-form that children find much easier to deal with than our alphabet?

Neel: The Japanese borrowed their written language from the Chinese but it does not correspond to the spoken language. A single explanation for the similar results for the Italians, Spanish and Japanese isn't going to be possible. There is now a phonetic version of Japanese (Katakana) but schools still teach the Chinese ideographs as well. There is a complete dissociation between the spoken and the written language.

Lachmann: Have you any data on the German-speaking Swiss, who speak and write different languages?

Childs: No.

Gurdon: The Japanese have three written languages while the Chinese have only one (Kanji), so the comparison you refer to would be much better discussed in terms of the Chinese. In the other two Japanese alphabets (Hiragana and Katakana), each symbol corresponds to a particular sound, as in ours.

Bodmer: Since in the Japanese language, where there is a total dissociation between the symbols and the sounds, the impression is that reading disability is rare, would it make sense to help these people in other populations by creating a hieroglyphic system like shorthand? How do blind people learn to read?

Childs: Blind people have great difficulty. It is harder to learn Braille than to learn ordinary reading.

Bodmer: Would they do better with a categorical alphabet which wasn't related to the phonetic variety?

Childs: That is what they have. The dots are not related to anything phonetic, are they?

Henry Harris: Braille is symbolic rather than phonetic.

Siniscalco: Has a twin study ever been made?

Childs: There has never been a really good twin study. The data from the most reliable among them show complete concordance for reading disability in single-ovum twins and much less for dizygotic twins. I have a feeling that a lot of the pairs came to the investigators because they were concordant.

Wolstenholme: What about music and the disabled reader?

Childs: I don't really know anything about that.

Cavalli-Sforza: There may be something wrong with the English language and with American schools. Most foreign post-doctoral students who come to my laboratory spell English better than the Americans.

Childs: Emphasis of certain kinds at school will be reflected in the performance of children. Spelling hasn't been fashionable for a long time but it is becoming important again.

Henry Harris: When you put a lot of effort into treating the severely handicapped children, do you get good results?

Childs: There is improvement but such people are likely to read slowly and spell badly for the rest of their lives.

Henry Harris: If this is a social problem, it won't be soluble simply by therapeutic exercises.

Childs: The social problem would be solved if the educational system were to accept that different people learn differently and that reading-disabled people can learn by auditory or visual means that don't involve reading. But schools will seldom allow children to take oral tests instead of written, or to tape lectures instead of reading texts. The handicap has to do with the expectations of society rather than with the disabled readers themselves.

Wolstenholme: I understand that the actress Susan Hampshire had a severe reading disability and took to acting in order to have a means of communication. She learned her roles from tapes.

Childs: That is a good example. Why shouldn't other people learn that way?

Scriver: The burden of the handicap is great in a society such as ours. You said that the principles learned from haemoglobinopathies, thalassaemia and phenylketonuria are pertinent to making a correct diagnosis of the reading disability and then designing remedies for it. There is another implication. Most infants in North America in the past 30 or 40 years were nourished not on human milk but on cow's milk. Recently it has been discovered that cow's milk is deficient in taurine (G.E. Gaull, personal communication). In some animals taurine is very important for the development of the nervous system, in particular the retina. It would be interesting to investigate whether reading disability is related to nutrition of the young infant with cow's milk.

Cavalli-Sforza: Have you done any work with EEG and evoked potentials?

Childs: Dr Malcolm Preston who works with us has been studying visual evoked responses. The subjects sit in a darkened room and are presented with visual stimuli, either light flashes or words, through a tachistoscope. Scalp electrodes are applied over the parietal and occipital areas. The responses to each stimulus are summed in a computer and emerge as a wave form with components which can be measured. In a series of normal subjects the response in the occipital areas to both light flashes and words were similar, which is perhaps not surprising, but in the parietal lobes responses to the word stimuli were much greater than to the light flashes. Over the left parietal lobe the response was greater than over the right parietal lobe, which coincides with the presence of the left angular gyrus where language is supposed to reside.

When Dr Preston tested people with reading disability the difference between the light flashes and the word responses was greatly diminished, more particularly over the left side. But because the method produces only group differences, and is unreliable for single individuals, test results have not been useful in defining subtypes. Dr Preston now has a far more elaborate programme which he hopes will elicit reliable individual differences.

Weatherall: Is there any other evidence of lateralization problems in the group with reading disability?

Childs: Two organic causes of specific reading disability have been suggested: (1) that there may be some structural variation in or around the angular gyrus, and (2) that there is incomplete lateralization. But there is insufficient evidence to favour either view.

Neel: You come down on the side of multifactorial inheritance, yet you have a ratio of 10 affected males to one affected female. It is true that in some traits attributed to multifactorial inheritance there are differences in the frequency in the two sexes and one talks about a difference in the threshold for phenotypic expression. But since the name of the game these days is heterogeneity, have you considered the possibility that there is a mixture of sex-linked inheritance and other more complicated modes? And have you picked out of the pedigrees those most consistent with sex-linked inheritance and then looked for a different pattern of dyslexia?

Childs: I used the word multifactorial because it is clear that the cultural conditions are important in this, but I hope I didn't imply that I came down exclusively on the side of multigenic inheritance. I thought that the data suggested that there could be some single gene types among the most severely disabled readers. We haven't found much to suggest sex linkage.

Bodmer: Ankylosing spondylitis provides a very close analogy to that sex difference. The sex ratio for this disease is somewhere between 4 and 10:1, males to females, where it is also HLA-B27-associated. But within the families the sex ratio is much less. In that way one picks up many of the less severe cases, whereas when you ascertain from the general population you are only picking out the more severe cases. In a sense when comparing family and population data you are not looking with the same bias towards severity in relation to sex.

On being an individual, or: the man in the red hat

CHARLES R. SCRIVER

Faculty of Science and Faculty of Medicine, McGill University, Montreal

Abstract If health is a state of equilibrium between intrinsic (genetic) functions and extrinsic (environmental) factors, then disease can be defined as a state of disequilibrium. Medicine has traditionally emphasized extrinsic factors in the origin of human diseases; medical genetics is concerned with the intrinsic factors (mutations) that either yield disease in the universal environment or constitute states of risk for individuals in particular (or universal) environments. Genetic screening is a process that defines specific risks for particular individuals. Screening is an ineffective activity if there is no participation by clients. Newborn (and homozygote) screening, on the basis of an experience involving 35 million infants, is usually considered as a 'successful' enterprise. But adult screening, usually for heterozygosity, is quantitatively a much more important activity for disease prevention and is much less 'successful' in its execution, judging from current experience. Comprehension of risk and perceived importance of biological individuality by potential participants and advocates are part of the problem. A major revision in the education of medical personnel and citizens is indicated if medical genetics is to achieve its goals.

Route 66 is a famous highway in America. Dreamers and travellers for decades have considered its possibilities and experienced its realities. Ciba Foundation Symposium No. 66 is a route map of sorts: it has charted human biology. A vista along the way is my contribution to this symposium. I have chosen a lookout on the reality of human biological individuality from where I can see the possibility of an evolving relationship between genetics and medicine.

Health maintenance is a fashionable topic today (Hamburg & Brown 1978). There is good reason for it to be. The World Health Organization has proclaimed the highest attainable standard of health to be the right of all persons (WHO 1960). Later, Justice Emmett Hall, for example, acknowledged that statement in his Royal Commission report on Health

Services in Canada (Hall 1964-65). Today our health system has assumed gargantuan proportions—with costs accordingly—and Canadians are forced to speak of 'health maintenance' in order to contain the costs of disease. We share that problem with other nations (Graubard & Knowles 1977). Our foe is disease and because his chief herald is thought to be our life style, we have mounted a campaign against the herald (Lalonde 1974). Of course, the commonsense theme in the life style campaign is needed, and it will yield some benefits, but before the war advances much farther it would be wise to examine the enemy and to consider the strategy that we have initiated to save our citizens. The collective has assumed that the strategy will benefit most individuals. Not yet asked are the questions: Who is the enemy? Does the collective reflect the individual?

ON BEING DIFFERENT

Knowledge of human genetics and the understanding it brings ought to benefit both the person and society. The person benefits through self-knowledge about his or her biological individuality; society benefits because the true basis of individuality is revealed. But that is a Utopian view. The more probable, collective perception of human individuality is subtly revealed in a poster I call *The Man in the Red Hat* by the Belgian artist Folon.

The artist conveys an ambivalent message. He has depicted 50 men; 49 wear blue hats; one wears a red hat. Three are looking at the odd fellow while the majority look blankly at us. We perceive that the man in the red hat is unmistakably different from his associates; he is recognizable in the crowd. However the three that look at him appear to be uneasy with their distinctive neighbour and the remainder seem to be avoiding him.

A human or medical geneticist can find a message or two in Folon's picture. First, it implies that to be different is, perhaps, to be blighted: conformity is more comfortable. Anyone who has evaluated the impact upon self-image of newly acquired knowledge about personal heterozygosity knows about the blighted-image phenomenon. Second, the picture suggests how recognition of individuality could be useful. Suppose the odd fellow is known to be at specific risk for some unpleasantness that, as yet, is unknown to himself. Because his individuality is recognizable, we can warn him without bothering anyone else. But if we had no clue to his identity, how could we help him? What could we do: warn the crowd as if its members were all at equal risk—or, if that were too daunting—give up? Aficionados of detective stories understand the value of a clue; so do geneticists in this case. The clues concern biological individuality, a phenomenon about which some of us, unfortunately, know little while others don't want to know. Yet a recognition of individuality is a vital strategic component that will be needed to win the campaign for health maintenance of the collective.

THE WAR OF MEDICINE: CAMPAIGNS

The traditional medical paradigm still teaches us to wage war on disease. A historical view reveals illnesses of the collective: epidemics and decimations. Fortunately, our campaigns against scourges were victorious. Sanitation and public health benefited society greatly, after which the causes of disease—the enemy—could be identified, and cures mobilized. Thus, *breakthroughs* were made against *killer* diseases: polio was *vanquished;* plague and smallpox *crushed; entrenched* infections *overcome;* nutritional diseases *eradicated.* We used the language of World War I to describe our campaigns and victories. With so many enemies driven from the field, the nature of contemporary disease changed. Now it is the degenerative processes, malignancies, hereditary ailments and birth defects that characterize the disease burden of modern society. Yet we keep the language of familiar experience. We *bomb* cancer with cobalt; and we *eradicate* the *killer* Tay-Sachs disease. But who now is the enemy? And where is the magic bullet that will vanquish him?

The medical paradigm makes soldiers of physicians. The traditional physician resembled an honourable knight. His campaign was conducted with great esteem and the contest between himself and the enemy was personalized, indeed immortalized in the painting *The Doctor* by Sir Luke Fildes. In the era of epidemic disease, the physician's campaign resembled that of Henry V at Agincourt: the vast enemy collapsed before a happy few. But today the knight is an anachronism and military battle is an immense and elaborate process, vastly expensive, impersonal, belittling and cruel (Keegan 1976). While still intended as a humane endeavour, the modern practice of medicine has many analogies with the modern field campaign. It is dependent on expensive technology; it is often impersonal and it can belittle the patient. Moreover the medical war is largely directed by medical generals from positions of power. They make policy, control budgets and design attacks on the high cost of health care. Yet even they hearken back wistfully to crusades and knightly endeavours. Moral views are offered about the conduct of the war, to which we should conform. A new life style is encouraged—indeed, recommended (Lalonde 1974; Knowles 1977). The new–old campaign must help to win the war against disease—which, because it persists in the face of earlier victory, has the taint of sin. But what other modes of battle can we expect as long as the prevailing paradigm seeks an extrinsic enemy?

Juvenal's words on Roman life, two millennia ago, have a familiar ring. Even then, premature heart attacks were immoral events related to life style.

What a grossly ravening maw
That man must have who dines off whole roast boar—a beast
Ordained for convivial feasting! But you'll pay the price
All too too soon my friend, when you undress and waddle
Into the bath! your belly full of undigested
Peacock-meat—a lightning heart attack, with no time to make your
final will.

(Juvenal, *The Sixteen Satires,* trans. by Peter Green. Penguin 1967)

I wonder about campaigns, either ancient or modern, that make natural disease immoral. But what alternative strategy is there? Perhaps it can be found in the genetic paradigm of health and disease.

A GENETIC PARADIGM OF HEALTH AND ILLNESS

Genetics recognizes a relationship between the organism, with its various biological functions, and the numerous environmental forces acting upon it (Childs 1977; Scriver 1976; Scriver *et al.* 1978*b*). When nature and nurture are in equilibrium, a state of fitness is maintained which, in human terms, is a state of health. Biological functions in their final analysis involve matter and energy, for which the ultimate source is the environment, and the interaction between organism and environment is likely to be very complex. In man, it is usually controlled by a number of gene products acting either in parallel or in series. When the extrinsic forces change, or when one or more of the genes controlling the intrinsic components of the relationship are modified by mutation, disequilibrium in the interaction between nature and nurture can result.

Medicine emphasizes the importance of extrinsic events in the origins of disease; genetics asks that intrinsic factors be recognized as well. Accordingly, disequilibrium, or disease, can be placed on a spectrum. At one end are diseases that result from particular environmental forces acting upon the universal genotype; at the other, disease occurs when particular genotypes are expressed in the universal environment. In the middle of the spectrum are conditions in which specific extrinsic and intrinsic factors interact to cause disequilibrium.

The model asks the physician to consider the relationships carefully. For example a fracture disables and its cause is largely extrinsic. Fractures belong at the 'environmental' end of the spectrum. Yet there are some persons who fracture their bones frequently because they have the Mendelian condition osteogenesis imperfecta. The physician who fails to recognize the intrinsic factor fails to interpret the debility or to advise the patient of potential risks to his offspring. Phenylketonuria belongs at the 'genetic' end of the spectrum; mutations of various types impair phenylalanine-4-monooxygenase in the affected homozygote, and mental retardation is the outcome in the absence of treatment. But disease occurs only because phenylalanine is an essential nutrient. When the extrinsic component in the relationship between the homozygote and the environment is regulated by dietary treatment the effect of mutation can be relaxed and mental retardation prevented.

Sometimes we observe a change in the relative importance of extrinsic and intrinsic origins of disease in the course of time. Rickets was endemic in the industrial societies of Europe during the 18th and 19th centuries. It was a disease of life style, reflecting atmospheric pollution and urbanization. It belonged at the environmental end of the spectrum. The discovery of vitamin D restored the extrinsic factor to a proper relationship and the prevalence of environmental rickets abated. Cases of rickets seen in the same clinics now have high heritability. Intrinsic factors that modify calcium or phosphorus homeostasis are largely responsible for rickets today and, more often than not, rickets is resistant, in some degree, to vitamin D therapy.

Such case studies of osteogenesis imperfecta, phenylketonuria and familial vitamin-D-resistant rickets are unlikely to make a great impression on the medical generals responsible for collective health maintenance. Nonetheless, recognition of an intrinsic basis for specific disease can be very important to the affected individual. Moreover, we have shown that it can also be cost-effective when the appropriate knowledge about a 'genetic' disease is applied (Clow *et al.* 1973). The phenylketonuria experience—in particular that of screening, diagnosis and treatment—is clearly a prototype for disease prevention in the new mode (Scriver 1974, 1977).

There are many lessons to be learned from the phenylketonuria experience, among which is the message of genetic heterogeneity. Phenylketonuria is a collection of mutations, and precise titration of the relationship between each patient and his environment is required to achieve a satisfactory outcome from treatment in each case.

Recognition of genetic heterogeneity is also relevant to more common—meaning less esoteric—disease. Take the example of premature myocardial infarction. About one in a million citizens will die prematurely because that person is a homozygote or genetic compound for type II hypercholesterolaemia. And a rare citizen may die prematurely because a bizarre diet exposes that person to cholesterol in great excess, as was the case with a chicken sexer known to me who was paid a bonus wage in eggs for his vital job and died in his 30s of atherosclerosis. These persons are extreme examples of, respectively, intrinsic and extrinsic causes of premature myocardial infarction. But we know that most coronary heart disease is not classed with either of these two extreme examples. The Seattle study (Goldstein *et al.* 1973*a, b*) reveals that about half of those with premature heart disease are hyperlipidaemic and 10% of these are heterozygous for type II hypercholesterolaemia. To possess such an allele is to be at risk for premature myocardial infarction; the event will arrive 11 years earlier than anticipated for the general population. A modified life style may remove some environmental insults, but it will not change the inherited risk in this population.

Only 0.2% of the population carries the type II hypercholesterolaemia allele, but these individuals become 5% of those with premature coronary heart disease. This streaming effect is not because they are immoral people consuming large amounts of wild boar and peacock meat. It is because they have a deviant process of absorptive endocytosis in the plasma membrane of their cells, and because endogenous cholesterol metabolism is disregulated (Goldstein & Brown 1977). Such persons cannot reset their cholesterol metabolism to normal levels by prudent living. Their hope lies ultimately in two activities: the development of a pharmacological agent that will control intracellular cholesterol metabolism and be independent of absorptive endocytosis for its cellular uptake; and a screening test to recognize those who need the drug.

These examples of disease reveal why it matters whether we can recognize men in red hats. They indicate that most individuals, except monozygotic twins, have a different and relative state of health because each of us exists in a unique relationship to the universal environment. That is a biological view of man that affirms long-held and enlightened political views.

MAPS OF MAN

There have been many maps of the human event. Cave drawings on the walls of ceremonial caves at Lascaux and Altamira reveal man in miserable effigy by comparison with magnificent animals. By contrast, Egyptian drawings show man in command of the environment, bending it to his will; giving reason for deification of the human fact. Vesalius challenged subsequent views that survived through the Middle Ages. In the spirit of Renaissance he dissected the matter of man and revealed details that became the foundation of medicine and human biology for centuries to follow. First microscopy, then physiology and biochemistry, took us into the interior of cells. The inter- and intracellular facts of human existence became known. And now we have genetic maps of man. Mendel's factors, named genes by Johannsen, are being assigned to the chromosomes and parts of chromosomes. Loci are no longer abstractions; they are actual places—specific genetic addresses— in keeping with the Latin word we use.

The nature of our genes is being revealed in many ways. For example, study of the inborn errors of metabolism has led to the discovery of genetic heterogeneity at many loci. Examination of the products of the α- and β-globin loci, mapped recently to chromosomes 16 and 11 respectively, has exposed at least 71 different mutant alleles at the α locus and 141 at the β locus. But is such heterogeneity pertinent to the understanding of the healthy person? Harris and his colleagues at the Galton Laboratory have shown us that it is. Their study of the products of 104 genes, by electrophoretic methods, uncovered extensive genetic variation at 32% of human loci (Harris 1976). Those loci are polymorphic, meaning that the frequency of the variant allele exceeds 1%; accordingly, the frequency of the particular heterozygote in the population exceeds 2%. Harris estimates the average heterozygosity per structural gene locus in man to be at least 6%. Interpretation of diversity is a different activity. The biological significance of diversity is caught up in the debate about neutralist versus selectionist evolution (Harris *et al.* 1977); its medical significance is sometines more easily encompassed.

If mutation—revealed as polymorphism—accounts for our individuality, we then realize that some of our intrinsic diversity may account for diseases that affect ourselves or our offspring. But we also realize that in diversity lies adaptability for the species. Each person is indeed adjusted to the universal environment in a different way and each has a different and relative state of health—for better or for worse.

GENETIC SCREENING: USE OF THE MAPS

Genetic screening, in the medical context, is a search for persons within a population who, because of their genetic constitution, may be at risk for disease in themselves or their descendants, either in the universal or in a particular environment (Committee for the Study of Inborn Errors etc. 1975). Thus, screening has three primary objectives. First to detect those at risk for their own health and for whom medical intervention might offset harmful effects of gene expression; second, to detect those at risk for passing on potentially harmful genes to their offspring, who might benefit from counselling about reproductive options; third, to enumerate the frequency of variation in a population and to investigate the biological significance and genetic epidemiology of the variation.

In the aggregate, the objective of genetic screening is to define and identify specific, intrinsic risk factors for particular persons. The paradigm is rather different from the public health mode of disease prevention. It recognizes that life style as a mode of health maintenance will not benefit all persons equally. It accepts that the enemy in the continuing war on disease may well be ourselves (Scriver *et al.* 1978*b*).

Genetic screening has been widely applied to the newborn for the prevention of mental retardation in conditions such as phenylketonuria. Thirty-five million infants have been screened and about 3000 probands have been identified and treated for phenylketonuria; the majority have benefited from the experience. Phenylketonuria screening has not been an uncontroversial activity nor has it always been well done everywhere (Scriver 1977). Yet, there is enough experience to show that voluntary screening is as effective as compulsory screening and that the quality of results is quite dependent on the organization and execution of the screening process (Committee for the Study of Inborn Errors etc. 1975; Clow *et al.* 1973; Scriver *et al.* 1977). These lessons are pertinent to future developments that might encompass heterozygotes.

There are over 2500 recognized Mendelian variants in man (McKusick 1978), many of which are silent in the heterozygote. Their biological and medical significance may become apparent only when paired with another allele at the same locus in an offspring or when expressed in a particular environment into which the heterozygote has entered.

Knowledge of genotype may allow a course of action that can prevent disease. If prevalent heterozygosity is the origin of some or, perhaps, much disease in modern society, it follows that screening for heterozygosity may become the major medical activity by which we can define specific risks to health for particular citizens. If that sublime state of genetic epidemiology were to come about, can we anticipate how citizens will feel when their individuality is declared; or how physicians and other health professionals will take to the new paradigm?

EDUCATION OF CITIZENS—AND PHYSICIANS

Rather few citizens or physicians show evidence that they perceive the real extent of human heterozygosity, or the particular benefits and risks to personal and collective health that it may confer. The evidence comes largely from surveys of knowledge and attitudes and from experience with Tay-Sachs and haemoglobin-S carrier screening. Data from the surveys and programmes reveal that physicians by their actions give genetic screening rather low priority (Beck *et al.* 1974). On the other hand, their practical behaviour is in contrast to their personal attitudes; in general, physicians favour voluntary genetic screening and the provision of other services (Rosenstock *et al.* 1975). The discordance between attitudes and behaviour is understandable; physicians tend not to perceive the existence of genetic factors in the actual medical and counselling problems they encounter, nor do they always possess the correct knowledge about genetic phenomena. A simple explanation exists for the gap between attitudes and performance of physicians in the area of medical genetics: few of those now in practice have had more than a rudimentary exposure to the discipline. Childs (1974) has examined this problem in detail and offers solutions. If this lacuna in medical education is soon filled, the provision will be timely; it is certainly needed.

The education of citizens is another matter. Most carriers when informed of their particular heterozygosity feel blighted initially. Black heterozygotes felt stigmatized by haemoglobin-S screening (Culliton 1972) and knew little or nothing of the evolutionary selective advantages in their heterozygosity (R. Murray, personal communication, 1977). Jews felt anxious about their Tay-Sachs heterozygosity, even though they perceived the medical advantages in the new self-knowledge (Clow & Scriver 1977). Again there is a gap between experience and theory.

Surveys of high-school biology students (Scriver *et al.* 1978*a*; Hickman *et al.* 1978) exposed to fine teachers and modern texts found that while they are aware of silent heterozygosity and its presence in mankind, they were far less likely to see themselves involved by it. The nature of biology education—in America at least—probably explains the deficits in the education of citizens in human genetics. The problem goes back to the Scopes Trial (Grabiner & Miller 1974) and continues in the present conflict between creationism and evolutionism in the school curriculum (Nelkin 1976). Current textbooks which determine curriculum content and classroom teaching contain relatively little human biology and less genetics (Scriver *et al.* 1978*a*). Furthermore the genetic material has a low human profile. On the other hand, student preferences strongly favour more emphasis in the curriculum on evolution, human biology and genetics. Small wonder, then, that students lack a good basic knowledge of genetics and are confused about the difference between a gene and a chromosome, for example. Nor is the fact that they are uncertain about how to use Mendel's laws in practical human situations surprising, even though they hold attitudes strongly favourable towards genetic screening and services in society (Scriver *et al.* 1978*a*; Hickman *et al.* 1978).

These findings describe contemporary students; the situation is likely to be worse for their parents and the evidence suggests that this is indeed so. In fact there is, in all likelihood, a national illiteracy in genetics. That may explain why citizens are not easily recruited into heterozygote screening programmes (Beck *et al.* 1974; Childs *et al.* 1976*a, b*) and why over-all participation rates are usually less than 20% of the eligible population. It also explains why citizens do not interpret the test result with objectivity when their personal heterozygosity is revealed to them.

The challenge ahead for the educators of citizens and health professionals is clear. It has been said that the importance of learning 'lies not in what meaning it makes of the world, but what it makes of the learner' (Allen, cited by Kennedy 1978). The goal in genetics education must surely be to have all of its students become 'more knowledgeable, independent and mature in their approaches to the problems of the modern world' (Kennedy 1978). To that end, an effort has begun to improve human genetics education and the textbooks that sustain it (Scriver *et al.* 1978*a*; Hickman *et al.* 1978).

POSSIBILITIES AND REALITIES

To improve individual and collective well-being, we must first utilize, to the optimum, the health services we have. Then we must improve them through the application of new knowledge (Comroe 1978). In the winter of our discontent about health maintenance and disease care, we have appealed to the individual's own responsibility for disease prevention (Knowles 1977). But if—as I have tried to show—the enemy of health is as likely to be within ourselves as without, then we must know ourselves in order to fulfil our responsibility. We must place extraordinary value on learning about our own biology and the genes that make us human. Awareness of our human dimensions through personal action and collective policy, could improve health maintenance, disease prevention and medical care. As I see it, this new dimension in public affairs would enhance and clarify the view of individuality given to us by the poet Housman:

From far, from eve and morning
And yon twelve-winded sky
The stuff of life to knit me
Blew hither, here am I.

ACKNOWLEDGEMENTS

I thank Barton Childs, Carol Clow, F. Clarke Fraser, Harry Harris and Claude Laberge in particular for formative discussion and work together. I have been supported in my work by the Medical Research Council of Canada, the Quebec Network of Genetic Medicine, The Department of National Health and Welfare (Canada) and the National Genetics Foundation (New York).

References

BECK, E., BLAICHMAN, S., SCRIVER, C.R. & CLOW, C.L. (1974) Advocacy and compliance in genetic screening. Behaviour of physicians and clients in a voluntary Tay-Sachs testing program. *N. Engl. J. Med. 291,* 1166-1170

CHILDS, B. (1974) William Allan Memorial Award Address. A place for genetics in health education and vice versa. *Am. J. Hum. Genet. 26,* 120-135

CHILDS, B. (1977) Persistent echoes of the nature-nurture argument. *Am. J. Hum. Genet. 29,* 1-13

CHILDS, B., GORDIS, L., KABACK, M.M. & KAZAZIAN, H.H. JR. (1976*a*) Tay-Sachs screening: motives for participating and knowledge of genetics and probability. *Am. J. Hum. Genet. 28,* 537-549

CHILDS, B., GORDIS, L., KABACK, M.M. & KAZAZIAN, H.H. JR. (1976*b*) Tay-Sachs screening: social and psychological impact. *Am. J. Hum. Genet. 28,* 550-558

CLOW, C.L. & SCRIVER, C.R. (1977) Knowledge about and attitudes toward genetic screening among high-school students: the Tay-Sachs experience. *Pediatrics 59,* 86-91

CLOW, C.L., FRASER, F.C., LABERGE, C. & SCRIVER, C.R. (1973) On the application of knowledge to the patient with genetic disease. *Prog. Med. Genet. 9,* 159-214

Committee for the Study of Inborn Errors of Metabolism. National Research Council (1975) *Genetic Screening Programs, Principles and Research,* National Academy of Sciences, Washington, D.C.

COMROE, J.H. JR. (1978) The road from research to new diagnosis and therapy. *Science (Wash. D.C.) 200,* 931-937

CULLITON, B. (1972) Sickle cell anemia: the route from obscurity to prominence. *Science (Wash. D.C.) 178,* 138-142

GOLDSTEIN, J.L. & BROWN, M.S. (1977) The LDL-cholesterol pathway. *Annu. Rev. Biochem. 46,* 897-930

GOLDSTEIN, J.L., HAZZARD, W.R., SCHROTT, H.G., BIERMAN, E.L. & MOTULSKY, A.G. (1973*a*) Hyperlipidemia in coronary heart disease. I. Lipid levels in 500 survivors of myocardial infarcta. *J. Clin. Invest. 52,* 1533-1543

GOLDSTEIN, J.L., SCHROTT, H.G., HAZZARD, W.R., BIERMAN, E.L. & MOTULSKY, A.G. (1973*b*) Hyperlipidemia in coronary heart disease. II. Genetic analysis of lipid levels in 176 families and delineation of a new inherited disorder, combined hyperlipidemia. *J. Clin. Invest. 52,* 1544-1568

GRABINER, J.V. & MILLER, P.D. (1974) Effects of the Scopes trial. *Science (Wash. D.C.) 185,* 832-837

GRAUBARD, S.G. & KNOWLES, J.H. (eds.) (1977) Doing better and feeling worse: health in the United States. *Daedalus (Boston) 106,* 1-278

HALL, E.M. (1964-65) *Canada Royal Commission on Health Services,* 2 vols. Queen's Printer, Ottawa

HAMBURG, D.A. & BROWN, S.S. (1978) The science base and social context of health maintenance. An overview. *Science (Wash. D.C.) 200,* 847-849

HARRIS, H. (1976) Enzyme variants in human populations. *Johns Hopkins Med. J. 138,* 245-252

HARRIS, H., HOPKINSON, D.A. & EDWARDS, Y.H. (1977) Polymorphism and the subunit structure of enzymes: a contribution to the neutralist selectionist controversy. *Proc. Natl. Acad. Sci. U.S.A. 74,* 697-701

HICKMAN, F.M., KENNEDY, M.H. & MCINERNEY, J.D. (1978) Human genetics education: results of BSCS needs assessment survey. *Am. Biol. Teach. 40,* 285-303

KEEGAN, J. (1976) *The Face of Battle,* Viking Press, New York

KENNEDY, M.H. (1978) A machine which they worship for its own sake. *Am. Biol. Teach. 40,* 316ff

KNOWLES, J.H. (1977) Responsibility for health. *Science (Wash. D.C.) 198,* 1103

LALONDE, M. (1974) *A New Perspective on the Health of Canadians. A Working Document,* Goverment of Canada, Ottawa

MCKUSICK, V.A. (1978) *Mendelian Inheritance in Man. Catalogs of Autosomal Dominant, Autosomal Recessive and X-linked Phenotypes,* 5th edn., Johns Hopkins University Press, Baltimore

NELKIN, D. (1976) The Science-Textbook controversies. *Sci. Am. 234* (4), 33-39

ROSENSTOCK, I.M., CHILDS, B. & SIMOPOULOS, A.P. (1975) *Genetics Screening. A Study of the Knowledge and Attitudes of Physicians,* National Academy of Sciences, Washington, D.C.

SCRIVER, C.R. (1974) PKU and beyond: when do costs exceed benefits? *Pediatrics 54,* 616-619

SCRIVER, C.R. (1976) Realized and potential neutralization of mutant genes in man by nutritional selection. *Fed. Proc. 35,* 2286-2290

SCRIVER, C.R. (1977) Screening, counseling and treatment for phenylketonuria: lessons learned—a précis, in *Genetic Counseling* (Lubs, H.A. & de la Cruz, F., eds.) pp. 253-262, Raven Press, New York

SCRIVER, C.R., LABERGE, C. & CLOW, C.L. (1977) Genetic screening and allied services. Structure, process and objectives, in *Medico-Social Management of Inherited Metabolic Diseases* (Raine, D.M., ed.), University Park Press, Baltimore

SCRIVER, C.R., SCRIVER, D.E., CLOW, C. & SCHOK, M. (1978*a*) The education of citizens: human genetics. *Am. Biol. Teach. 40,* 280-284

SCRIVER, C.R., LABERGE, C., CLOW, C.L. & FRASER, F.C. (1978*b*) Genetics and medicine, an evolving relationship. *Science (Wash. D.C.) 200,* 946-952

World Health Organization (1960) Constitution. WHO, Geneva, p.1

Discussion

Cavalli-Sforza: As long as there are differences between individuals from the same family one cannot really say whether phenylketonuria is genetic or not. What is the present status of evidence of genetic heterogeneity in phenylketonuria?

Scriver: There is clear evidence for heterogeneity at more than one locus. One locus controls phenylalanine 4-monooxygenase activity while another locus or set of loci determines dihydropteridine reductase activity and the synthesis of tetrahydrobiopterin, the essential co-factor used stoichiometrically with phenylanine in the conversion of the latter into tyrosine. The reductase-deficient type of phenylketonuria involves more than phenylalanine oxidation; it also perturbs the synthesis of 5-hydroxytryptophan and catecholamines in the brain. This particular form of phenylketonuria is not easily recognized. Simple dietary treatment with phenylalanine restriction cannot benefit the individual with this form of the disease. Accordingly, we have an example of genetic heterogeneity involving two loci in the interpretation of the phenylketonuria phenotype that might be diagnosed by newborn screening.

At the phenylalanine hydroxylase locus itself there is evidence for several different alleles causing another set of phenylketonurias. One type is transient for reasons which we don't know. It starts out with profound hyperphenylalaninaemia after birth and requires dietary treatment. Then there is transition back to adequate enzyme activity with a fall in blood phenylalanine concentration and a relaxation of dietary tolerance for phenylalanine. Another type is 'more relaxed' than the classical type, with respect to dietary phenylalanine intake. We have seen this form in three siblings. It is important to recognize this particular form because to treat it as if it were classical phenylketonuria would lead to phenylalanine deficiency and subsequent harm. Recent studies also indicate different levels or types of residual hydroxylase activity in the livers of patients with the different forms of phenylketonuria. So I think there is evidence for heterogeneity at a single locus as well as heterogeneity at different loci.

Bodmer: You emphasized the possible role of heterozygotes for comparatively rare recessives. What is the relative contribution of these sorts of heterozygotes to contributions from polymorphic genes? Some of those we study, such as HLA, may have major effects on disease susceptibility.

Scriver: That effect might be considerable too. You mentioned ankylosing spondylitis earlier (p. 376). Garrod talked about idiosyncratic responses to drugs. If the frequency of glucose-6-phosphate dehydrogenase deficiency in some populations reaches the dimensions of a polymorphism, then the frequency of aberrant responses to anti-malarial drugs in that population is directly related to the presence of the variant allele.

I mentioned familial hypercholesterolaemia in my paper because although the allele is not polymorphic in the general population the gene becomes polymorphic in the strict sense of the word in the population with premature coronary heart disease where its prevalence approaches 0.025. We might ask ourselves here whether we are learning enough from the current general genetic screening experience. For example, when students leave school, we don't do cholesterol typing with pedigree studies or whatever is required to tell them whether they are at risk and what medication they should take to reduce the risk of a premature heart attack. We have no experience yet with such screening, counselling, or the treatment modality. One of the yields of this symposium could be a start at thinking about such tests and their fuller implications.

Harry Harris: Unless you have the medication it is rather difficult.

Scriver: That is why we don't do the screening now. We lack two things: the medication and a simple screening test.

Weatherall: We don't know enough about the natural history of hypercholesterolaemia to be able to give solid advice to persons at risk. Furthermore, we don't know how this population would react to this kind of knowledge.

Scriver: Those are further questions to be asked. A good example of doing screening before full awareness of its implications is at hand is found in the sickle cell screening experience in the United States. It was initiated largely for political reasons. Prenatal diagnosis of the homozygous fetus was not possible at the time and probability decisions had to be taken by families instead of binary decisions about risks for at-risk couples. This led to profound disruption within some of the black communities. That is quite a different experience from the development and acceptance of Tay-Sachs screening in the Jewish community, where the various components of a complete prevention programme were available at the outset.

Recently we have asked Greek and Italian communities in Montreal whether they want thalassaemia screening now that prenatal diagnosis is available and it is possible to detect heterozygotes for β-thalassaemia. Our studies show that the people are ready for a screening and prevention programme. They know that the resources are available and we are about to start a programme at their request. We will continue to measure the kind of attitudes and responses we encounter.

Weatherall: This is an extraordinarily important question. There have certainly been pressures in both the USA and the UK for prenatal screening for sickle cell anaemia, yet not one good study has yet been reported of the natural history of sickle cell anaemia. We are thinking about prenatal diagnosis programmes before we even know the natural history of the illness.

Henry Harris: What are you going to do with the textbooks? Are you going to substitute human examples for all the examples taken from other species?

Scriver: We are going to do something on the basis of our findings. The teachers and the students want examples in their own image of the larger messages that are being provided in their biology textbooks. For example, a picture of a white gorilla is interesting but the students would rather have albinism presented to them in human terms.

Henry Harris: Harvey didn't work out the circulation of the blood on humans; and what we know about *Drosophila* chromosomes still makes our knowledge of human chromosomes look trivial. Besides, the classical experiments were done on *Drosophila*. I am all for a picture of a human albino, if you want one, but I wouldn't really recommend textbooks that left out the primary experiments that determined the nature of the subject.

Scriver: There is no reason why both kinds of example shouldn't be used.

Brenner: A much more severe conceptual problem is that we are still not very clear how to go from the molecular level to complex phenotypes. Many of our explanations may well be misleading. One might wait to revise the textbooks because the whole of biology may change. No one can confidently say that they know how the genes determine structure of the organism, except in principle.

Polani: Another practical aspect is that the task of counselling will be infinitely easier, and the outcome more useful, when people understand more. If one has to give biology lessons to every person one counsels, it gets very difficult and very unproductive. They cannot learn in one easy (or difficult) lesson when they hear information which affects them emotionally as well.

Crick: We need to find out what most people actually know. I had to brief a charming lady, who wanted to visit the Salk Institute, on the elements of molecular biology. I was astonished to find that she thought that molecules were made of cells. She made a very favourable impression at the Salk but Leslie Orgel discovered that she thought worms were generated spontaneously in apples. And at parties in Cambridge, which after all is a university town, you can meet people who do not know that the sun is a star. If you are going to revise the textbooks you should first study things like that. How can you advise people if they don't know the difference between a gene and a chromosome? It is these very simple things that we would like everybody to know. We know too much ourselves.

Scriver: We are not going to throw away the whole of biology just because we want to focus on human biology. Average citizens would like to have biology interpreted in language that they understand. They understand, for example, that one person looks different from the next unless they are identical twins, that somebody in the family died of a heart attack, and so on. Yet when we asked our students whether they had experienced sickle cell disease, 5% said they had sickle cell disease in their family, which is absolutely impossible in Montreal. One hopes to dispel such fantasies in the same way that we hope citizens will know about worms and the sun. What one wants to do is to bring information and experience together, so that the biology that people understand from their textbooks becomes also the biology of themselves.

Henry Harris: My position would be that you should simply give the best examples. If they happen to be human, that is a bonus.

Cavalli-Sforza: You implied that the worries of people who are found on screening to be carriers of Tay-Sachs disease, for example, could be avoided if they were well informed.

Scriver: Our own data (Clow & Scriver 1977) show that students have a fair understanding of how to use the information. Carriers have a much better knowledge about the Tay-Sachs gene and the disease than do non-carriers. The experience of being told that you are a carrier is a learning experience. One learns that this new information can be used productively. One learns that it isn't necessary to change one's marriage partner to avoid having a child with Tay-Sachs disease. Our follow-up studies clearly reveal that screened students are equipped to use the new knowledge. The screening experience in the school is a practical experiment in biology that they could not obtain from the pages of a textbook. The carriers discovered in themselves something that is a tangible risk. It was the first time in their lives that this had been made apparent to them. I would like to see this particular insight generalized, not to a level that causes useless anxiety about everything, but to an intelligent human use of that information in designing, applying and utilizing the social structures of medical care services.

Nossal: You mentioned society's new concern about costs and your talk was pervaded with a kind of infrastructure of cost–benefit. But another angle is what one might term the 'compliance' angle. One health risk that people know a lot about is smoking, yet the total cigarette consumption of western societies is still rising, although there may have been a slight beneficial shift in the curve, related to the lower tar content.

The two examples of screening in the United States are each rather special cases. Tay-Sachs disease occurs among a population of high intelligence and high socioeconomic stratum, and sickle cell trait carries this rather strange and special racist element. So what has happenend about screening for sickle cell trait in Greek villages where the trait is so common that nearly everyone has a relative or friend who has died of sickle cell anaemia? Three or four years ago it appeared that the screening programme in such villages had made absolutely no difference to mating patterns, to family numbers or to intervention in terms of abortion. Is that still true?

Scriver: David Weatherall can probably answer that. I would say they shouldn't have been asked to participate in the first instance. What could they have gained by learning that they are carriers in the absence of prenatal diagnosis or treatment? Let me give you an analogous answer to that question. I live in Quebec, where there are many devout practising Catholics. Tay-Sachs disease occurs as often in three large French-Canadian demes as it does in the whole of the Quebec Jewish population. The three demes have now been screened and carrier couples identified. The traditional culture of the French-

Canadian population might be expected to say 'thou shalt not screen, because abortion or prenatal diagnosis is not acceptable'. But the affected and at-risk individuals in these families said differently. They asked for screening and where indicated they followed through with prenatal diagnosis. Healthy children have thus been born because the option of abortion was available. A paradox perhaps, but a real situation. I don't know how the scenario will unfold in the Greek villages but I suspect that when the technology allows them to make binary decisions instead of probability decisions, these families will do the same as the French-Canadians.

Weatherall: I believe that the results of the Greek study were disappointing. Preliminary follow-up studies showed that virtually no difference had been made to the breeding patterns and Dr Stamatoyannopoulos was convinced that the programme had caused a considerable amount of anxiety and social disruption in the community.

Reference

CLOW, C.C. & SCRIVER, C.R. (1977) Knowledge about and attitudes toward genetic screening among high-school students: the Tay-Sachs experience. *Pediatrics 59,* 86-90

Molecular and genetic organization: the future

WALTER BODMER

Genetics Laboratory, Department of Biochemistry, University of Oxford

I am very diffident about summarizing this symposium, as everyone tells me that it is an impossible task. When asked to be a summarizer of a meeting, the one thing you shouldn't do, I believe, is actually to summarize the whole of it. So I thought that I should present you with a few notions, a few speculations, that in part follow on naturally from the earlier discussions and which also take account of the fact that 'possibilities', which refers to the future, is in the title of this symposium.

We have gone full circle from culture back to culture and I want to start a little bit after the culture, namely with the basic advances. One of the things that Sydney Brenner particularly emphasized at the beginning was the dramatic advance in our knowledge of genetic structure at the nucleotide level, namely all these really remarkable discoveries of overlapping genes, gene inserts and discontinuous genes. Pontecorvo, interestingly, gave some of us a little note from his book of some 20 years ago which reminded us that he had already then clearly predicted gene splicing and overlapping. These are really very dramatic developments that have a lot of implications for our understanding, eventually at all levels.

One aspect that we spent some time discussing (I, perhaps, more than anyone) was the evolutionary implications of these discoveries. Evolutionary discussions often arouse great interest because they tend to be inconclusive and so are often the basis for much argument. The molecular biological approach to understanding evolutionary questions is perhaps more important for understanding the evolutionary interrelationships within the genome of any one organism than for the more conventional problem of looking at evolutionary relationships between organisms. This is, for example, illustrated by the complement cascade. The question of what is the history of a genome is an extremely interesting and important one, that can only really be approached at the molecular level. At the molecular level, also, Francis Crick reminded us that the study of DNA intervening sequences, if they truly are not subject to selection, which I personally would doubt, might give us a handy tool in population genetics.

Our discussion has suggested new insights into the molecular genetic bases for the differences between prokaryotes and eukaryotes. The change from prokaryote to eukaryote is, perhaps, one of the major evolutionary transitions. It allowed the development of extreme flexibility in the control of expression of genes, which may underly all the complex processing of messenger RNA that we see in higher organisms and which seems to depend on the existence of the nuclear membrane. I sometimes wonder whether fungi might not sit a little more in the middle between prokaryotes and eukaryotes than is usually assumed, and whether they have the same extent of gene duplication, which I like to emphasize in terms of gene clustering.

Another problem worth thinking about is whether the genetic structure of the organism moulds the phenotype or whether the phenotype moulds the genetic structure. Is it the nature of the genetic organization which determines how the organism works rather than the other way round? For example, having got genes which can be duplicated, does the organism take advantage of that, rather than duplicating the genes because that is advantageous? In more conventional population genetic terms, for example, a comparable question is, how does the mating structure influence what goes on in the population?

Another problem that intrigues me, and which I think those who study protein structure need to consider more, is the saltatory evolution of new proteins from old ones. If you can take a nucleotide sequence, frame-shift it and produce a new protein which may have little relationship in terms of amino acid sequence to the previous one, can you put that to use? Can you take a more or less arbitrary amino acid sequence and have it do something, however inefficiently, and then gradually improve its efficiency? I think this relates to the question of genetic structure moulding the phenotype. If a gene is duplicated and then expressed after a frame-shift, can this new and apparently unrelated product be functionally useful—even, perhaps, in a way that is related to the function of the original gene product? Can the organism take advantage of what the genetics has given it rather than what the phenotype determines?

Moving away from the more speculative evolutionary questions, I think we come very naturally to the basic question that permeated some of the earlier discussions, namely of how to control gene expression. We don't really know what 'promoters' are in higher organisms, and we don't know much about switching genes on and off—both of which are most important questions for differentiation and development. I am sure we can expect to see very rapid and important advances here. A part of the answer is obviously going to lie in understanding, in much more detail, the nucleotide sequences of the genes of higher organisms. As I understand it, at the present rate of nucleotide sequencing, which I gather is about 100 per day, it would take perhaps 100 man or woman years to sequence the HLA region. I suppose, therefore, if you put 100 people to work on it, it would take one year! This does point to a problem, namely the need to analyse and understand these sequences. Outside biology great advances have been made in the design of microprocessors and computers, and in automation, and one wonders what these advances will contribute to our area of biology especially in the automation of sequencing and the analysis of the sequences produced. As has already been emphasized, we are going to produce lots of nucleotide sequences and we are going to have to be able to analyse them. I am sure that a more incisive and intelligent analysis of nucleotide sequences is going to become extremely important in the future.

Many of you might not agree with some of my conclusions about the number of gene clusters and the basic number of genetic functions. But I think that whether one looks at the genomic complexity in the way I tend to present it, in terms of gene clusters, or in terms of smaller amounts of functional DNA, one is inevitably led to the conclusion that the number of functional genetic units is not by any means as large as one might have been led to expect by the amount of DNA. The compromise between amount of DNA and number of genes that follows if one thinks in terms of gene clusters and functions and so regards the number of gene functions rather than the number of genes as being reduced, is reassuring. But the whole question of the number of basic genetic functions is an extremely important one. The actual numerology is very relevant for some of the questions that we were discussing towards the end of the symposium. If it is true that the number of basic genetic functions is of the order of thousands rather than hundreds of thousands, and complexity is achieved by making use of combinations of combinations, this has a number of fundamental implications. I think, first, that it has important implications for what the concept of a gene is in terms of human syndromes, and so in terms of the genetic diseases that Charles Scriver was talking about. He pointed out that McKusick's catalogue lists some 2500 mutants. Many of these may to some extent be repeats. On the other hand, if the number of basic genetic functions is really in the thousands, then the proportion of these functions already represented by mutants in the catalogue may be quite high. In other words, we may not be that far from having some knowledge of a fairly high proportion of the number of basic genetic functions. If that is true, it is an extremely important conclusion. A lot of us, I am sure, certainly including myself, have until recently assumed that we have only seen the absolute tip of the mutant iceberg and that we are not even within three orders of magnitude of finding representative mutants of all the genes; and that, maybe, is totally wrong if we think in terms of basic genetic functions. Perhaps we are much nearer than we thought we were to having a sample of mutants that is representative of many of the basic genetic functions we need to understand. Undoubtedly in that case some of the possibilities that Charles Scriver was mentioning, of screening mutants and making use of them, would be greatly simplified. It means, to my mind, that the chance of finding a gene which affects any particular attribute is much greater than one might at one time have thought.

This leads me into another area. As a geneticist, I am often embarrassed that the everyday questions that one is asked can rarely be answered. Why does little Johnny look like his grandmother? What is the chance if my mother had osteoarthritis that I will have it? And similar questions, many of which we know something about but can't answer precisely. The reason we can't answer them is because they deal with complex phenotypes whose genetic basis is poorly understood. Sydney Brenner at the start of the symposium and several times subsequently emphasized the problem of the definition of the complex phenotype, which I think is very fundamental. Those of us who have spent some time at the other end of the scale from the molecular end, looking at quantitative inheritance, are very conscious of this problem. Barton Childs provided an excellent example of the way in which one can try to dissect a complex phenotype from the top downwards rather than from the bottom upwards. This dissection is needed in the study of a whole variety of phenotypes, at whatever level, including many of the major chronic diseases. I think that submission to a model of quantitative inheritance based on variance analysis is a surrender of despair to complexity which shouldn't and needn't be made. It is even less necessary if the number of basic genetic functions is not so large, so that we really do have a chance of finding something out about most, or at least many of them. I have often thought about studying the inheritance of facial features. All of us, I am sure, look at faces and recognize certain elements in common and I wonder whether the 'Identikit' that the police use could be used for some sort of classification of facial features for genetic studies. Why does it seem that some people look like the way they behave—are there really any relationships between facial features and behaviour? How long will it be before we can approach the understanding of this sort of complexity? Charles Weissmann told us about the revisionist genetics of the 1970s. What will the revisionist genetics of the 1990s be like? Will it actually help us in any way to solve the problem of understanding the genetic bases of some of these extremely complex attributes?

I want to turn now to a somewhat different topic. There are often overtones in scientific discussions such as we have had, and there have been such overtones in the latter part of this symposium, concerning the need to be accountable and the need to be more applied. A group like this is all too familiar with the arguments for basic research, which is needed to solve the unknown problems, and the fact that there is no way in which we can have strategic or directed programmes to solve difficult problems if we don't yet know what to do. I said to someone that it is like trying to put a man on the moon but not knowing in which direction to shoot, and got the answer that it is even worse than that: we don't actually know whether there is a moon! However, I am beginning to wonder whether we shouldn't be just a little less dogmatic in that direction. There are some remarkable advances that underlie the subject of this symposium—the development of somatic cell genetics, recombinant DNA techniques and cloning of DNA sequences, rapid DNA sequencing, the ability to make monoclonal antibodies, and the manipulation of early embryos. We have to be careful not to overlook the possibilities for applications of these rather dramatic advances. In the case of cancer, for example, which has permeated our discussions, I think there are some definite possible applications. Cesar Milstein used the rather nice term 'cellular fingerprint' and surely the clear-cut identification of different types of cells in solid tumours is not a trivial matter, but may be achieved using monoclonal antibodies. Will, for example, the much-discussed idea of targeting a cytotoxic drug onto the cancer cell by an antibody become a reality in the not too distant future? Even if we don't find that Holy Grail of an absolutely unique attribute of the cancer cell, nevertheless by combining different properties such as the particular origin of the cell and its special ability to multiply, each of which may occur in different normal tissues but which may not normally occur together, we may end up by having the possibility of a unique attack on the tumour cell which leaves the normal cell alone.

Another question concerns the familial cancers, which were briefly mentioned by Paul Polani. Where are the responsible genes and what do they do? Can we understand their function at the cellular level? The remarkable specific chromosome rearrangements that George Klein mentioned raise the obvious question of what are the particular genes that are involved in these changes. Why is it that such changes are such an advantage for the development of the tumour? There is a whole lot of activity, which Gus Nossal would be able to say much more about than I could, in the area of tropical diseases, involving, for example, studies on the specificity of the protozoan antigens using monoclonal antibodies, and immune responses to them. There is relevant research on other chronic diseases such as the role of immune responses in rheumatoid arthritis, on hypercholesterolaemia predisposing to heart disease, and on the possible role of slow virus infections in senile dementia. It seems to me that there are now a number of areas where we are not quite so far as is sometimes said to be the case from a useful understanding of the problem and where there is just a hint, at least, of the direction in which to go.

We have heard a lot from Charles Scriver on the importance of human individuality. Whenever I give talks about genetics to schools, or to the general public, one of the major facts that seems worth emphasizing is the genetic individuality of all human beings. This fact is, I think, one of the most striking contributions of genetic studies to our culture and our knowledge about ourselves, and it is one to which Harry Harris has made a major contribution. This reminds us that we mustn't forget the cultural importance of understanding the biological and genetic basis of our own culture, which takes us back to some of the points that Luca Cavalli-Sforza was making at the beginning of the symposium. Our social problems may well be much more difficult to solve than the medical ones. They combine the biology with the complex structure of our society, about which we know even less than we know about the biology. The recent revolution in biology came from the introduction of the physical sciences, and indeed of physical scientists, into the biological sphere. It seems to me that, for dealing with the social questions, present and future, the revolution in sociology and economics which may be needed will come from the contributions of the biologists to the social and economic sphere. Maybe, unfortunately, this will have to depend on the slower process of migration of people rather than of ideas. Luca Cavalli-Sforza told us about the migration of farmers rather than their ideas. Maybe, similarly, it will have to be the biologists who go into sociology and economics and who will help us to solve those problems of the future (mostly not genetical), many of which I am sure will still be with us in 20 years' time.

Final general discussion

Neel: Looking to the future, one important topic which we have scarcely touched on is the whole area of mutation. The explosion of information about the extent of genetic repair mechanisms equals the explosion of information in the other fields that we have discussed. It now looks as if 80–90% of all of the primary would-be mutations are repaired by some device or another. I submit that it is not only of theoretical but also of very practical importance that we should understand these repair mechanisms and how to influence them one way or the other.

Recently I undertook to estimate the impact of mutation on human disease (Neel 1978). Most of the efforts to do this have centred their attention on dominant and sex-linked disorders and chromosomal defects, plus a cautious effort to make some allowance for the mutational contribution to multifactorial disorders and congenital defects. By this approach, for example, UNSCEAR (1977) estimated that about 2% of newborn infants were destined to develop genetic diseases maintained by mutational pressure. This estimate fails to take adequate account of all the 'null' diseases now being recognized, diseases due to the lack of an enzyme or a receptor or transport protein. It is difficult to believe that these are maintained by heterozygote advantage—they must reflect mutation pressure. In *Drosophila,* the ratio of mutation resulting in 'null' traits to electrophoretic traits is 5:1 (Mukai & Cockerham 1977). Even higher ratios have been observed in human and mouse cell lines (Milstein *et al.* 1974; Arthur *et al.* 1975; Adetugbo *et al.* 1977). By an indirect approach we have recently estimated that electrophoretic variants arise in Amerindians with a frequency of 1.6×10^{-5} per locus per generation (Neel & Rothman 1978). Let's make the conservative assumption, guided by the data just cited, that in humans nulls for any particular protein arise with a frequency of 2×10^{-5} per generation. These nulls of course can result from the mutation of operator-regulator loci as well as structural loci. Let's assume there are 5000 proteins whose absence results in serious disease, and that the mutations responsible for absence of these proteins have no heterozygote advantage and function as alleles. Then about 10% of all newly fertilized eggs should be homozygous for one or more of these alleles. What proportion of these reach term is unknown. It could be that, as in the case of chromosomal abnormalities, intrauterine loss is very high.

This is a rather soft estimate, but clearly recessives maintained by mutation pressure do make a major contribution to disease, and clearly the total impact of chromosomal and 'point' mutation to disease is greater than we have been accustomed to thinking. Coming back, now, to the subject of genetic repair mechanism, it would appear that if we could increase the efficiency of repair from 80 to 90%, or 90 to 95%, this in the long run would have a very significant effect on disease. Surely this is as important a field for future research as any we have been discussing. In particular, should the carriers of the genes responsible for defects in genetic repair mechanisms, such as were discussed by Dr Polani, be singled out for studies on how to influence repair mechanisms, on the thesis they may be the source of a disproportionate share of 'spontaneous' mutation?

Polani: Should we also be thinking about those mutants that may exist in the population and are perhaps resistant to carcinogenesis? I have in mind a parallel situation with that in *Micrococcus radiodurans.* People with psoriasis, for example, don't seem to get cancer very often. This is anecdotal, but a dermatologist in Australia (Kocsard 1976) has shown that solar keratosis, an allegedly pre-malignant condition, is quite common in the population at large but extremely rare in people with psoriasis. Perhaps we should put psoriasis cells through the radiation-sensitivity mill and test them with other mutagens as well.

Lachmann: Psoriasis patients treated with arsenic develop squamous carcinoma of the skin.

Polani: On the other hand they have been treated with tar for a long time and without much skin cancer, but one could argue that they will get their cancer anyway. Something similar may be happening with psoralen and 'black light'. The question is not whether they do or do not get cancers, but whether they get them quickly or slowly. Of course, nobody is immune from cancer, I am sure, but some may be more resistant than others.

Henry Harris: DNA synthesis in psoriatic cells does seem to be aberrant in some respects. I don't know whether these cells are aberrant in their DNA repair mechanisms. Your general point may well also apply to atherosclerosis, in which the primary lesion is probably a monoclonal proliferation of aberrant smooth muscle cells. If you want to cut down atherosclerosis, one way might be to cut down the mutagenic load. That would be a more intelligent approach than cutting down fat in the diet. Since there are individuals who repair DNA very badly, there might be individuals who repair DNA a good deal better than most, if we could find them.

Weatherall: We seem to have accepted that all the studies suggesting a high incidence of cancer in these families are really solid, but there are all sorts of problems about ascertainment. The family histories also show an increased incidence of other disorders like diabetes. How good is the current technology for picking out heterozygous carriers for the DNA repair defects? Could one do this more systematically?

Polani: One might do this prospectively. Or one might start from cancer subjects, testing for heterozygosity at the loci under suspicion. However, one can only pick up some heterozygotes, and at a very heavy price at the moment. New tests are being developed which should be productive and might even be useful for the type of more direct enquiry one has in mind.

Bodmer: One of the problems with psoriasis is the high correlation with HLA, which doesn't show any obvious pointer in the direction of the repair problem.

Polani: Quite true of course. I used psoriasis because it was the only example I could think of. Nevertheless we should be looking for these things, for these relative resistances to cancers, not necessarily in psoriatics. After all one could say that people with psoriasis don't get solar keratosis simply because they can exfoliate their sun-damaged skin very efficiently. This could be very useful to the psoriatic patients, but a trivial explanation for the relative lack of solar keratosis in psoriatics.

Siniscalco: A point of general importance concerns eugenics versus euphenics. Administrators will expect a book with a title such as this one to help them when they have to assign priorities in deciding what types of scientific activity to support. Since funds for such activity are more and more limited, shouldn't we discuss the crude question of where the money should go? Should it be spent on massive screening for rare genetic defects, or on understanding relationships between the genotype and susceptibility to the everyday diseases, or on efforts to correct 'euphenically' (as opposed to 'eugenically') genetic defects within the diseased tissues, within the individual and within the ecological environment? As a medical man, I am interested in curing rather than in eliminating the genetically handicapped newborn; as a geneticist, I wouldn't dare to decide what is 'good' or 'bad' for the generations to come. Of course, I agree with the idea of massive screening in given populations which are known to be at risk for a specific birth defect. I agree also that everyone has the right to be helped if a particular genetic problem arises in the family. But screening the general population for as many genetic defects as are known is a formidable and a risky task. Funds may be exhausted for little reward in many cases. Human resources are limited, too, so that if systematic screening has to be extended everywhere, there will be hardly anyone left to deal with the more difficult, but most important, research needed for a true medical solution of genetic diseases —that is, to make everyone's life bearable, no matter what genetic make-up nature has bestowed on us.

Childs: Most screening programmes can show that they are cost-beneficial. Far from taking people away from anything else, and taking more money, they conserve both.

Bodmer: It is also possible to automate them.

Childs: Screening is also every-day medical care. It is simply one of the ways that one ministers to the ills of the population. If the techniques were not cost-beneficial your kind of question might arise, but if they are cost-beneficial, it doesn't arise.

Harry Harris: I can't quite understand your antithesis between screening and euphenics. If you screen for phenylketonuria and treat those who are found with a low phenylalanine diet, that end of it is euphenics, isn't it?

Bodmer: The antithesis would be that if you only start treating a disease when it is first conventionally diagnosed, then you wouldn't need to screen; but that is often too late for treatment to be effective.

Brenner: Are you contrasting this with the elimination of those genes from the population? The word eugenics was mentioned.

Siniscalco: Yes.

Bodmer: The standard answer, as Charles Scriver pointed out, it that all of us are heterozygous for some bad genes and if you screen all those out, you get rid of all (or nearly all) of us. The scheme of eugenics for this purpose is totally unrealistic. It might possibly work for certain abnormal genes, but it can't have any major impact.

Lachmann: Surely the effect of eugenics is to preserve bad genes, not to eliminate them? If heterozygotes are advised to mate only with people who are not heterozygous for the same gene, then the loss of the mutant gene in the homozygote is prevented.

Crick: But if it is used together with abortion, that is not true.

Polani: The point just made relates to another form of screening. We screen in a very different way when we screen for neural tube defects in the general population, as it is now possible to do. This introduces another dimension in screening from that which pertains to the protection of an individual: here one is screening for something which, if detected, some people would find difficult, undesirable, unethical or counter to their religious beliefs, as it may logically lead to abortion. So one has to think of this type of screening in very different terms from the more conventional form of screening with which we are more familiar. Charles Scriver introduced this very clearly.

Henry Harris: I propose that we should remain aggressively crusader in our attitudes to cancer and atherosclerosis, so as to give people more options on the way they choose to die. We might, within the foreseeable future, actually be able to do something about these conditions, and I should like to have the option of dying of old age.

The other thing of very great importance is that we must seek to induce clever schoolchildren to take up biological science. In the 19th century nearly everybody admitted to hospital died of some form of suppuration. In England, a suppuration foundation was set up, and pus was collected from Land's End to John o' Groats. Its chemical composition was analysed, its crystals examined and so on. And while all this busy organized work was going on, there was a young French chemist working on diseases of silkworms, and that made all the difference.

Scriver: I have somewhat the same concern but in a different context. One problem is that knowledge isn't on the national agendas of most nations. However, the French-speaking part of Canada is rather exceptional in this regard. Our provincial government has invested in the idea that new knowledge can benefit health care. In the Quebec Network of Genetic Medicine (Scriver *et al.* 1978) part of the annual budget is committed to research and development. From that came the development of the first national programme of screening in the newborn period for a classic medical illness, namely congenital hypothyroidism. That model could be applied in many other places. As Dr Childs has shown, what one gets is largely determined by the social and medical care structures. One of the messages from this symposium is surely that knowledge is important to society.

Harry Harris: We all extrapolate from what we know and develop great schemes for the future, but we ought to remember the completely unexpected discoveries that may come. We have this recent dramatic example of the non-translated inserts in the middle of structural genes. Nobody predicted anything like this. Now it has been found, it is likely to open up quite new and previously unanticipated developments.

Siniscalco: That is exactly what may be jeopardized by massive world-wide routine screening for genetic differences, which requires a lot of money.

Scriver: The programme in Quebec costs the equivalent of $7 per live birth for screening, diagnosis, counselling, treatment through life, etc. The cost-saving in our programme exceeds the basic cost of Genetic Network. Some of the savings emerge in government funds directed to the Quebec Medical Research Council. That Council is now training a large number of fledgling scientists and it is supporting established investigators. A cost accountant could say that the money saved through the prevention of disease is now generating new knowledge.

Crick: I think you are at cross-purposes. Marcello's point is the investment for the future. The implication was that with a limited amount of talent, some of that talent should be used for the future and some of it now. It is not a complete answer to say something is cost-beneficial now; you also have to say what is the cost-benefit for future work. If you cut off future work you may have a good cost-benefit now but you will have a very poor return in the future. Clearly some balance is needed.

Bodmer: When screening works it becomes part of the health care system, not part of research. Health care spending is an absolute ocean compared to the drop that is spent on basic research. When screening goes into health care it is cost-effective there, not on the research side. One must always argue for the support of adequate basic research, so I don't see the relevance of your argument, Francis.

Crick: I agree with you, but the concern was that that wouldn't happen, that the money would be diverted from fundamental research and put into other things.

Bodmer: It comes out of a different pot.

Childs: But screening conserves manpower for the future. People are prevented from becoming sick, so they don't need care and other people are available for other services.

Siniscalco: My concern is based on another personal bias, that the solution to the genetic problem can never be narrowed down to a particular gene. It is always the entire genome which matters. If you screen for phenylketonuria before birth you don't actually know what that gene would have been in that particular genetic background. I wouldn't be sure of the rightness of spending large amounts of money world-wide to do one thing which may be the wrong thing. I have to learn what the correlation between a variety of genes and their environment means before taking steps like that.

Childs: That sounds like a straw man of some kind. That issue has been proved as far as phenylketonuria is concerned.

Siniscalco: Our ignorance is about what the genetic make-up means with respect to the environment within which the individual lives.

Brenner: Like François Jacob, I believe that complex biological systems may have structures so conditioned by their history that the simple logic we have been accustomed to in the natural sciences may fail in generating acceptable explanations. We may well have to acquire some more conceptual equipment, though no one knows exactly what this might be. Our experimental activity is going to generate lists, and the amount of information we will have will be enormous. We are going to see hundreds of thousands, perhaps millions, of sequences. Work with monoclonal antibodies will give us maps of cell surfaces and maps of cellular distribution in tissues. How will we be able to assimilate this information and integrate it? Not only would that be good for science but it will also be essential for those people who might wish to use the information in practice. Perhaps we could discuss these questions.

Ruddle: About a year ago some of us on a genetics advisory committee for NIH made some of the same statements. We are on the brink of a large information explosion, especially with regard to recombinant DNA-generated sequences, and there will be an enormous effort to piece it all together and analyse the pieces in various ways. We thought that one way of aiding that process, simplifying it to some extent, would be to institute libraries which would store and distribute the sequences themselves, and information related to the sequences, and which would also develop computer methods for handling the information. That recommendation is still on the table. A decision hasn't been made. There was not much enthusiasm when an advertisement or letter was put into *Science* giving the aims of this programme. It was a little premature perhaps.

Crick: It is going to be too late unless it is done soon.

Jacob: The same problem is going to happen with the monoclonal antibody. That is very important.

Ruddle: Yes, that is another angle. We already have some experience along these lines with a cell bank for human mutants. This has been cost-effective and has served to stimulate research which might otherwise not have been done.

Gurdon: If we just discuss genes as opposed to other properties of cells, then if it turns out that there are in fact 5000 genes and that the sequences that precede the genes have certain functions, this won't be so bad as you make out. Sequences need not be catalogued as an endless string of bases; particular sequences will have functions (e.g. a promoter), which can be used for classification.

Crick: I don't believe a word of it, John! First of all, if there are 5000 genes, a gene to Walter has 3×10^6 base pairs! Secondly, there are other animals, to say nothing of plants, so an awful lot of taxonomic information is going to come along. There is absolutely no doubt about the large amount of sequence data. And large numbers of people will want to use those data, although I profoundly hope that generalizations will emerge so that we don't always have to look at all the data.

Gurdon: I didn't mean that this information would not exist, and I assume that it will be possible for it to be stored somewhere. What I mean is that the problem is likely to become easier to comprehend than might be imagined at present. We don't all want to know the sequence of bases in every structural gene.

Crick: No, but my experience is that even if you don't want to know it for one reason, somebody else wants to know it for another reason. If only for evolutionary studies, Walter will certainly want to know it. If you want to see the evolution of the haemoglobin locus, think of the number of animals you could look at, how far back you can go, and in what detail you would have to look at them for just that one thing.

Bodmer: I am sure that this is a soluble problem. It is a question of realizing the problem and doing something about it.

Crick: I agree with Sydney that a number of other problems are related in that they involve storing a lot of data or a lot of things. François mentioned the practical problem not of storing the information but of having the actual clones and so on as a tool, which is certainly yet to come. Then let's take the simple problem, which is going to be a real problem by the year 2000, of identifying every single human protein. People are starting to do that, and that is just one species, without even thinking about where they are in the body. When we know all the components, we won't need all of them all the time, but a good subset will be needed in any one problem. We then have to study the interacting functions, and it is perfectly clear from the number of systems, including the nervous system, that new methods will have to be developed for handling these data by computers. We hope that new generalizations are going to come out. The more people who study neuroanatomy, for example, the more regular it seems compared to what was thought 30 years ago. But the functions of these structures are still almost totally obscure. We have heard a little about human behaviour at this symposium. We are going to see a big explosion in that direction. We have to realize that we cannot solve some of these problems with the small amounts of data that we had in the past.

Jacob: I think we are entering an era of what could be called 'molecular anatomy' and then we'll go to comparative anatomy before we go to physiology. Look at the brain peptides, for instance. Nobody can predict anything about these. It could turn out that sleep, anxiety, pain and all this mess is due to a number of peptides about which we can predict absolutely nothing. We shall just have to isolate the molecules, purify them, produce the sequences and try in the year 2000 or so to find the evolutionary road which led to those structures.

Brenner: I often have the gnawing feeling that there may well be nothing more to be said than the declaration of what exists. Compare, for example, the arrangements of the histone genes in *Drosophila* and in the sea urchin. Here we have five genes, whose protein products are highly conserved in evolution. In both organisms, the group of genes is tandemly repeated in the genome. The general patterns are the same. Nevertheless, the order of the genes and their orientations are very different and there is no simple way of converting one set into the other. I find this worrying; why is the unit so variable? It looks as though the moment we go beyond the cistron, literally anything can be found, and it is conceivable that higher level organization of sequences does not matter, so that everything will be found.

Scriver: We need to find one very important gene, synthesize the product and then make it widely available. That gene product is going to be called optimin—which I hope will be one of the enkephalins.

Bodmer: This is a terribly important, challenging and interesting problem. It may be a new problem for the molecular biologists but it is not a new problem in other fields, including population genetics, where we have already been dealing with information on a large scale, sorting it, analysing it and providing it for other people. It is not an insurmountable problem.

Crick: It has to be tackled. Computer storage and handling is going up at such a rate that we should be able to tackle even greater things than you have had to handle in the past.

Neel: Francis referred to the people who want to plot all the human proteins. We are currently working on this subject with the Andersons (1977). There is an enormous problem in data reduction. However, for 20 years the military and others have been working on the analysis of complex patterns resulting from monitoring by satellite. We believe that the programs developed for this purpose have a high transfer value for the reading of two-dimensional gels. If this proves to be the case, then automated gel reading may not be far away.

Henry Harris: When one serves, as I have done for many years, on a number of grant-giving bodies, one has the clear impression that the bottleneck is not money, but talent. It is a good day when one gets two or three intelligent and original applications that look feasible. The real problem is to get talented people to turn their attention to important problems.

Brenner: Suppose it is true that we may have to collect a vast amount of information about a given organism in order to understand it. Does this not raise questions about the choice of the organism we have to work on? It has been argued that there is only one interesting higher organism and that is human beings.

Bodmer: Surely the mistake would be to study, for example, only *Drosophila* and other similar insects. It must be all right to choose model organisms at the different levels that matter and concentrate on those. The choice between *Drosophila* and humans doesn't need to be made. You can use both.

Brenner: It is a choice one has to exercise. Maybe all of these are rather uninteresting dead ends in evolution and won't tell us anything about the things that really matter.

Henry Harris: Wouldn't it be plausible to say that you choose the material that you think might answer the problem you are interested in?

Brenner: That is easy to say but how do you define the problem to choose the material?

Henry Harris: You don't define it. You must play your hunches—as we always have.

Brenner: That leaves lots of people in trouble.

Crick: There is really a good case for not choosing here, there and all over the place. In the last century there was a tendency, because of their interest in evolutionary and taxonomic matters, for people to work on all sorts of organisms. What we do now is really quite sensible. People concentrate on a few organisms. They concentrate, for example, on *E. coli,* on *Drosophila,* on the mouse and on humans. There isn't much wrong with the present arrangement. In most fields people are not doing one thing here and one thing there and so on. Clearly you choose what you think is the best material. Nevertheless when you have chosen an organism, as *Drosophila* was chosen for rapid breeding and all the rest of it, you then find that it has unexpected dividends. It can be used for problems you didn't think of. So there is not a real problem here. You must have a selection of organisms at different levels. In addition we should have one or two plants.

Weissmann: So when you decide it is going to be humans, the question is, who is it going to be!

References

ADETUGBO, K., MILSTEIN, C. & SECHER, D. S. (1977) Molecular analysis of spontaneous somatic mutants. *Nature (Lond.) 265,* 299-304

ANDERSON, L. & ANDERSON, N. G. (1977) High resolution two-dimensional electrophoresis of human plasma proteins. *Proc. Natl. Acad. Sci. U.S.A. 74,* 5421-5425

ARTHUR, E., STEEL, C. M., EVANS, H. J., POVEY, S., WATSON, B. & HARRIS, H. (1975) Genetic studies on human lymphoblastoid cell lines: isozyme and cytogenetic heterogeneity in a cell line, with evidence for localization of the Pep A locus in man. *Ann. Hum. Genet. 39,* 33-42

KOCSARD, E. (1976) The rarity of solar keratoses in psoriatic patients: preliminary report. *Australas. J. Dermatol. 17,* 65-66

MILSTEIN, C., ADETUGBO, K., COWAN, N. J. & SECHER, D. S. (1974) Clonal variants of meloma cells, in *Progress in Immunology II,* vol. 1, *Immunochemical Aspects* (Brent, L. & Holborow, J., eds.), pp. 157-168, North-Holland, Amsterdam

MUKAI, T. & COCKERHAM, C. C. (1977) Spontaneous mutation rates at enzyme loci in *Drosophila melanogaster. Proc. Natl. Acad. Sci U.S.A. 74,* 2514-2517

NEEL, J. V. (1978) Mutation and disease in man. *Can. J. Genet. Cytol.,* in press

NEEL, J. V. & ROTHMAN, E. D. (1978) Some indirect estimates of mutation rates in tribal Amerindians. *Proc. Natl. Acad. Sci. U.S.A.,* in press

SCRIVER, C. R., LABERGE, C., CLOW, C. L. & CLARKE FRASER, F. (1978) Genetics and medicine: an evolving relationship. *Science (Wash. D.C.) 200,* 946-952

UNSCEAR (1977) *Sources and Effects of Ionizing Radiation,* pp. 725, United Nations, New York

Chairman's closing remarks

SYDNEY BRENNER

MRC Laboratory of Molecular Biology, University Medical School, Cambridge

In closing the meeting I do not think I can say much more than has been said many times already. Molecular biology, cell biology and developmental biology are all going to go through an enormously exciting period. The meeting has emphasized how rapidly we can obtain information using all the new techniques of construction, analysis, and assay. The questions that make us pause are those discussed by François Jacob in his essay in *Science* (1977). These are questions about the very nature of the objects we are studying. Is it true that if we just go on doing more of what we have done before, all questions will be answered?

If natural complex systems resemble any man-made structure we know that they would be full of compromises, patches and even useless appendages. Perhaps we should be hesitant to read into what we observe something of the logic of the system itself, if indeed it has any logic to speak of, apart from the fact that it survives. This is, of course, no ground for discussing the enormous gain that we may expect in our knowledge of the genetics of higher organisms, and, in particular, in the genetics of man. There will be practical dividends and even if this is only a small fraction of the accretion within the subject, that, in itself, would justify the effort.

I gain further reassurance from what I think is a general feature of science. Some problems are never solved; they simply vanish. Perhaps future biologists reading this symposium volume will just not understand why we looked on some questions as deep problems, and what worried us.

We now live and work in an era when the connections between science and society have become complex. Scientists should not promise society too much. In the past, we have been over-enthusiastic in making promises because of the special social contract we have with society. Society has defined the context in which our work has to be carried out and we have been compelled to play the game. Our promises have been made too easily, and although many of the predictions we make about applications are right, it is a lesson of the past that we are nearly always wrong about the time they will take.

I make these remarks because Sir Gordon Wolstenholme, in whose honour this conference was held, now mediates (in GMAG—the Genetic Manipulation Advisory Group) between some of us and society. I think we should now question the basis of the relationship between science and society. The evolution of this relationship has perhaps set too rigidly the stage on which many of us have to play our parts, often as fairly unskilled actors. There have been facile promises coupled with too rigid demands. But worst of all has been the transformation of scientific activity into the coinage of the realm. *A priori* cost-benefit analyses raise such wide questions that I, for one, do not know whether there is any reasonable way of dealing with them.

We have had a very exciting meeting and our symposium will be a landmark even if it only records our confused perception of the future of genetics and human biology.

Reference

JACOB, F. (1977) Evolution and tinkering. *Science (Wash. D.C.) 196,* 1161-1166

Index of authors

Entries in **bold** *type refer to papers, other entries to contributions to discussions*

Bodmer, W.F. 16, 19, 20, 21, 22, 41, 42, 43, 59, 74, 75, 77, 128, 129, 136, 137, 138, 141, 142, 143, 176, 178, 179, 182, 201, 202, 203, **205**, 223, 224, 225, 226, 227, 228, 229, 247, 248, 270, 271, 277, 279, 280, 331, 351, 353, 354, 374, 389, **395**, 403, 404, 406, 408, 409, 410

Brenner, S. **1**, 20, 21, 23, 42, 43, 45, 59, 60, 61, 74, 75, 77, 79, 128, 129, 135, 136, 137, 138, 139, 140, 141, 144, 174, 180, 183, 184, 200, 203, 204, 224, 225, 227, 228, 229, 275, 277, 278, 279, 280, 281, 327, 332, 391, 404, 406, 409, 410, **413**

Cavalli-Sforza, L.L. **5**, 16, 17, 18, 19, 20, 21, 22, 43, 144, 246, 374, 375, 388, 392

Childs, B. **359**, 374, 375, 376, 404, 406

Clegg, J.B. **147**

Crick, F.H.C. 17, 19, 21, 22, 39, 40, 41, 58, 60, 128, 129, 137, 141, 142, 143, 144, 145, 202, 203, 224, 225, 226, 227, 228, 229, 277, 278, 280, 291, 404, 406, 407, 408, 409, 410

DeRobertis, E.M. **63**

Finucci, J.M. **359**

Fincham, J.R.S. 41, 79, 137, 142, 184, 200, 228, 277, 278, 280, 281

Gardner, R.L. 75, 138, 140, 174, 175, 272, 330

Gurdon, J.B. 60, **63**, 73, 74, 75, 76, 77, 78, 79, 135, 136, 137, 138, 139, 140, 141, 142, 143, 145, 176, 200, 201, 273, 374, 407, 408

Harris, Harry 19, 20, 39, 179, **187**, 200, 201, 202, 226, 271, 328, 332, 390, 404, 405

Harris, Henry 20, 39, 43, 44, 61, 73, 74, 79, 129, 131, 136, 137, 139, 140, 141, 142, 143, 144, 180, 199, 200, 202, 225, 226, 246, 248, 279, 280, 307, **311**, 327, 328, 329, 330, 331, 332, 351, 374, 375, 390, 391, 402, 405, 409, 410

Hobart, M.J. **231**

Jacob, F. 23, 43, 60, 76, 139, 273, 275, 329, 330, 407, 408

Klein, G. 76, 130, 142, 271, 275, **335**, 349, 350, 351, 353, 355, 356

Koprowski, H. 77, 78, 131, 246, 266, 349, 350, 352

Lachmann, P.J. 20, 180, **231**, 245, 246, 247, 248, 249, 354, 374, 402, 404

Melton, D.A. **63**

Meyer, F. **47**

Milstein, C. 43, 79, 137, 182, 200, 202, 203, 223, 224, 225, 226, **251**, 270, 271, 272, 273, 279, 332

Müller, W. **47**

Neel, J.V. 43, 130, 176, 179, 184, 248, 277, 278, 279, 374, 376, 401, 409

Nossal, Sir Gustav 76, 131, 132, 175, 226, 245, 249, 269, 350, 392

Pasvol, G. **147**

Peto, R. 17, 18, 42, 43, 78, 127, 128, 129,

130, 138, 201, 202, 272, 280, 308, 309, 329, 350, 356
Pickering, Sir George 20
Polani, P. 41, **81**, 127, 128, 129, 130, 131, 132, 138, 141, 175, 177, 178, 203, 275, 280, 308, 355, 356, 391, 402, 403, 404
Pontecorvo, G. 59, 60, 128, 181, 228, 275, 308

Ruddle, F.H. 40, 59, 61, 75, 76, 138, 182, 183, 200, 203, 228, 272, 328, 354, 355, 407

Scriver, C.R. 132, 328, 375, **377**, 380, 389, 390, 391, 392, 405, 409
Secher, D.S. **251**
Siniscalco, M. 22, 131, 135, 136, 139, 140, 142, 143, 177, 184, 202, 247, **283**, 308, 309, 355, 374, 403, 405, 406
Springer, T. **251**

Taniguchi, T. **47**

Walker, P.M.B. **25**, 39, 40, 41, 43, 45, 127, 144, 145, 280
Weatherall, D.J. **147**, 174, 175, 176, 177, 178, 179, 180, 181, 182, 183, 246, 376, 390, 393, 403
Weber, H. **47**
Weissmann, C. 42, 44, 45, **47**, 59, 60, 61, 74, 77, 79, 245, 279, 410
Wolstenholme, Sir Gordon 373, 374, 375
Wood, W.G. **147**

Indexes compiled by William Hill

Subject index

α-amanitin
68, 78
aconitase
193
adenosine deaminase deficiency
101
adenovirus
216
leader sequences in 33
alpha fetoprotein in ataxia telangiectasia
99, 101
aflatoxin
95
agriculture
12, 14, 19, 21, 22
alcohol dehydrogenase
192
aldolase
192
alkaline phosphatase
192
alkylating agents
95
alleles
218
control functions of 218, 223
amethopterin (methotrexate)
34, 40, 297
amino acids in haemoglobin
179
amino-fluorene derivatives
95
amylase
192
ankylosing spondylitis
211, 376, 389
antibody
chains 252
genes in myeloma hybrid cells 253
antibody-producing lines
derivation by cell fusion 254
antibody production
in hybrid myeloma 257
antibodies, pure
from impure antigens 256
antibody response
251
antigens
Forssman 260
in HLA system 208
antigens, differentiation
258
antigens, impure
pure antibodies from 256
antigens, surface
see surface antigens
asparaginase
297
ataxia telangiectasia
98-103, 113, 296
alpha fetoprotein in 99
basic defect in 99
cancer and 98
carriers 117
chromosome anomalies in 130
chromosome breaks in 99
chromosome 14 involvement in 102
clinical aspects of 98
immune deficiency in 116
incidence of 98
neurological defect in 99, 101, 131
X-radiation and 100
atherosclerosis
405

bacterial genes, mutations in
179
bacteriophage Qβ RNA
carrying A-G transition 51
ribosome binding capacity 53
site-directed mutagenesis in 48
with A-G substitution 49
with G-A substitution 51
with mutations in coat cistron initiator region 52
bacteriophage ΦX174
nucleotide sequences in 216
behaviour
genetics of 359
benzanthracene derivatives
95
β globin complementary DNA
53
biological individuality
378
biological transmission
6
blood group antigens
247

Bloom cells
303, 305
Bloom syndrome
103-107, 113, 129, 308
basic defect in 106
cell transformation in 130
chromosome anomalies in 104, 105, 106
chromosome exchange in 129
DNA repair defect in 106
immune deficiency in 116
immunological defect in 104
incidence of 103
sister chromatid exchange in 105, 107, 130
bone marrow transplantation
in sickle cell disease 168
Burkitt's lymphoma
102, 103, 339
aetiology of 341
age distribution of 356
banding patterns in 340
chromosome 14 in 341
cytogenic features of 340
geographical pathology of 342, 353
immune response in 356
immunotherapy 299
interaction of EB virus with B cells 335
location of 356
malaria and 354
recurrence of 350
stages of development 343
T cells and 350

caffeine
94
cancer
analysis by cell fusion 311-333
as dominant trait 353
ataxia telangiectasia and 98
attitudes to 405
chromosomal changes in 355
complement deficiency and 245
DNA repair defect and 97
familial 399
Fanconi anaemia and 109
gene and chromosome repair defects in 86
gene mapping and 283-309
genetics and 399
induction of 130
mutational hypotheses 115
of immunoglobulin-producing cells 252
production stages 309
step theories 115
viruses in 115
xeroderma pigmentosum and 96, 115
cancer cells
abnormality of membrane glycoprotein in 312, 317, 325
chromosome mutations in 298
chromosomes in 321
compared with normal cells 320
growth of 297
membrane proteins in 313
origin of 116
response to insulin 328
selective killing of 299
sister chromatid exchange in 307
trisomic alterations in 298
virus-induced 335-357
cancer therapy
296
carbohydrates
monoclonal antibodies against 270, 271
carcinogenesis
402
cell fusion
137
analysis of malignancy with 311-333
derivation of antibody-producing lines by 254
experiments on haemoglobins 148
techniques 82, 89, 264
cell hybrids
82, 136, 139
complementation for sister chromatid exchange in 299
composition of membranes 312
disruption of linkage groups 291
G6PD 292
gene mapping and 284
human—mouse 330
influenza 266
melanoma 268
monoclonal antibodies from 267
mouse—chick 331
mouse tumour 312
myeloma 253, 254, 262, 266, 272
normal and malignant 311, 352
production of 285
spleen 263, 268
stability of 272
tumorigenicity 352
cell proliferation
85
cell surface antigens
251-276
cellular fingerprints
399
centric fusions
83, 85
chemotherapy for cancer
296
chromatid segregation
86
chromosomal DNA
60
β-globin-specific 44
chromosomes
assignments in multilocus enzymes 193, 202
carcinogenic agents affecting 349
changes in T cell lymphoma 343
fusion of 84
genes assigned to 287
heterochromatin in 27
in cancer cells 298, 321, 355
in hereditary persistence of fetal haemoglobin 161

in juvenile myeloid leukaemia 177
insertion of foreign DNA in 75
of mice and men 273
size of 31
chromosomes, acrocentric 83
chromosome anomalies
in ataxia telangiectasia 130
in Bloom syndrome 104, 105, 106, 129
in Fanconi anaemia 108
incidence of 83
chromosome breaks 36, 114, 196, 278
in ataxia telangiectasia 99
in Fanconi syndrome 108, 109, 110
radiation causing 144
chromosome instability disorders 81-133
chromosome repair defects
malignancy and 86, 114
chromosome 8, 354
chromosome 11 determinants
antibodies against 270
chromosome 14
in ataxia telangiectasia 102, 131
in Burkitt's lymphoma 341
peculiarity of 131
chromosome 15, asymmetry of 127
coeliac disease 211
complement
in HLA system 209, 215, 227
in malaria 354
loci coding for 233
complement receptors
EB virus receptors and 349
in lymphocytes 354
complement system 231-250
clinical associations of deficiency 240
deficiencies 239
deficiency and cancer 245
enzymes in 236
factors 232
functional homologies 248
gene duplications and 236
inflammation and 235
isolated component deficiencies 239
loci 238
pathways 233, 236, 239
polymorphism in 238, 246
terminology of 231
unassigned loci 235
complement 3 convertase 236
concanavalin A binding 313, 315, 320, 329
congenital word blindness
see specific reading disability
conservation of linkage 203
convergent evolution 345
coronary heart disease 382
creatine kinase 192
cultural change, relevance of 5-23
role of selection in 9
cultural traits 10
culture 2
adaptive significance of traits 10
conservation of 8
diffusion of 13
evolutionary aspects of 5, 20
homogeneity 7, 8
interaction with genotype 16
retention of traits 10
stability of 7
transmission 6, 8
darwinian fitness 9
Daudi tumour cells 320
dengue shock syndrome 246, 248
DNA
see also under types, satellite DNA, etc.
amount in genome 220
analysis of small segments of 67
content of genome 277
differing configuration 67
functions of 22
gene number and 397
highly repetitive 41
in spacers 27
injection in oocytes 64
insertion into plasmids 67, 70
methylation of 73, 78
modifications of 48
organization of 25
origins of replication in 42
photons absorbed by 90
radiation damage 280, 291, 292
SV40 76, 77
transcription after injection 77
variability 27
DNA, cloned injection of 60
DNA damage
cancer and 115
radiation and 280, 291, 292
u.v. 89
DNA, EcoRI-resistant Pβ G 54, 55
DNA, foreign
insertion into chromosomes 75
DNA, non-coding 30
DNA repair
caffeine affecting 94
disruption of linkage groups as 291

DNA repair defects
81-133
cancer and 97
carriers of 117, 403
in Bloom syndrome 106
in Fanconi anaemia 111
in xeroderma pigmentosum 89, 93
DNA replication
in Bloom syndrome 106
mutations and 86
DNA sequences
1
expansion mechanisms 42
genetic functions and 219
homogeneity of 31
in antibodies 252
inserted 29, 37
insertion in plasmids 70
in spacers 31
middle repetitive 28, 29
of Y chromosomes 43
promoter 58
range of properties 30
recombinant 407
regional functions 2
regulation and 144
regulatory molecules 79
repetitive 29, 33, 34
representative 27
variability 26
DNA sequences, non-coding
evolution and 39, 42
genes and 25-45
properties of 30
role of 37
DNA synthesis
defect in 93
in psoriasis 402
DNA transcription
76
in injected oocytes 65
diabetes
insulin-resistant 98
juvenile onset 211
differentiated functions
262
differentiation antigens
258
monoclonal antibodies to 260
quantitative expression of 261
dihydrofolate reductase gene
34, 36, 40
dimethylbenzanthracene
349
dimethylsulphoxide
59
disease
changes in 379
intrinsic and extrinsic origins 381
mutations and 402
domains
279
domestication of plants and animals
12

EcoRI resistant PBG DNA
54, 55
electrophoresis
187, 188
endonucleases
92
endolase
192
enzymes
complement system and 236
enzymes, multilocus
see multilocus enzymes
Epstein-Barr (EB) virus
102, 130, 353
biology of 336
in aetiology of Burkitt's lymphoma 342
interaction with B lymphocytes 335
isolation of 336
oncogenic in monkeys 338
phenotypic changes in transformation 337
receptors 349
related diseases 338
relation between receptors and complement 349
transformed lymphoblastoid cell lines 337, 354
E. coli
gene clusters in 225, 226
mutation rate 280
ethane methane sulphonate
278
eugenics and euphenics
403
eukaryotes
228
change from prokaryotes 396
evolution
39
allelism in 218
convergent 345
expansion and deletion in 224, 225
gene duplication in 197, 206, 213, 224
gene overlapping and splicing and 395
histone genes in 409
HLA system and 228
introns and 44
multilocus enzymes in 201, 202
of haemoglobins 214
of immunoglobins 214
of new gene functions 212
excision XP cells
93, 94
exonuclease
92
expansion and deletion in evolution
224, 225

Fanconi anaemia
108-113
cancer and 109
chromosome instability in 108, 109, 110
DNA repair defect 111
immune deficiency in 116
molecular nature of defect 110
sister chromatid exchange in 112
transformability of cells in 110
fetal erythrocytes
survival of 176
fetal haemoglobin
174
hereditary persistence of 160-164, 168, 179, 181
increase in production 164

in pregnancy 177
in sickle cell disease 164
production of 164, 170, 177
fibroblasts
330
fluorescence banding
82
food production
12
Forssman antigen
260
Friend's erythroleukaemia
329

genes
amount of DNA and 219
assigned to chromosomes 287
geographical factors 13, 17
jumping 281
nature of 383
non-coding DNA sequences and 25-45
overlapping 395
re-expression of 139
regulation of 25
relation to protein structure 195
return to organisms 60
spread of 17
structural 187
gene clusters
397
complex organization of 215
evolution of 212, 217
expression 214
HLA system and 205-229
in *E.coli* 225, 226
radiation and 280
regulation 225
significance of 206
gene coding
for myoglobin 197
for RNA 214
gene control
polymorphism for 220
gene dosage method
297
gene duplication
complement system and 236
evolution and 197, 206, 213, 224
expression of 218
followed by mutation 195
in multilocus enzymes 194
origin of 195
sister chromatid exchange and 202
tandem 196, 203, 220
gene expression
136
in clusters 214
regulation of 33
gene frequencies in HLA system
207, 209
gene inactivation
272
gene injection, genetic analysis by
64
gene mapping
cancer and 283-309
cancer therapy and 296
cell hybrids and 284
progress of 283
gene multiplication
200
gene number
DNA and 397
gene products
2
gene repair defects
malignancy and 86, 114
gene repression
272
gene transfer in thalassaemia
183
genetics
education in 384
relation to medicine 377, 380
genetic analysis by gene injection
64
genetic changes, spontaneous
219
genetic counselling
391, 398
genetic engineering
2
genetic heterogeneity
381, 388
genetic organization
future possibilities 395
of higher organisms 220
genetic repair mechanism
402
genetic screening
381, 383, 390, 392, 404, 405, 406
genius
16
genomes
187
DNA in 220, 227
manipulation of 2
mapping 286
genotypes
drift 17
interaction with culture 16
glomerulonephritis
240, 241
glucose transport systems
323, 324, 328, 332
glucose-6-phosphate deficiency
179
glutamate-oxalate transaminase
193
glycoprotein
abnormality in cancer cell membrane 312, 317, 325
as glucose transport system 323, 324, 328
in cell membranes 328
mannose-rich 330
monoclonal antibodies against 271
purification in cancer cell membranes 325
glyoxylase
228
gonococcal sepsis
240
guanine analogues
298

haemochromatosis
211
haemoglobin
α-chain variants 153
amino acid deletion in 179

haemoglobin, *continued*
β-chain variants 153
cell fusion experiments 148
chains in 180
developmental biology of 169
evolution and 214
excessive chain production 181
fetal *see fetal haemoglobin*
globin chain synthesis 168
loci 150, 197, 408
molecular mechanisms of structural variants 155
number of globin genes 148
pathophysiology of structural variants 155
structural variants 152
structure of 147
switch from fetal to adult 169, 174, 176
unequal crossing over in 154, 174
with elongated chains 153
haemoglobin expression
74
dimethylsulphoxide inducing 59
haemoglobin genetics
147-185
future developments and problems 167
organization of 147
haemoglobin genome
150
haemoglobin polymorphisms
166
haemoglobin production
152
haemoglobin variants
179, 181, 224
health
disease and 380
maintenance of 377, 386, 405
HeLa cells
143, 320, 327
hepatitis
211
hereditary angio-oedema
241, 242
heterochromatic regions, mass of
143
heterochromatin
27
enlarged regions of 36
function of 32
separation of 143
hexoseaminidase
214
histocompatibility systems
205
see also HLA system
histones
202
histone genes
215, 409
HLA system
206
antigens 208
complement components 209, 215, 227
definition 206
development of 209
disease and 211
evolution and 228
gene clusters and 205-229
gene frequencies 207, 209
immune response genes 211
loci in 207, 273
mapping 210
molecular structure of gene products 211
recombination frequencies 209
hybrid cells
see cell hybrids
hypercholesterolaemia
382, 389
hyperpigmentation
108
hypothyroidism
405
hypoxanthine phosphoribosyltransferase
61

immortal strand hypothesis
86, 107, 128
immune complex disease
240, 241, 242, 246, 249
immune deficiency
116
immune response
351
in Burkitt's lymphoma 356
immune response genes in HLA system
211
immunization
269
immunoglobulins
137
allelic products 219
chains 226
in Bloom syndrome 104
in evolution 214
structural variants 182
tetraploidization 201
immunoglobin-producing cells, cancer of
252
immunotherapy for cancer
296
individuality
377, 378, 400
infectious disease
243
infectious mononucleosis
338
inflammation
235
influenza virus
268
insertion sequences
217
insulin
binding sites 325
cancer cell response to 328
receptors 323
introns
44, 45

keratoacanthomas
88
kuru
21

lactate dehydrogenase
188, 204, 214
forms of 195
gene duplication 198
tissue differences in expression 192

language
7
learning disabilities
aetiology of 360
definition of 360
familial properties 361
genetics of 359-376
leukaemia
B cell 270
chronic myeloid 296
Fanconi anaemia and 109
juvenile myeloid 177
Lewis blood groups
234, 247
linkage
interaction with selection 213
linkage disequilibrium
227
linkage groups
291
lymphocytes
complement receptors 354
in Fanconi anaemia 109
B lymphocytes
in infectious mononucleosis 330
interaction with EB virus 335
leukaemias of 270
neoplasia 350
T lymphocytes
226, 261, 263, 270
Burkitt's lymphoma and 350
infectious mononucleosis and 339
neoplasia 341, 343, 350
lymphomas
344
see also Burkitt's lymphoma

macrophages
226
spreading factor 236
maize, jumping gene
281
malaria
166, 167, 354
malate dehydrogenase
193
malignancy
see cancer
malignant melanoma
88
mutation and 308
Maloney-virus-induced lymphoma cells
318
man, mapping of
382
medicine
as war 379
relation to genetics 377, 380
melanoma cells
hybrid 268
meningococcal meningitis
240
mental retardation
361, 365
methotrexate
34, 40, 297
methylcholanthrene-induced sarcoma
319
microcephaly
88
migration
15, 18, 19
molecular anatomy
408
molecular organization
385
monoclonal antibody
407
against glycoprotein 271
cell surface and 407
cell surface antigens and 251-276
classification 260
profiles of 267
mouse tumours
312
multilocus enzymes
187-204
chromosomal assignments 193
chromosomes in 202
classification 189
evolution and 201, 202
function and structure 194
gene duplication 194
gene multiplication 200
incidence of 188
mitochondrial forms 193
molecular size 191
protein in 201
substrates 200
subunit structure 190
tetraploidization in 197, 200, 203
tissue differences in expression 192
multiple heterozygotes
284
multiple sclerosis
211, 242
mutagens
278
mutagenesis
cloned β globin complementary DNA 53
in phage Qβ RNA 48
mutation
383, 397, 401
cell banks 407
DNA content and 277
DNA replication and 86
disease and 402
following gene duplication 195
frequency 279
in bacterial genes 179
in bacteriophages 278
in cancer cells 115, 298
in DNA repair defects 114
in *E.coli* 280
ratios 401
replication error 279
screening 398
myasthenia gravis
211
myeloma hybrid cells
253, 254, 262, 266, 272
antibody production 257
myeloma proteins
252
myoglobin
gene coding 197

natural selection
9, 15, 41
neuraminidase
332

Neurospora
277
nuclear RNA
34
nuclei
transplantation of 76
nucleoids, uncoiling of
91
nucleotide sequences
216, 397
nutrition, reading disability and
375

oncornaviruses
243, 245
oocytes
analysis of small segments of DNA in 67
and ordinary cells 77, 78
DNA injection experiments 64
DNA transcription in 65
genetic realities and possibilities 71
genetics in 63-80
X-chromosome activation in 138
osteogenesis imperfecta
381

PG19 melanoma
330
pain
11
phenylketonuria
381, 384, 388, 404, 406
phosphoglyceromutase
192
plasmids
DNA insertion in 67, 70
pleasure
11
polymorphism for gene control
220
polyoma
216
polyoma virus
78, 319
Prader-Willi syndrome
84
progeria
131
prokaryotes
228, 396
promoters
75
control of activity 73
promoter DNA sequences
58
properdin
232
protein
identification of 408
in cancer cell surface 313
in multilocus enzymes 201
plot of 409
structure 195, 396
psoriasis
211, 402, 403
pyruvate kinase
192

rabies virus
268
radiation
ataxia telangiectasia and 100
chromosomal damage and 144, 291, 292
DNA damage from 280, 291, 292
gene clusters and 280
recombination
at original site 61
suppression 32, 41
religion
20
replication error mutations
279
reversed genetics
47-61
reward mechanisms
11
rheumatoid arthritis
211
RNA
evolutionary new 45
gene coding for 214
leader sequences and 33
synthesis of mutation 49
transcription of 70
RNA polymerase
77
transcription 74
RNA synthesis
68
mRNA structure
150, 151
plasmid RNA
70
rickets
381
Rous sarcoma virus
349

SV40
216, 351
DNA 76, 77
sacral aplasia/hypoplasia
275
satellite DNA
27, 28, 85
control by 32
evolution of 43
homogeneity of sequences 31
length of 144
sequences 127
structure of 43
transcription 75
sex chromosomes
135
Shope papilloma
350
sickle cell disease
162, 167, 181, 246
fetal haemoglobin production in 164
management of 168
screening for 390, 392
sister chromatids
distinguishing 86
sister chromatid exchange
39, 41, 128
complementation in murine—human hybrid cells 299
gene duplication and 202
in Bloom syndrome 105, 107, 130, 309
in cancer cells 307
in Fanconi anaemia 112
in solid tumours 308
in xeroderma pigmentosum 95

skin cancer
88
spacers
28
DNA in 27, 31
specific reading disability
aetiology 376
classification 362
definition 361, 362
familial transmission 364
geographical and ethnic differences 363, 373
infant nutrition and 375
neurological basis 371, 375
phenotypic heterogeneity 364
secular changes 362
sex differences 369, 376
social implications 371, 375
spelling mistakes 367
sub-types of 365
spelling
367, 370, 374
spina bifida
273, 275
spleen cell hybrids
263, 268
stem cells
128
structure-function relationships
47
subacute sclerosing panencephalitis
243
super-genes
213
superoxide dismutase
191, 192
surface antigens
285
surface antigenic proteins
285
systemic lupus erythematosus
239, 241, 242

Tay-Sachs screening
390, 392
teratocarcinomas
329
tetraploidization
197, 200
in immunoglobulins 201
in multilocus enzymes 197, 200, 203
thalassaemias
152, 155-160, 167
α 156, 158, 177
β 156, 162, 164, 167, 168, 176, 184
β° 157
$\delta\beta$ 158, 164, 165, 170
chains 155, 157, 158
gene transfer in 183
management of 168, 183
mapping 184
molecular basis of 159
screening for 390
traits
frequency of 7
transcription process
66
translocation
200
triple X females
142
trisomy 15
cancer and 355
lymphomas and 344
trisomy 21
84
tryptophan synthetase
217
tumours
see cancer
tumour-specific antigens
291
vesicular stomatitis
243
viral transformation
335-357
viruses in cancer
115

wheat germ agglutinin binding
313, 315, 320, 322, 329

X chromosome
activation in 136, 137, 139
dosage of genes in 140
inactivation 140, 142
purification 135
reactivation 138
xeroderma pigmentosum
87-98
basic defect 88
cancer and 96
cell transformation in 130
clinical aspects of 87
DNA repair defect in 89, 93
enzymic efficiency of cells in 91
neurological involvement 88, 93, 96, 131, 132
sister chromatid exchange in 95